BRAVE NEW BRAIN

*Conquering Mental Illness
in the Era of the Genome*

NANCY C. ANDREASEN

OXFORD
UNIVERSITY PRESS

2001

OXFORD
UNIVERSITY PRESS

Oxford · New York
Athens   Auckland   Bangkok   Bogotá   Buenos Aires   Cape Town
Chennai   Dar es Salaam   Delhi   Florence   Hong Kong   Istanbul   Karachi
Kolkata   Kuala Lumpur   Madrid   Melbourne   Mexico City   Mumbai   Nairobi
Paris   São Paulo   Shanghai   Singapore   Taipei   Tokyo   Toronto   Warsaw

with associated companies in
Berlin   Ibadan

Published by Oxford University Press, Inc.
198 Madison Avenue, New York, New York 10016

Oxford is a registered trademark of Oxford University Press

Library of Congress Cataloging-in-Publication Data
Andreasen, Nancy C.
Brave new brain: conquering mental illness in the era of the genome/Nancy C. Andreasen
p. cm
Includes bibliographical references and index.
ISBN 0-19-514509-7
1. Mental illness. 2. Mental illness—Genetic aspects. 3. Human Genome. I. Title.
RC455.4.G4 A53 2001
616.89'042—dc21   00-050141

9 8 7 6 5 4 3 2 1
Printed in the United States of America
on acid-free paper

IN MEMORY OF GEORGE, WHO HELPED ME BEGIN THE JOURNEY.

AND TO TERRY, WHO HELPS ME CONTINUE.

CONTENTS

Part IV

BRAVE NEW BRAIN

In the early 1980s I wrote a book, *The Broken Brain: the Biological Revolution in Psychiatry*. It described a major paradigm shift that was occurring in American psychiatry: the movement from a psychodynamic model to a biomedical and neurobiological model. It was written for laypeople, especially those who suffer from mental illnesses and their families, to help them understand how the brain works and how it becomes "broken" in mental illnesses. I also wanted to reduce the stigma associated with mental illnesses by making it clear that they are brain diseases that cause enormous human suffering. I wanted people to understand that the human sufferers should be accorded the same compassion and respect that we accord people with other illnesses such as cancer or diabetes. *The Broken Brain* was generally a success. It is still in print and is still selling, perhaps because its prediction of a paradigm shift turned out to be true and perhaps because its social message was so important.

As time passed, the scientific basis of modern psychiatry continued to advance. By the end of the final decade of the last century, known as the Decade of the Brain, so much had happened that it was time to write a new and different book to describe our growing knowledge about causes and treatments of mental illnesses in the twenty-first century. Hence *Brave New Brain*.

We now have a wealth of powerful new technologies that illuminate the causes and mechanisms of mental illnesses on many different levels. These include the tools of molecular genetics and molecular biology, which are being used to map the genome and identify the genetic basis of many different kinds of illnesses, including those of the mind and brain. In addition, neuroimaging techniques now permit us to visualize and measure the living brain. Psychiatrists may not be able to read minds, as many people used to believe, but they *can* watch the mind think and feel by using the tools of neuroimaging. The terrain of the brain is being mapped in parallel with the mapping of the genome. The convergence of these two domains of knowledge is one of the most exciting things that is happening in medicine and mental health at the moment. Their convergence has already changed how we think about both the causes and treatment of mental illnesses.

Brave New Brain tells this new story. It is the story of a voyage of discovery, a story about what scientists and clinicians are learning when mind and molecule meet. At present, no book exists to explain it to lay readers interested in following the explosion in knowledge that will result from the meeting of mind and molecule. The lay public should be prepared to share in the excitement of the discoveries that will unfold over the next several decades. The title of this book, which comes from some famous lines from Shakespeare's *The Tempest*, was chosen to convey the sense of enthusiasm and optimism currently felt by clinicians and scientists who work with mental illnesses: "Oh brave new world, that hath such people in it."

Voyages of discovery create both perils and opportunities. *Brave New Brain* contains messages of both warning and hope.

One warning is that we must not lose the human face of psychiatry as it becomes progressively more scientific. We must recognize the perils of a variety of false dichotomies: mind versus body, medications versus psychotherapy, or genes versus environment. Second, in the area of public education, we still have a great deal of work to do in order to ensure that people with mental illness receive appropriate treatment. As described in the first chapter, mental illnesses are among the most disabling and costly of all illnesses. Some illnesses, such as depression, appear to increasing in incidence, as do some consequences such as suicide. Although both lives and money could be saved because many mental illnesses can be easily and effectively treated, health care coverage for them is usually not on parity with other illnesses. This situation must be remedied. A third warning is that we must use our powerful new scientific tools both wisely and well, for the ability to manipulate the basis of life itself—the genome—and the basis of humanity—the mind—creates a heavy burden of responsibility for all of us. The brave new world created by the tools of science must be the humane and enlightened one of Shakespeare, not the totalitarian and self-indulgent world envisioned through the misuse of science in Huxley's *Brave New World*.

The message of hope is even more important. Mental illnesses are a scourge that afflicts the minds, brains, and spirits of billions of people throughout the world. We can be comforted by the fact that our social perceptions about them have finally emerged from the Dark Ages. Furthermore, the scientific study of mental illnesses is now occurring in the era of the genome and the golden age of neuroscience. The powerful tools of molecular genetics, molecular biology, neurobiology, and neuroimaging have only been used to understand the causes of mental illnesses for a few years. As described in this book, we have already learned a great deal about mind and molecule, but the best is yet to come. Progress in Alzheimer's Disease is occurring rapidly. Slow but steady progress, perhaps punctuated by occasional spectacular breakthroughs, is also likely to occur for schizophrenia, mood disorders, and anxiety disorders. The short-term goal for those of us actively engaged in mental illness research is to understand the causes so that we can find better treatments. The long-term goal is to figure out how strike early and prevent at least some of them from occurring at all. The goal of the twenty-first century is to find a "penicillin for mental illness." We would like to fight schizophrenia or dementia as effectively as we can currently fight infectious diseases such as neurosyphilis or pneumonia. We hope to discover a brave new world in which mental illnesses, now painfully common, become infrequent and easily treated.

The book is in four sections. The first introduces readers to the major topics: the personal and economic burden of mental illnesses, the internal human experiences of those who suffer from them, and the various ways that they are often misunderstood using oversimplified through false dichotomies.

The second section contains three chapters that provide a mini-tutorial on neuroscience and molecular genetics. They introduce readers to the workings of the brain/mind and the workings of DNA and genes. Some people may find these sections hard going. They may want to skip ahead and return to them later after reading the third section. The second section is an important reference resource, however, and so people should be sure to skim and dip in it, using it to learn more about mind and molecule, but not feeling guilty about failing to absorb it all. After all, those of us in science have spent a lifetime trying to understand the complexity of the mind and the genome.

The third section contains five chapters that focus on the definition and scientific advances in four major groups of mental illnesses: schizophrenia, dementias, mood disorders, and anxiety disorders. It begins with a chapter that describes the history and conceptual framework of the scientific study of mental illnesses, providing a "report card" on where we stand at present. The remaining four chapters fill in the details for the specific illnesses, beginning with a patient's story, and continuing to describe the defining symptoms, how the specific disorders affect people over the course of their lives, their social and neurobiological mechanisms, and their treatments.

The last section contains a single chapter, which ponders the social, moral, and economic implications of our growing knowledge of mental illnesses.

✻   ✻   ✻

I wish to thank many people have helped me complete this book. Oxford University Press, and especially Fiona Stevens, had faith in my choice to write a serious book that did not compromise content and that assumed people really do want to learn something about mind, brain, genes, and mental illnesses. Susan Schultz made many helpful editorial suggestions, and others also offered useful comments, including Raymond Crowe, Elliott Gershon, Jack Gorman, Sergio Paradiso, and Steven Hyman. Many thanks are due to those who helped with the preparation of the manuscript and the illustrations: Luann Godlove, Shirley Harland, Brian Wilson, Ron Pierson, Vince Magnotta, and Helen Keefe. Of course only I am responsible for whatever errors remain in the text.

I also wish to thank the many patients whom I have seen over the years. They have taught me more about mental illnesses than can ever be captured in any book. They have also taught me—and many others—about the courage and dignity of those who suffer and strive to overcome their suffering, as well as the bravery and love shown by so many of their family members.

The case histories in this book are all based on real people, but significant details have been changed in order to protect their privacy and anonymity.

PART I

BROKEN
BRAINS
AND
TROUBLED
MINDS

CHAPTER 1

BRAVE NEW BRAIN

Confronting the Burden of Mental Illness

O, wonder!

How many goodly creatures are there here!

How beauteous mankind is! O brave new world,

That has such people in't!

—William Shakespeare

The Tempest, v,i, 182–186

Human beings are wondrous, goodly, and beautiful creatures, as Miranda observed in Shakespeare's magical final play, *The Tempest.* This play was his farewell to London theater. Shakespeare wrote it in his late forties and then walked away forever, retiring to a quiet life in Stratford-on-Avon. We do not know why. In this last play that he would ever write, he must have wanted to give the world a message that he considered very important. Just as *Romeo and Juliet* is a great play for teenagers, *The Tempest* is a great play for grown-ups. It is about lofty and fundamental themes, such as conquering evil with goodness and ignorance with wisdom. It is about love and hope. Because it is both wise and affirmative, it is my favorite play.

Miranda speaks these lines when she sees other human beings for the first time, after a tempest wrecks a ship and the survivors struggle ashore. She has grown up on an isolated island, surrounded only by nonhuman creatures such as the ethereal spirit Ariel and the primal Caliban. Her father, Prospero, is the only human being she has ever seen. By a twist of fate, the survivors include Prospero's brother, Antonio, who betrayed him and banished him from Milan many years ago. Also among them is a handsome young man, Ferdinand, with whom Miranda falls in love. (Even at almost fifty, Shakespeare still understood the nature of being in love.) Miranda suddenly envisions a brave new world, filled with beautiful and goodly people. The play is about the reconciliation between the estranged brothers and the love that develops between Miranda and Ferdinand.

Despite the fundamental optimism of his final play, Shakespeare also knew that human beings can be very troubled creatures. *The Tempest* recognizes that there are many dark forces around us and within us: misunderstanding, betrayal, cruelty, hatred, and evil. We murder one another. We lie to one another. Our loved ones develop illnesses, suffer, and die. We ourselves also become ill, suffer, and die. *The Tempest* is a play that confronts such darkness with a seasoned and realistic eye . . . and counters it with light. Optimism based on a foundation of realism is the only true path to a "brave new world."

Like *The Tempest*, this book looks at pain and suffering and expresses the conviction that they can be conquered through enlightenment and knowledge. This book is about building a "brave new *brain*." It is about one group of illnesses that human flesh is heir to: the illnesses that arise from the brain and are expressed through the mind. Mental illnesses. It is about the people who develop them, the friends and relatives who share their suffering, the physicians who treat them, and the scientists who study them so that causes can be discovered and better treatments can be found. Ultimately, it is about how the powerful tools of genetics and neuroscience will be combined during the next several decades to build healthier, better, braver brains and minds. To achieve that goal, however, we must first confront the facts of illness, pain, and suffering. As in *The Tempest*, our optimism must be built on the solid foundation of reality, not on the ephemeral foundation of naiveté.

Mental illnesses are often ignored, misunderstood, or stigmatized. Confronting any serious illness makes us feel charged with emotion and fear. It makes those of us who have a capacity for empathy or introspection recognize that we too are vulnerable, and that we too could suffer the same fate, as could any of our loved ones. We speak the names of illnesses—"cancer" . . . "heart attack"—in a hushed and respectful voice. Mental illnesses probably produce the most intense reaction of all, since they are the least well understood among the many human illnesses. Our intuitive reaction, when confronted on the sidewalk with a mumbling and disheveled person suffering from a mental illness, is to look away. Even when a close friend has a problem that requires hospitalization, we are reluctant to visit her. (The excuse is often, "I don't want to embarrass her." Or, "I wouldn't know what to talk about.")

There are many important reasons why we cannot afford to ignore mental illnesses.

First, they are very common. Almost anyone who picks up this book has a friend with mental illness, or a family member, or suffers from one himself or herself. Mental illnesses are among the most common diseases that afflict human beings. Schizophrenia affects 1% of the population, manic-depression another 1%, major depression another 10–20%, and Alzheimer's disease 15% of people over 65. And that is only mentioning the most severe illnesses.

Second, they are incredibly costly, both economically and psychologically. Worldwide, the cost runs to billions of dollars. In 1990 the World Health Organization did a survey of the cost of medical illness throughout the world. The results were recently summarized in a book called *The Global Burden of Disease*. If asked, many people would guess that the greatest costs are from cancer or heart disease. Wrong. Mental illnesses cost us more than any other general class of disease. There are many ways to summarize the economic burden of disease, but they consistently lead to the conclusion that mental illnesses should be given a high priority for treatment and research because of the many ways that they are costly to society.

Table 1–1 shows the figures for the costs due to disability for people who are between 15 and 44 years old. The costs are expressed in a unit of measurement developed by the Harvard researchers who wrote *The Global Burden of Disease*, known as Disability-Adjusted Life Years (DALYs). This is a composite measure of time lost due to premature mortality and the time lived with the disability. A loss of one DALY is equivalent to the loss of one year for one person. Among people in the prime of life, depression costs society more than any other disease, and four mental illnesses are in the top ten. Self-inflicted injuries (usually suicide as a consequence of a mental illness) are also in the top ten. In this age-group mental illnesses cause us to lose millions of years of potentially productive life.

TABLE 1–1

The Ten Leading Causes of Disability in the World

Type of Disability	Cost (in DALYs)
Unipolar major depression	42,972
Tuberculosis	19,673
Road traffic accidents	19,625
Alcohol use	14,848
Self-inflicted injuries	14,645
Manic-depressive illness	13,189
War	13,134
Violence	12,955
Schizophrenia	12,542
Iron deficiency anemia	12,511

Mental illnesses are not just economically costly. They also take a cruel psychological toll and are unfortunately often fatal. Suicide claims approximately 10% of people with schizophrenia and 10% of people with depression. Suicide rates are inexorably rising in our most valued national and international asset: our children. Losing a child by suicide may be the most painful experience a person can have. But observing how schizophrenia invades the personality and mental skills of an adolescent or young adult also causes nearly unbearable pain to both the young person and his family. Watching a parent or a spouse die a slow death from Alzheimer's disease is heartbreaking.

In fact, if we confront reality honestly, we realize that mental illnesses stand out from other human diseases as both special and frightening. They affect the most important organ in our bodies and the most important capacities that we have. They affect the brain and its product, the mind. Modern medicine has taught us that we do not die when our heart stops or when we stop breathing. We die when our brains die, when they stop producing the characteristic electrical rhythms that indicate that our nerve cells are firing. We feel ourselves truly alive when our brains are most active, when we play (or even excitedly watch) a basketball game, when we read an interesting book, when we listen to a particularly engaging song or a symphony. What we fear most is not being paralyzed in an accident or even a sudden death by a heart attack, although either would be cruel. We fear *losing our minds*.

Mental illnesses are that 600–pound gorilla, hidden in the closet, that we fear to confront. But we must confront them. They are important now, and they will only become more important as the next several decades pass by. The global burden of mental illness will continue to increase, until or unless we proactively identify ways to improve treatments or implement preventive measures. Several demographic trends create this situation. First, our country is aging, with the consequence that the number of elderly will increase and will develop the most important mental illness that affects the elderly: Alzheimer's disease. Second, the baby boomers are aging. This particular cohort already has higher levels of depressive and anxiety disorders than previous generations. As this large group moves into their sixties, they will also swell the ranks of those with Alzheimer's disease. Unless we do something, our children will be left holding a heavy bag of responsibility and suffering.

The reality of mental illness is painful on many fronts. No wonder we would like to ignore it.

As in *The Tempest*, however, we can counter the reality of suffering with the reality of hope. During the last two decades of the twentieth century, both psychiatrists and their patients have steadily recognized that mental illnesses are diseases of the brain that can be understood and treated using established scientific tools. The last decade of the twentieth century was designated by Congress as "The Decade of the Brain." We are at present in the midst of a golden age of biomedical research. We are currently engaged in two of the most important endeavors in the history of science and medicine. We are simultaneously mapping the human brain and the human genome. Each of these is a daunting task. The brain contains billions of neurons—most estimates are around 10^{12}. The human genome contains far fewer genes. Most estimates are around 80,000, or maybe even fewer. Not all are active ("expressed") in all parts of the body—only around 20–30,000 in the liver, for example. But most are active in the brain. The mapping of the brain is made possible by a variety of new technologies that permit us to understand things on a large scale, a scale that neuroscientists refer to as the "level of systems," by which they mean functions of the mind such as memory and attention. The mapping of the genome is made possible by spectacular advances in the technology of molecular genetics and molecular biology, which work on a very small scale at the level of the molecule. The achievements of these two endeavors are described in detail in chapters 4 through 6 of this book.

The achievements occurring on these two levels will meet one another some time within the next decade or perhaps two. When they do, the payoff will be impressive. We will understand how the cells in our brains go bad when their molecules go bad, and we will understand how this is expressed at the level of systems such as attention and memory so that human beings develop diseases such as schizophrenia and depression.

Once mind and molecule meet, prevention is possible. Improvements in treatment are certain.

This book is a travel guide to the future, written to help the curious understand how better treatments and preventive measures will be created or identified. Part 2, "From Mind to Molecule" (chapters 4–6), introduces readers to the scientific foundations of modern psychiatry: the study of brain and mind, and the study of genetics and molecular biology. Part 3, "The Burden of Mental Illnesses" (chapters 7–11), gives readers cutting-edge information about what we have learned so far about the diagnosis, mechanisms, and treatment of four major groups of mental illnesses: schizophrenia, mood disorders, dementias, and anxiety disorders.

The readers of this book will be sitting in the theater during the first decades of the twenty-first century, watching as the hope of conquering mental illnesses gradually turns to reality. Perhaps for the first time in history, we can be optimistic about mental illnesses, which have previously been a dreaded scourge. It is an exciting time.

The lights are going down. The first act in this drama begins with a real story about real people, people who suddenly find themselves dealing with mental illness, in a world that is still ill-prepared to look at it and see it.

Let's begin with the painful reality. Let's see how mental illness can affect the lives of ordinary people like you and me.

CHAPTER 2

A WAKING NIGHTMARE
Mental Illness
and Ordinary People

> O the mind, mind has mountains; cliffs of fall
> Frightful, sheer, no-man-fathomed. Hold them cheap
> May who ne'er hung there. Nor does long our small
> Durance deal with that steep or deep.
> —Gerard Manley Hopkins
> *No Worst, There Is None*

Mary was concerned. She had known Jim for eleven years and had been married to him for six. He was as solid as a rock. She could always count on him. Now, suddenly, he seemed to be falling apart, for no apparent reason. Her rock was turning into a pile of pebbles right before her astonished—and increasingly frightened—eyes. What could be going on?

It started about four months ago.

Jesus, what is wrong with me??

Jim fixed a vacant stare on the Alka Seltzer as the bubbles swarmed about the tablet before rising in the glass.

I can't do it. I just CAN'T do this. I can't work all day and try to study all night. Why did I let myself get into this trap? Mary would be better off without me.

The knot in his stomach tightened. He reached for his toothbrush and knocked the Alka Seltzer glass shattering to the floor. He sank down to the cool tile floor and held a smooth piece of glass to his forehead as if to chill the worried thoughts that kept boiling to the surface. He tried to remember when he began feeling so terrible. He tried to think of a reason for the pained feeling in his stomach that seemed to poison his concentration and sap away his usual spirit. He could think of no reason. His wife Mary was the joy of his life. She'd been contentedly supporting his return to business school so that he could move on from his construction work to a better life. He was only one semester away from taking his place in the world of boardrooms and laptops. He ought to have it made. So why did everything look so gray and ominous? Why couldn't he shake the sick visceral feeling that hit him like a rock the moment he woke up in the morning?

As he sat, he analyzed the feeling. It was the same sick fearfulness he'd felt in grade school when he thought he'd stepped on the wrong bus going home. It was that surreal "something is wrong with this picture" feeling that comes in a panicky wave. Back then it would evaporate as soon as he saw a familiar face on the bus or recognized the right route, but this time it stuck. The mornings were barely tolerable, hence the Alka Seltzer. It subsided a bit during the day but the feeling never, never left. It had tortured his sleep for weeks so that he now felt like the walking dead.

How can I keep on going this way? It's all my fault for trying to change who I am. This is my penance for trying to be something other than a construction worker like the rest of my family. People like me should stay in their place . . .

This was a time when he should have been feeling great. He was about to realize a dream that he had had for many years—to finish his master's degree in business and get a high-paying job with a company, preferably in a marketing and sales division. Both of them had worked, planned, and sacrificed to achieve this goal. They had put off starting a family so that she could keep her job as a secretary and help pay the expenses of going back to school. Mary's biological clock was ticking—she was thirty-one and he was thirty-four—so they couldn't wait too much longer. They had held off from buying a house or even from starting to buy furniture or a good set of dishes, since they knew that they would be moving. Jim had kept working half-time in the family construction business—installing roofs on houses. It was hard work, and somewhat dangerous. Soon he would be able to move into the white-collar world, with all its benefits. He had actually been looking forward to wearing a suit and tie every day!

It apparently began with a bout of the flu. He was tired and had a low-grade fever and muscular aches. He also lost his appetite. He missed classes for a few days and spent most of the time in bed. The fever and aches went away, and so he went back to work and school, but he didn't feel much better. In fact, he felt worse in some ways. The loss of appetite turned into nausea. When he got up in the morning, he felt sick to his stomach. This lessened as the day went on, but he never felt much like eating. Jim and Mary joked about his "morning sickness" and the possibility that he rather than she might be "expecting." When he had lost about five pounds and continued to feel worse, even vomiting a couple of times when the nausea got too bad, the joke stopped being quite so funny. He was also having other problems. He would wake up in the middle of the night, full of anxiety about the things he had to accomplish the next day, and he was unable to get back to sleep for two or three hours. Then he was exhausted the next day. He had big projects in two of his classes, and he was beginning to wonder if he would be able to finish them. He felt frightened and full of self-doubt—not a way he was accustomed to feeling. Usually he approached big projects with confidence, and usually he was the top student in his classes. Why was this happening to him now? He started to feel nervous about everything. Sometimes his hands would even get shaky. He was beginning to think he was a real loser, and Mary was having trouble convincing him otherwise. The situation came to a crisis when he announced that he had decided not to apply for a job next year—that he would just return to working full-time in the roofing business. He might even drop all his classes and not finish his MBA.

They had health insurance with an HMO through Mary's job. Both of them were in good health, so they never used it, except for Mary's annual appointment with the gynecologist to get her birth control pills. They decided that Jim should probably see a doctor. He was obviously sick with something.

The doctor was a young, blond man with a small, droopy moustache and a wiry build, wearing the obligatory white coat over an open-necked casual shirt and khakis with Birkenstock shoes. His name was Dr. Mc-Nerny. He listened to Jim's story—bout of the flu, tired afterward, nausea, some vomiting, shakes, trouble sleeping. He then asked lots of questions—first about how much Jim drank. Jim and Mary usually had a beer before going to bed at night. It was their time to unwind together, talk, and review the events of the day. They also occasionally had a bottle of wine with dinner on Saturday night—maybe once a month. Jim had a hard time convincing the doctor that he didn't drink more than that. Then the doctor asked about coffee and other drinks containing caffeine. Jim had to admit that he drank quite a bit of coffee—five or six cups a day. He also drank a couple of cans of Coke or Pepsi. Then the doctor went on to diet. Jim and Mary were very proud of their record there. Jim's dad had had bypass surgery, and so they were very careful to follow a "heart-smart" diet—meat only a couple of times a week, lots of fresh fruits and veggies, pasta, rice, and olive oil instead of butter or margarine. Although Jim loved ice cream (and would have preferred that as their bedtime snack), Mary almost never allowed him to eat it. And then there was exercise, smoking, and use of illegal drugs. Again, they had a good record. They had actually met one another running, and running or biking together was part of their regular routine. Neither of them had ever smoked or used illegal drugs. There were also lots of questions about what kinds of illnesses Jim's mom, dad, and other family members had experienced—heart disease, diabetes, cancer, kidney disease, and so on.

The doctor listened carefully to Jim's heart and lungs, checked out his balance and coordination, and looked inside his eyes and mouth, eventually seeming to cover most parts of his body. Everything was fine. When Jim was weighed, however, he observed that he had lost a total of eight pounds over the past six weeks. Dr. McNerny looked a bit worried when he heard that. He asked if Jim had been drinking a lot of fluids and peeing a lot.

After completing his physical exam, Dr. McNerny explained that he would like to do some blood and urine tests. Lots of things could explain Jim's problems, although most of them were unlikely. He was wondering about thyroid disease or diabetes; the weight loss also made him worry about cancer. However, he thought the most likely explanation for Jim's problem was too much caffeine. He suggested that Jim switch to herbal tea (not even decaf, since it contains *some* caffeine) and caffeine-free soft drinks.

Jim worried about the possibility of cancer too, after leaving the office. In fact, the possibility made him *really* worried. A few days later Dr. McNerny's assistant called to say that all the blood and urine tests had been normal and that Jim was considered to have a clean bill of health.

Despite following the doctor's advice about caffeine, however, Jim continued to get worse. He felt nauseated all day, and he could barely get through brushing his teeth in the morning without vomiting. In fact, he was starting to heave up his breakfast three or four mornings a week. He felt lousy. On the one hand, he was tired and apathetic and discouraged. On the other hand, he was wound up like a tightly coiled spring. He was having periods lasting for hours when he felt so anxious that he couldn't think clearly or concentrate. He could barely drag himself to class or to work. He began to consider dropping all his courses and delaying graduation. He was near a definitive decision to forget about the job market. He had no energy to pull together a new resume, check out available openings, and send in applications. Furthermore, he didn't think anyone would want to hire him.

Mary decided to take some action herself. Always a feisty take-charge type, she figured it was time to do research on her own about the nature of Jim's problem. She started checking into the various Internet sites that provided information about health. She discovered some very interesting things. She learned that Jim's symptoms seemed to match pretty closely with a problem called "panic disorder" or "generalized anxiety disorder," which were very common mental illnesses. In fact, for his age range and his family history, the chances of having thyroid disease, cancer, or diabetes were exceedingly slim. On the other hand, lots of people his age apparently had problems with anxiety. There were new drugs available to treat these problems that were supposed to be very effective. She downloaded an article from *The Journal of the American Medical Association* and gave it to Jim to read. She also downloaded the description of generalized anxiety disorder and panic disorder from the American Psychiatric Association's *Diagnostic and Statistical Manual* (DSM). Jim was also surprised at how closely his symptoms matched the description. Mary pushed very hard and finally got him to call Dr. McNerny back for another appointment. The HMO didn't think it was necessary—Jim had already been given a clean bill of health. But Jim finally convinced them to refer him to a psychiatrist in order to get an evaluation for panic or anxiety disorder. However, he had to wait another three weeks for the appointment. He didn't think he could make it. He was so nervous he could hardly stand it.

This time he saw another young man—in fact, this doctor looked even younger than Dr. McNerny, and he wasn't wearing a white coat. It turned out his name was Mr. Morgan, and that he wasn't a doctor at all. He had a master's degree in psychology. Mr. Morgan explained that the HMO didn't have a psychiatrist. If absolutely necessary, they referred patients elsewhere, but that was only for the really "bad cases," and it was unlikely that Jim was such a "bad case." Mr. Morgan was the person in the HMO who diagnosed and treated mental disorders, under the supervision of doctors like Dr. McNerny. He had an air of confidence and competence, and so Jim had no qualms about entrusting Mr. Morgan with his care.

Mr. Morgan explained that he had already reviewed the records from Dr. McNerny, so he didn't need to repeat all the questions about drinking, drugs, and the like. Instead, he began by asking a bit about Jim's schoolwork and career goals. Jim explained that his career was in a real crisis. He was afraid that he would have to drop out of school and give up his plans to interview for a prestigious MBA job somewhere. Mr. Morgan smiled sympathetically and began to ask a laundry list of questions in a rapid-fire manner. Jim recognized them immediately as the DSM criteria for panic and anxiety disorder that Mary had downloaded. Jim already knew that he fit both of them loosely and neither of them perfectly, but he dutifully answered each question. After about fifteen minutes of questioning, Mr. Morgan excused himself and returned carrying a prescription. He explained that it was for a medication called Celexa, which was similar to Prozac. The dose was twenty milligrams, and Jim was to start out taking it that evening. It might take up to four weeks to notice any effect, however. Jim's heart sank. He was hoping for more rapid relief. He had never heard of Celexa, but he recognized Prozac as the drug that makes you "better than well." Well, maybe it would work. It was certainly worth a try.

Jim had accumulated a long list of questions while he waited three weeks for the appointment, and he was desperate to get the answers.

"What do you think is wrong with me?" Jim asked.

"Oh, it's a chemical imbalance in your brain, probably a tendency you were born with," Mr. Morgan replied.

"How long will this last?"

"You could have the problem for the rest of your life," Mr. Morgan said, "but don't worry about it, because the medications usually work very well."

"Do you mean that I might have to take medications for a long time?" Jim asked fearfully. Jim was really health conscious. He hated to take any kind of pills. He was also very self-sufficient, and he didn't like the idea of being on anything "for life."

Mr. Morgan, who appeared to be even younger than Jim, sat up very straight and looked at him sternly before replying. "Now, if you had diabetes, you wouldn't object to taking medications, would you? Well, this is like diabetes. But it's a brain disease instead."

"But couldn't it be treated with psychotherapy instead of drugs?" Jim asked.

"Well, maybe," Mr. Morgan replied. "But that would take a lot longer, cost a lot more, and probably not work as well. You were perfectly fine until you developed this problem. You don't have any neuroses or hangups, so there probably is not much to work on with psychotherapy." Jim had to agree with that!

"So what is your diagnosis?" Jim wondered.

"Well, you don't fit panic disorder perfectly. You don't have typical 'attacks' of panic. Those periods of anxiety that you describe last too long—for hours instead of five to ten minutes. But I'm still assuming that panic disorder is the problem. You're just having atypical attacks. You don't meet the criteria for generalized anxiety disorder because that requires six months of symptoms. You have only been having problems for about two and a half months."

For Jim, those two and a half months had been an eternity.

"You also have some obsessive-compulsive traits. You are orderly, organized, and like to be in control. You also have some symptoms that resemble depression. That is one of the reasons I chose Celexa. It works pretty well for people with comorbid problems. Comorbidity of anxiety, depression, and OCD is very common these days, but it is harder to treat successfully."

That didn't instill much confidence in Jim.

"Is this really the best drug? Will it really work?" Jim asked.

Mr. Morgan was starting to get restless. He had obviously used up his allotted time and needed to get on to the next patient.

"It should work like a charm. These new drugs target specific chemical systems in the brain that are messed up. You have a serotonin deficiency. Celexa is what is known as a Selective Serotonin Reuptake Inhibitor, or SSRI. It corrects your deficient serotonin by increasing the amount available. It may take up to four weeks, but I'm pretty sure that you'll feel better. Any more questions?"

The final question conveyed the message that there had been enough already.

So Jim walked out, carrying his prescription, feeling only slightly more hopeful. He didn't like the idea of having a brain disease, even though he knew that Mr. Morgan was trying to reassure him with that information. He also didn't like the possibility of a "life sentence." But he filled the prescription, went home, reported the events to Mary, and took his first dose right after dinner, not expecting to feel much, since the drug was supposed to take four weeks to take effect.

Mary felt pretty concerned about Mr. Morgan's comments too. But she kept her worries to herself. Jim had enough of his own. She knew she had to be encouraging and hopeful. Most of the information on the Web suggested that the new medications worked wonders.

Within just a few hours after taking his first dose, however, Jim was crawling the walls. Rather than signs of relief, Jim began to feel much more anxious and agitated. He figured that he needed to stick with the regimen, however, and he wanted to get well quickly, so he took his second pill for the day before going to bed that night. It was a disaster. He was awake the entire night, tossing and turning, ruminating and fretting. Was he losing his mind? Was he going to be sick forever? Was he going to require institutionalization? He could almost feel that serotonin draining away from his brain, depriving him of his will, drive, personality, and intelligence. What brought on this serotonin deficiency? What had he done? What could he do? What would his professors think when they saw how sick he was? How could he ever go to a job interview? How could he ever produce a good resume? How could he manage to do all the work of looking up which companies had openings and sending in his application? If he ever got an interview, the interviewers would obviously discover that he was a charlatan with no business talents at all. He couldn't even begin to design a model marketing plan for a product. What would his parents think? How could he let them down like this? How could he let Mary down? Would they have to give up their hopes for the future? Maybe he should just leave her. After all, she should not have to deal with a husband with a mental illness. She was still young, and she could find another, better husband. Maybe he should kill himself, since the future looked so hopeless.

Jim dutifully took the medication for another two days. He was sleepless every night, tormented by fears, doubts, and worries. He was constantly nauseated and vomited up everything he ate. Normally stoic, he was having spells when he burst into tears and cried inconsolably. Seeing him like this hurt Mary almost unbearably. He was suffering so much, and neither of them could understand why. Finally, it was Mary who had to call Mr. Morgan and Dr. McNerny. She had already done her Web research, and she had learned that such side effects *could* occur with an SSRI, although most people seemed to tolerate the drug very well. After all, half the population seemed to be taking some kind of SSRI! But Jim seemed to be an exception. She noticed that there were other drugs that worked on anxiety, such as Xanax, and they worked more rapidly and produced almost immediate relief. She and Jim were ready for a little immediate relief. Mr. Morgan listened to the story. He told her that Jim should be calling himself. He also said that Dr. McNerny would have to make the final decision, since Xanax and the other "tranquilizers" could be habit-forming. (What about that business of having to be on Celexa for life? Wasn't that habit-forming as well?) Mary finally conveyed the idea that Jim was really, really sick, and that she was really, really frightened. She had never seen him like this, and she had known him for 11 years.

Before Dr. McNerny called back, Mary had a long talk with Jim. Both of them knew his current medication was making him worse—an unexpected and unwanted outcome. Mary knew she had no business trying to call the shots with his medical care, but she desperately wanted to help him get better. From what she could tell, one of the tranquilizers would give some immediate relief. Maybe he could get some sleep and keep some food down. Then they could regroup and figure out what the next step might be.

When Dr. McNerny called, Jim managed to speak with him and explain how badly he was doing. He was close to tears, and Dr. McNerny could probably tell. Dr. McNerny agreed to phone in a prescription for Xanax and to see Jim again in another week, just to make sure that there was nothing "really serious" that was wrong with him. Apparently the severe anxiety, despondency, vomiting, and insomnia were not really serious.

For Jim and Mary the problem was serious enough that they decided to get some help from friends and family. Up to this time they had not discussed Jim's problem with anyone, but they were feeling very alone and in need of help.

Somewhat surprisingly, they received different advice from different people.

Professor Ernst Vogel, Jim's mentor and favorite teacher at the business school, was really kind. He had noticed that Jim was "not himself" and wondered what was wrong. He seemed relieved that it was a problem with self-doubt and anxiety, rather than the other possibilities he had been considering—taking drugs, drinking too much, marital problems. Jim and Mary were one of the nicest young couples that he knew, and he would have hated to see them break up. Further, Jim was one of the best students that he had ever had, and he was destined to go far. Jim was personable, smart, sincere, honest, creative, and hard-working—like an ideal son. Professor Vogel was very prominent in his field, and he knew that he would be able to help Jim get a good job, based on his own high recommendations. He was stunned to learn that Jim was getting panicky about job interviews and had stopped working on seeking open positions and preparing applications. Professor Vogel thought Jim should see a psychotherapist so that he could work through whatever was causing the problem. His own wife had had some problems with anxiety and depression and had gotten enormous benefit from psychotherapy. No medications at all had been necessary. The therapy lasted for about six months, and she had been fine ever since. That sounded pretty good to Jim, who jotted down the name and phone number of the therapist. Ol' Ernst made him feel better about himself than anyone he had seen so far. Maybe this problem could be solved after all. Maybe he could take control and return to his old self. Maybe therapy could show him how to do that.

They also turned to Jim's parents, Max and Helen. Mary's parents were both dead, and so Jim's were the main resource. Mary and Helen were very close, and Mary felt comfortable in confiding in her. Helen was a "with-it" lady who had also worked as an office employee for much of her life. However, she was pretty conservative, and very much a no-non-sense type, so Mary wasn't sure how she would react to learning that her nearly perfect son had turned into a basket case . . . for reasons that no one except Mr. Morgan had been able to explain.

Over a cup of coffee, Helen listened intently to Mary's story. Then she reached out and held Mary's hand and said, "I'm going to tell you some-thing I never planned to discuss with you, but I think you need to know about it now. I had a very similar experience myself, right around the time that Jim was born."

Helen then recounted an episode when her emotions ricocheted out of control in her late twenties, and she experienced intense anxiety and despondency. The symptoms were very similar, but the diagnosis was dif-ferent. She saw a psychiatrist, who said she was suffering from endoge-nous depression and treated her with a tricyclic antidepressant, nortripty-lene. The medication calmed her almost immediately and relieved the depression after about two weeks. She went from feeling out of control to feeling like herself again. It was like a miracle. She was able to discontinue the medication by tapering it slowly after about six months. She had another episode when Jim's sister was born and again received successful treatment. Apart from these two episodes she had been fine. She felt slightly ashamed that she had had these problems, so she hadn't discussed them with the kids. Apparently depression does run in families, so Jim might have exactly the same thing. If so, then exactly the same treatment might work.

Then Helen leaned forward and said quietly, "You know, people with depression sometimes become suicidal. Ask Jim if he has thought about this. I know it will be hard. But you need to know. If he has a depression rather than an anxiety or panic disorder, his life could be at risk. I don't want to worry you, but you need to find out. If he is depressed, he should be treated for it. Celexa, which apparently helps a lot of depressed people, backfired with him. I have read quite a few books about depression over the years, because of my own experience, and it seems that some of the older drugs work on other chemical systems besides serotonin and can be more effective for an episode like mine—and perhaps Jim's—that occurs very severely and for no apparent reason, what they used to call endoge-nous depression. We'll all have to think about this some more. But for sure we want our Jim to get well and be himself again."

Helen reached out her arms to Mary and gave her a big hug, as both of them choked back tears. "Don't worry, we'll get him back again. I know we can," Helen said.

Mary knew Helen was with it. But, wow! Helen didn't even need to consult the Web! She got it from books and personal experience.

Mary returned home, full of food for thought. Before Jim got home, she had already checked out the *Diagnostic and Statistical Manual* Website. BINGO! Major depression was a perfect match. She didn't know about suicidal thoughts, but he had every other symptom for sure: depressed mood, decreased energy, weight loss, insomnia, physical restlessness and agitation, fatigue and loss of energy, feelings of worthlessness, and trouble thinking and concentrating. She thought she had a firm handle on the problem and was prepared when Jim walked in the door. It would be back to the HMO, and referral to a psychiatrist who was expert in treating depression and familiar with all the different kinds of medications available. Humming "Yellow Submarine," she formulated her attack on the HMO as she cleaned romaine lettuce for the salad and ran a sundried tomato sauce through the cuisinart.

"I've decided I need to see a psychotherapist and get to the bottom of my problems," Jim said over dinner, looking as if he had finally worked his way to a decision after three months of agonizing self-doubt and indecisiveness. Mary hardly knew what to say, having reached a totally different conclusion herself. So she just asked a question.

"What made you decide that?"

Jim described his meeting with Prof. Vogel, whom he adored. Prof. Vogel had treated Jim more like a son than a student during the past three years that Jim had spent working on his MBA. Mary didn't disagree with Prof. Vogel's suggestion, although she wasn't exactly sure what "bottom" there might be to get to. Jim was usually so normal and well adjusted. But there was plenty about the mind that Mary didn't understand, and maybe psychotherapy would help. Anything that would help would be welcome. However, she did need to share the new insights that she had gotten by talking with Helen.

"Psychotherapy sounds like a good idea," she said. "Maybe it will help you feel more confident, more like your old self. Let's figure out how to get an appointment. I also got some really helpful ideas from your mom."

Mary started describing her conversation with Helen. An inquiry about suicidal thoughts revealed that Jim had been preoccupied with this possibility for a couple of months, right down to going to Fin and Feather, the local sporting goods shop, to see about buying a handgun.

"Jim, please. Don't ever do that. I love you so much. I couldn't bear to lose you. Please, I need you. I just want to get our life back to normal."

Jim, her exuberant but solid rock of a husband, started quietly crying and mumbling something about being worthless and no good for her.

Mary then told Helen's story. Jim seemed surprised, since his mother was also like a rock, but somewhat comforted as well. If his mom could get through two depressions and look and feel so great, then he probably could too. Mary also extracted a promise that there would be no guns and no suicides. They had been waiting for years for their chance to have a good job, a nice home, and kids . . . together, as Mary and Jim. Jim admitted that he was also feeling a little better on the Xanax, so there seemed to be some hope already, even though he hated taking *anything*, and especially anything that might be habit-forming. They decided to attack the problem on both fronts—both medications and therapy. Since Mary had more free time—and more patience in dealing with the healthcare bureaucracy—she would assume responsibility for setting up the necessary appointments.

On Monday, Mary assaulted the HMO over her lunch hour. It took a bunch of phone calls back and forth, but the mention of suicide apparently scared someone, and so the HMO finally authorized a referral to a psychiatrist. However, they would not authorize psychotherapy. That would have to be paid "out of pocket," unless Jim would accept fifteen minutes every other week with Mr. Morgan and Mr. Morgan was willing to provide it. But six sessions of brief psychotherapy within the HMO was all that they covered. Mary scheduled an appointment for Jim with the psychiatrist authorized by the HMO—for a date another interminably long two weeks away. She thought about the situation after hanging up the phone, called back, and told the receptionist about Jim's suicidal thoughts. The receptionist was sympathetic and got Jim an emergency appointment at the end of the week, on Friday afternoon.

Mary then checked out the therapist recommended by Prof. Vogel. The receptionist indicated that this was an independent private group practice, consisting of three psychologists and a psychiatrist. Jim would have to see their psychiatrist at least once at some point, but the weekly therapy would be done by one of the psychologists. Dr. Emily Brill, the therapist recommended by Prof. Vogel, did have an opening on Wednesday afternoon and could see Jim then for an initial evaluation. It would cost $200, and each session thereafter would cost $100. Mary took a deep breath. That would buy a lot of pasta or furniture. Over time it would buy a down payment on a house. But it was what Jim wanted. And she wanted Jim to get well. She scheduled the Wednesday appointment.

Jim saw Dr. Brill on Wednesday. She was a petite, graying woman who appeared to be in her early fifties and who had striking olive skin and dark, penetrating eyes. Although she didn't look like his mom, she had a similar no-nonsense manner, and he liked her immediately. She began by reviewing the history of "his problem"—the feelings of anxiety and depression, the sleepless nights, the nausea and vomiting. She listened intently as he described how the various medications had affected him— the disaster with the SSRI and the recent improvements with Xanax. She spent an hour with him and asked a lot of other questions—about his parents, about Mary, about his childhood and adolescence, about how he usually reacted to challenges, about how he got along with friends and superiors, about his hopes for the future. At the end she said, "I think this is mostly situational anxiety and depression, probably brought on by all the changes that you will soon be making in your life—finding a new kind of job in a white-collar world, perhaps moving to a new city and leaving friends and family, and starting a family of your own. It is a lot to take on all at once. You are probably more threatened by all these changes than you realize. I need to see you a few more times before I can say that for sure, though. I'd like to try to help you understand why this happened right now, and to figure out how to prevent it from happening again, the next time you confront a big challenge."

That all sounded pretty good to Jim. But he was worried about the cost. He mentioned that.

"It's a long-term investment in your health," Dr. Brill replied. "Your health is the most important asset that you have. You have to take care of it."

That made sense to Jim. After all, it was just like business . . . even though the investment was pretty steep. But maybe he could just see Dr. Brill, without also seeing the group psychiatrist. That would be another cost, and a duplicate one, since he was already scheduled to see the doctor covered by the HMO.

"No, I'm sorry," Dr. Brill said ruefully. "Our group has an agreement. If you see me, you have to see our group psychiatrist. We think this is important for continuity of care. The psychiatrist will make the decisions about medications, while I'll handle the psychotherapy, but we'll work together as a team."

Jim was starting to feel caught in the middle, since he knew that he probably had to see the HMO psychiatrist, and he might get different recommendations from him. He also wondered what Dr. Brill thought about Celexa, Xanax, or even the nortriptylene that his mother had taken.

Dr. Brill smiled. "Here's what I really think. These medications are kind of like a crutch. They help you get through hard times, and you probably will need to take some for a while. But you have to learn to walk without them eventually. That is what psychotherapy will help you learn to do. As for which medications you should take, I really don't know anything about that. Dr. Hauptman, our group psychiatrist, will decide that. When you leave, our receptionist will schedule a follow-up appointment with me, and also one with Dr. Hauptman. In the meantime, you will have to figure out for yourself what you should do about the mental health care provided by your HMO."

Jim stopped at the front desk on his way out. The next appointment with Dr. Brill would be in another two weeks, and probably at weekly intervals after that. The appointment with Dr. Hauptman could not be scheduled for another three weeks. Jim wasn't sure whether he was falling between the cracks, or being given some time to make up his mind about what he really wanted to do. There sure seemed to be a lot of different points of view about what was wrong with him, and about how to treat it. It was also hard to figure out how to get treatment and how to pay for it. He wondered how people who were *really* confused or not very well educated ever managed to get any help for their problems.

Jim saw Dr. Walker, the psychiatrist recommended by the HMO, on Friday. Dr. Walker was another graying, fiftyish sort who looked lively and intelligent and also very professional. He had copies of recent medical journals, such as the *American Journal of Psychiatry* and *The New England Journal of Medicine*, sitting on his desk in a rumpled and obviously read condition, suggesting that he was staying up-to-date with medical advances. He also spent an hour with Jim, beginning with the comment that he had carefully reviewed the records sent over by the HMO, so he already knew quite a bit about him.

"You've been having a rough time," Dr. Walker said. "Take about five or ten minutes and summarize it all in your own words, beginning with when you first noticed something wrong."

By now a bit tired of his story, Jim nonetheless reviewed it again. This time he added the information about his mother's history, her response to medication, and his own thoughts about suicide. He hated to mention it, but Mary had told him that he had to. He also described how Mary had looked up DSM criteria for depression, and how he had to agree—it was a perfect match, although the match seemed to have been made in hell rather than heaven.

"Your marriage sounds like a match made in heaven though," Dr. Walker observed. "You are lucky to have such a smart and supportive wife. She is right about the two of you needing one another. Promise me too that you will not take your life. If you ever get slightly close, call me immediately."

"I agree that you have a serious depression," Dr. Walker continued. "And your mother is right—it is what we used to call 'endogenous depression' because it seems to grow from within, rather than being caused by some obvious precipitating factor. In this case, we may be lucky, because we may have a clue in your mother's good response to one of the tricyclic antidepressants. That is what I am going to prescribe for you, along with Xanax as a backup to help you get through the next couple of weeks. The tricyclics are older drugs, and some doctors use them less often these days because of their side effects, but they are really 'tried-and-true' for the kind of problem you have been having. In this case, the side effects are advantages, especially in a healthy young man like you. Because the tricyclics are slightly sedating, they act almost immediately on the insomnia and anxiety, and they kick in and knock out the more basic depressive symptoms within a couple of weeks. Jim, don't worry about using a crutch for a while to get through this episode of depression. And don't worry about becoming addicted to medications to help your mood. When the right time comes, we'll taper you off of them, and you'll be fine. You may not get better than well, but you'll get well. And given the kind of person you are, that is plenty good enough."

"What about getting to the bottom of the problem with psychotherapy?" Jim asked.

"Psychotherapy helps a lot of people," Dr. Walker replied. "I am not sure that you are one of the ones who needs it, however. I know this will sound strange, since you feel like a real mess right now, but you seem to be a person who is pretty normal and psychologically healthy. You have a wonderful marriage, lots of friends, a close relationship with your parents, and lots of success in getting along with both superiors like Prof. Vogel and people your own age. I'd suggest that you see how you feel over the next couple of weeks and then make the decision yourself. If you decide that you want psychotherapy, then you can continue with Dr. Brill and her group, who are very good. Dr. Hauptman and I could probably work something out. If you want to stay within the HMO program, and if I think psychotherapy is needed, then I may also be able to provide some psychotherapy. But first let's get rid of this depression that is making you feel so miserable. Then you'll be in a much better position to decide if you need further help."

✴   ✴   ✴

Jim and Mary were fortunate. Their waking nightmare came rapidly to an end at this point. Over the next few weeks Jim steadily improved. He began sleeping better, and finally very well. He got his appetite back and returned to his normal weight. In three weeks he felt like his old self again. It was absolutely amazing, given how lousy he had been feeling only a month earlier. He was on medication, of course, which he thought of as his necessary and temporary crutch. He didn't like having to take it, but at least it was working. He decided to cancel the appointment with Dr. Brill, even though he liked her a lot. He just didn't feel he needed the additional help. Mary, his mom and dad, and Prof. Vogel seemed to be giving him enough psychological support. He finished his classes, applied for jobs, sailed through his interviews, and got five or six good offers. He chose a dream position in marketing with 3M in Minneapolis, close enough to his family in western Illinois so that his folks would be able to see their grandchildren often. He tapered off his medications after six months.

Five years later, he still had not had another episode of depression.

CHAPTER 3

BROKEN BRAINS, TROUBLED MINDS
Being Blinded by False Dichotomies

O chestnut tree, great rooted blossomer,
Are you the leaf, the blossom or the bole?
O body swayed to music, O brightening glance,
How can we know the dancer from the dance?
—William Butler Yeats
Among School Children

As we mentally navigate through life and try to figure out where we have been, where we are now, and where we are going, we use two very basic approaches. One is analysis, and the other is synthesis.

When we analyze, we quite literally break things down into their parts (ana = down, apart; lysis = break, destroy). The old poet Yeats, wandering through a schoolroom filled with young children and pondering their past and future and that of all human beings, ends his musings on the continuities and discontinuities of life by analyzing a simple chestnut tree. Where does it begin and end? Which part is its essence? Is it the trunk (bole) from which it grows, and from which the tree will renew itself each year in the rebirth of spring? Is it the flower, fragrant and beautiful but also ephemeral, which produces a seed and then dies? Or is it the leaf, the green umbrella that breathes for the tree and gives it energy to sustain its life? Obviously, it is not one of these, but all of them.

Analysis is a powerful tool. We human beings are probably the only creatures who can consciously analyze in this way. We use analysis to see structure and components in things. We can break the continuity of time into pieces and create concepts such as past, present, and future. We can define interfaces between ourselves and others and develop ideas such as family, tribe, and nation. We can divide the face of the earth or the universe into maps with boundary markers and regions, and give them names like Earth or Mercury. We can create units to weigh and measure things after we have broken them down into parts, such as meters, miles, and minutes. We can assign values to the things we see and measure, based on rarity (as in gold), beauty (as in art), or utility (as in food or water).

The more we analyze, the more we feel we understand. The more we analyze, the more we feel we can control. We forget that megabytes and millimeters and millennia have no intrinsic meaning and are merely human inventions. By trying too hard to understand everything, we may understand nothing. We analyze so much and so well that we may also destroy the vital essence and meaning of things by breaking them into pieces.

Synthesis is the counterapproach. We use it far too little. Synthesis puts the parts back together again and restores the wholeness of things (syn = with, together), (thesis = concepts, ideas). It permits us to see things in a pure and healthy state, before our analytic minds have broken and destroyed them. If we can achieve synthetic thinking, which is more difficult than analytic thinking, we can perceive things as they really are. Synthesis permits us to see things free of boundaries and barriers, as they exist in the natural world, as a divine creator made them. As the poet suggests, we cannot "know the dancer from the dance," and sometimes we should appreciate the whole rather than the parts.

*   *   *

Our understanding of mental illnesses has been handicapped by our human penchant for analysis. Over the years we have invented a set of false dichotomies and arbitrary categories. For creating a simple structure to assist us in analyzing and thinking, these dichotomies may be useful. But too often they have become reified, holding us back from perceiving clearly. Some of the great false dichotomies used in discussions about mental illnesses are mind and brain, drugs and psychotherapy, and genes and environment. Our human love for analysis has fooled us into thinking that these must be "either/or," when in fact the correct answer is often "both."

Mind versus Brain

When another person complains of a symptom such as "feeling anxious," the first question that many people ask is: Is it "all in his mind" or is it "real"? This usually translates into: Is it mental or physical? . . . due to mind or body? . . . due to mind or brain? The first dichotomy that is often addressed is "body" versus "brain"—an obvious false dichotomy, since the brain is part of the body. But even this erroneous distinction still pervades some peoples' thinking. The first steps in making Jim's diagnosis consisted of ruling out a "physical cause," such as diabetes or thyroid disease. Jim's anxiety did not have any of those obvious physical explanations, and so the "body" was ruled out as a source of the symptoms, even though many of them were in his body (e.g., vomiting, weight loss). His HMO gave him a clean bill of health because a "physical cause" was not found by laboratory tests. Getting the additional help for his mind/brain, which he badly needed, was therefore difficult.

When Jim went to see Mr. Morgan, the next dichotomy was mind versus brain. This is a more treacherous and tempting false dichotomy. Mr. Morgan quickly leaped to the "physical" end of the spectrum and assured Jim that he had a "brain disease" that would best be treated with medications. Jim, Mary, Prof. Vogel, and most of the others implicitly struggled with the question of whether Jim's problems were in his mind or brain, and whether they should therefore be given differential treatment. (Fortunately, by this time no one wondered if the symptoms were "real," except perhaps the HMO.)

The distinction between mind and brain is embedded in the language that we use every day. "Brain" refers to a physical organ, while "mind" refers to an abstract concept. Because it is not palpable, mind is sometimes considered less "real." On the other hand, because it is "only physical," the brain is sometimes considered less interesting or important. But just as the dancer and the dance are indistinguishable, so too are the brain and the mind. They are two different words that refer to the same thing/activity, and neither exists without the other in living human beings. What we call the mind is the product of activity occurring in the brain at the molecular, cellular, and anatomical levels.

In fact, human beings recognized the link between the mind and the brain as early as the Neolithic period. Skulls survive from Neolithic times that have been trephined, or drilled into. We surmise that the trephining was done as a medical treatment to release evil spirits, which were believed to be causing mental illness or epilepsy. In classical times the eminent physician Hippocrates observed:

> Men ought to know that from the brain, and from the brain only, arise our pleasures, joys, laughter and jests, as well as our sorrows, pains, griefs, and fears. Through it, in particular, we think, see, hear. . . .

The false dichotomy between mind and brain is a particularly nasty one because it is used to misunderstand and mistreat people in a variety of ways. It is used to divide illnesses into "physical" or "neurological" versus "mental." The former are treated with respect, while the latter are stigmatized. The stigma translates into social and economic discrimination. People applying for jobs or college often have to complete forms that ask if they have ever had a mental illness. The question is on applications for a license to drive a car, obtain health insurance, or join the military. Of course, information about health may be relevant. The problems are the false distinction between mental and other illnesses, and the potential for excluding people who have perfectly harmless mental illnesses. For example, I once heard a medical school dean propose that applicants should not only be denied admission if they had suffered from depression, but that they should also be turned down if they had a family history of depression. The openings were too precious to waste! Most health insurance places "caps" on coverage for mental illness or makes it difficult to obtain care, as Mary and Jim discovered. The distinction between mind and brain is the main rationale used to deny parity for health care for mental illnesses.

Given this stigma and discrimination, there should be little wonder that patients with mental illnesses want to stress that they have brain diseases. It legitimizes their suffering and enhances the likelihood that they will be treated with respect. As neuroscience advances, the evidence supporting this position grows steadily. Schizophrenia, manic-depressive illness (bipolar and unipolar mood disorders), the various dementias, and many of the anxiety disorders have been studied with the modern tools of neuroscience and molecular biology and have been shown to have some "physical" basis or component. The great risk is pushing this argument too far, however, and making the proverbial mistake of "throwing out the baby with the bath water." The risk is in forgetting about mind and focusing solely on brain.

Mental illnesses, although they have a physical component, are nonetheless also *mental*. Almost by definition, they are illnesses that affect the mind, however much they arise from the brain. They are illnesses that affect the most human abilities—remembering, conversing, thinking, feeling, interpreting information, reading social situations, bouncing back from a stress or rebuff. To ignore or minimize the importance of the *mind* in mental illnesses is almost to ensure that they will be misunderstood and mistreated from yet a different direction. If seen only as brain diseases, they are at serious risk for being dehumanized. Patients with mental illnesses risk being seen as generic "cases" that can be "managed" in a "one-size-fits-all" manner. Bodies can be treated fairly generically, but minds are individual and unique. Each person with a mental illness needs to have his or her symptoms evaluated and treated within the context of his or her own personal, social, economic, emotional, and intellectual resources.

In short, as we think about mental illnesses, we must synthesize and understand them as mind/brain disorders. We will be mindless if we address only the brain, and brainless if we address only the mind.

Drugs versus Psychotherapy

The tendency to divide mental illnesses into "brain diseases" versus "mind diseases" (sometimes expressed as "biological" versus "psychological") leads naturally to another false dichotomy. This false dichotomy often creates confusion in discussions about how mental illnesses should be treated.

The reasoning runs like this. If these illnesses are mental, then one should treat the mind with psychotherapy. If they are physical or "in the brain," then one should use physical treatments that affect the brain, such as medications.

The first problem with this line of reasoning is its basic premise—the mind versus brain dichotomy. But the mind versus brain (or mental versus physical, or biological versus psychological) distinction is very entrenched. Consequently, simply pointing out the false dichotomy is probably not sufficient to dissuade people from the foolishness of the drugs versus psychotherapy distinction.

A second problem with this particular false dichotomy is that it introduces an unfortunate oversimplification into the care of mental illnesses that we do not use for other illnesses. If a person suffers from diabetes, we do not ask: Should this person only take drugs, or should he also be counseled and supported to maintain a healthy lifestyle? Diabetes is treated with medications, but also with diet and exercise and (ideally) psychological support for the adjustments in lifestyle that are required. Psychosocial management is especially important in young people with diabetes, who have to inject insulin regularly, admit that they have a physical problem to their friends, and consume frequent and regular meals with a minimum of the junk food that their buddies can joyously gulp down. People who have coronary artery disease, or even a family history of it, are treated with psychosocial management—diet, exercise, and avoidance of stress—in addition to medications as needed. The first-line treatment for high blood pressure is learning to consume a low-salt diet. Medications are used when this does not work. We learn muscle strengthening and relaxation techniques to reduce back pain, taking analgesics only as needed. There are endless examples of combined therapies for other illnesses, and we don't pause for even a moment to raise false dichotomies when doctors suggest combinations of medications and other kinds of treatment. Why should we ask whether mental illnesses should be treated with medications *versus* psychotherapy?

A third problem with this line of reasoning is that it fails to recognize that drugs affect the mind, and psychotherapy affects the brain. Again, these points are so obvious that they should scarcely need to be made, and yet the dichotomy still seems to persist.

That drugs affect the mind (as well as the brain) *is* something that most people realize. If they don't, they have government-mandated notes on prescription bottles for allergy or pain medications, which point out that "these drugs may be sedating. Operating a car or machinery may be dangerous." "Mind functions" are things like attention, arousal, alertness, memory, and mood. Drugs of many kinds, used for mental illnesses and for other disorders (e.g., antihistamines for colds), affect these mental functions. Illegal drugs such as marijuana and speed are used precisely because they affect mind functions in ways that people find appealing. There are very few illegal drugs available that affect much of anything else.

So why do people make the drugs = brain (not mind) mistake at all? Why do we continue to have concerns about treating mental illnesses with medications? Why did Dr. Brill tell Jim that she thought the medications were "just a crutch," and that he needed to get to the root of his illness with psychotherapy? The fact that some "mind drugs" are illegal and that prescription drugs are sometimes abused probably contributes to the aura of taboo about using medications to treat mental illnesses. Paradoxically, the mistake may also arise because these drugs work "too well." As later chapters indicate, one of the major achievements of the past 50 years has been the development of medications that are effective in reducing or eliminating symptoms for three major groups of mental illnesses: schizophrenia, mood disorders, and anxiety disorders. A puritanical masochism may make us feel that human suffering should not be reduced so quickly—at least for diseases of the mind. We ought to rejoice that, as in the case of both Jim and his mother, the antidepressant medications worked miracles nearly as remarkable as those of antibiotics for infections. But instead we fret and worry that mental anguish should not be relieved so easily.

Because we revere the mind so much, we have the sense that its illnesses must be treated with "deeper" techniques, such as psychotherapy, which work *directly* on the mind and its functions, such as feelings and memories. There is nothing wrong with using psychotherapy to treat the mind, as long as the very real usefulness of medications is not devalued. Many people will benefit from support and counseling, and some will be helped greatly by extensive psychotherapy. The problem is not with using it, but with the failure to recognize that its effects are both mental and physical. Psychotherapy acts on both the mind and the brain. In fact, as we understand more and more about how the brain works and how it changes in response to experience, we are steadily recognizing that the effectiveness of psychotherapy is a consequence of the ability to affect "mind functions" such as emotion and memory by affecting "brain functions" such as the connections and communications between nerve cells.

Chapter 4 contains an extensive discussion of the concept of "brain plasticity." This concept, fundamental to a modern understanding of both mind and brain, stresses that our brains are in a state of constant dynamic change, which occurs as a consequence of the impact of experience on our mental functions and states. Psychotherapy affects specific mind/brain systems, such as learning, memory, and emotion. In the present golden age of neuroscience, we are learning how we learn—how the brain changes its structure and chemistry in order to store information or retrieve it, to respond to emotionally charged events, and to continually adapt to an ever-changing world. The essence of psychotherapy is to help people make changes in their feelings, thoughts, and behavior. This appears to occur through the multiplicity of psychotherapeutic techniques that are currently available—cognitive, psychodynamic, psychosocial, behavioral. But to the extent that these techniques are effective, they lead to changes in a "plastic brain," which learns new ways to respond and adapt that are then translated into changes in how the person feels, thinks, and behaves. Psychotherapy, sometimes denigrated as "just talk," is in its own way as "biological" as the use of drugs.

When psychoactive medications first became available to treat psychosis, depression, and anxiety in the 1950s, many people (including some psychiatrists) argued that psychotherapy might be "better" because it got to the "root of the problem." Because the medications were so obviously effective, this line of reasoning became increasingly difficult to defend. By the 1980s nearly all psychiatrists and most psychologists concurred that medications have a place in the treatment of many mental illnesses. The issues for 2001 and later are related to finding the right balance of medications and psychotherapy for each specific disorder and each specific person, and to continuing to develop *even better* medications and psychotherapies that closely match the needs of each patient and each illness.

When medications and psychotherapy are polarized and pitted against one another, unfortunate consequences result. Patients are confronted with conflicting advice and left in a state of confusion and doubt. They may be given only medications when they also need psychotherapy. They may be denied medications and given only psychotherapy. The best advice is not "either-or," but "either-or-both, as needed." Whatever the treatment, the most basic mechanisms are the same. Both affect mind functions by changing brain functions.

Genes versus Environment

When people discuss the causes of mental illnesses, they frequently fall into another false dichotomy. Are they due to genes or environment? This dichotomy gets tied in to the others. If something is caused by genes, it is physical, biological, a "real illness," less stigmatized, and "something the person can't really help." If something is caused by environmental factors, then it is mental, psychological, less real, easily criticized, and perhaps due to a moral weakness or an inability to cope with problems.

The first and most basic problem with this dichotomy is that very few things in human life—normal or abnormal—are due solely to genes, or solely to nongenetic factors. As we shall see in chapters 4 and 5, which discuss genes and the mind/brain in detail, even identical twins who possess the same genes differ from one another in many ways because of the shaping effects of nongenetic factors. There are a few relatively rare illnesses that are almost totally genetic and uninfluenced by environmental factors. These are most often fatal illnesses that arise in childhood, such as Tay Sachs disease, a disorder that is autosomal dominant and usually kills its victims within the first several years of life. However, the course of other disorders that are highly genetic, such as cystic fibrosis (a lung disease), is substantially moderated by nongenetic factors such as medications and lifestyle adaptations. At this time cystic fibrosis is usually a fatal disease, but its victims now live longer and better lives, thanks to environmental interventions.

Most diseases, including almost all mental illnesses, are caused by a mixture of genetic and environmental factors. Physicians and scientists call such illnesses "multifactorial." Many common illnesses, such as coronary artery disease, are familial and partly genetic. Having a father who died of a heart attack in his fifties or sixties increases his child's risk of suffering the same misfortune. But coronary artery disease is also caused partly by environmental factors, such as eating a high-cholesterol diet. We all know that rates are lower in places such as Japan, where fish replaces meat and milk as a source of protein. We also know that stress increases the risk for coronary artery disease and heart attacks, and that regular exercise reduces the risk. Some illnesses are heavily environmental, such as the various lung diseases (cancer and emphysema) caused by smoking. Other illnesses may be almost totally "environmental." Breaking a leg when hit by a careless driver who runs a red light is an example. The broken bone is a physical problem, caused by an environmental event over which one has no control, and a very real illness.

The equation between genetic and "real" or "something the person can't help" obviously leads to misperceptions and confused thinking about the nature of disease and the role of moral responsibility. A person who has a family history of heart disease is not exactly behaving in a morally responsible way when he munches on fries and a Big Mac, but usually only his wife or kids are likely to point this out. We do not berate the person dying of lung cancer for having been a smoker . . . or usually even mentally criticize her for "moral weakness." We certainly do not criticize the unfortunate victim of the careless motorist . . . nor even the person who has a car or skiing or plane accident and (perhaps) took a few too many risks.

Mental illnesses are multifactorial illnesses as well. Only one, Huntington's disease, is almost totally genetic. Some, such as schizophrenia and autism, are partly caused by genes, but also by nongenetic factors. Others, such as mood and anxiety disorders, may have a greater balance of environmental components because stress and life experiences can play an important role in their development. We often summarize the relative contribution of genes and nongenetic factors by looking at how often identical twins are "concordant" (i.e., manifest the same condition) in comparison with nonidentical twins. This is because identical twins have almost identical genes, while nonidentical twins share only 50% of their genes. The relationship is summarized by comparing the concordance rates in identical and nonidentical twins. Table 3–1 shows this relationship for several common illnesses, both mental and nonmental.

TABLE 3–1
Concordance Rates in Twins

Type of Illness	Identical Twins	Nonidentical Twins
Autism	60%	5%
Schizophrenia, bipolar disorder,		
Coronary artery disease	40%	10%
Depression	50%	15%
Breast cancer	30%	10%

Strikingly, autism is the most "genetic" illness in the table, with a con-cordance rate of 60% in identical twins and 5% in nonidentical twins. Schizophrenia, bipolar (manic-depressive) illness, and coronary artery dis-ease are very similar, with rates around 40% in identical and 10% in non-identical twins. Depression has relatively higher rates in both (50% versus 15%). Breast cancer is around 30% in identical twins and 10% in noniden-tical twins. As the table reveals, genes make some contribution to causing all these common illnesses, but they are far from telling the whole story, since people who have identical genes do not have an identical tendency to develop a specific disease. Nongenetic factors seem to play an even more important role in all of them.

Many people create a false dichotomy between genes and environ-ment because they do not understand the nature and behavior of genes, which are described in detail in chapter 5. The problem begins, as is so often the case, with the language that we use. We talk about genes versus environment when we should be saying "genetic" and "nongenetic" fac-tors, and not using the word "versus" at all. The word "environment" is too vague. It means everything ranging from things that influence the growth of the fetus in the uterus (such as the mother's nutrition or use of cigarettes), to infections with viruses, to mental stimulation such as occurs (we hope) when people attend college, to painful personal experiences such as the loss of a loved one. It is best to look at this topic from the point of view of the cell and the gene. The gene contains a code instruct-ing the body how to perform its various functions, beginning with cell division and continuing on to a complicated interaction among all the organs of the body, and between the body and the world around it. From the point of view of the gene, *everything except it* is nongenetic, and its point of view is probably the correct one. The gene is influenced by things as close by as the temperature or nutrition inside the cell, and by things as far away as the bottle of beer that its human owner might be drinking.

Further, like the brain, genes are "plastic." Genes are influenced by "the environment," and their behavior is changed by it. The genetic code is not the rigid dictator many people think it to be. Genes do not contain a static and unchanging set of instructions. Rather, they modify their influ-ences on the body in response to their own "environmental" or "non-genetic" experiences. So to speak, they notice that the cell's food supply is getting a bit too low, and so they "turn on" instructions to produce more food. They notice that a medication has entered the system and is shut-ting down the nerve receptors on the cell membrane, and so they turn on orders to tune up the "listening system" by growing additional receptors so the cell can overcome this environmental intrusion and break through the blockade. They notice that some neighboring cells are becoming too numerous and create instructions to destroy them. They even notice that they have made mistakes in duplicating themselves and fix them!

If it is difficult to recognize that the brain is plastic—that experiences in the mind cause changes in the brain—it is even more difficult to recognize that genes are also plastic and can be influenced by their environment. But they are. Gene plasticity frees us from genetic determinism. It offers us the hope that we will some day learn ways to modify the genetic instructions that may lead to diseases such as mental and other illnesses.

Understanding Mental Illnesses Using a Synthetic Model

Abandoning these false dichotomies gives us a much better grasp on how life actually works, if we can handle the more complex way of thinking that arises when the world consists of continuities without arbitrary dividing boundaries. It is a more richly textured and colored world than the black-and-white dichotomies of mind versus brain or genes versus environment. It is also more true.

If we think about mental illnesses using synthesis rather than analysis, the picture is something like the model shown in Figure 3–1. (Note: even this model has to use some analytic tools to make its point, by using divisive terms such as mind and brain.)

Why does a person develop a mental illness? Take Jim, for example. Why did he become depressed? What determined how it affected him? Whether he got a bad case or a mild one? How he should be treated?

Mental illnesses usually arise from multiple interacting causes, many of which we do not yet fully understand. They include influences that range from genes through many nongenetic factors. These are often *interacting* factors. They influence one another, and their combined interacting effects influence the brain. Jim probably carried a genetic predisposition to depression, transmitted from his mother. It was a predisposition to a more severe "endogenous" form, which is not obviously brought on by life experiences. Jim had no obvious triggers, because things were going well for him. But on the other hand, he did have some more subtle ones. His body (including his brain) had been stressed by a case of the flu. Perhaps a virus had some impact on the way his brain cells were functioning. (Viruses have a special affection for nerve cells.) Perhaps his low-grade fever changed the nerve-cell environment and affected gene expression. Perhaps the flu, by bringing a general malaise and restless sleep, affected

FIGURE 3-1

A Synthetic Model for the Development of Mental Illnesses

Causes

Multiple Interacting Factors Such As

Genes	Gene Expression	Viruses	Toxins	Nutrition	Birth Injury	Personal Experiences
↓	↓	↓	↓	↓	↓	↓

Brain Structure and Function

(e.g., brain development and degeneration, plastic changes in response to experience, brain chemistry, changes in response to medications, changes in response to psychotherapy)

↓↑

Mind Functions

(e.g., memory, emotion, language, attention, arousal, consciousness)

↓↑

The Unique Person in a Specific Social World

(i.e., individual behavior and response in a specific personal and social environment)

↓↑

A Specific Mental Illness

(e.g., schizophrenia, mood disorders, dementias, anxiety disorders)

stress hormones, which in turn affected his brain. Perhaps he had become run down from working two jobs, and this affected his stress hormones and ultimately his brain.

Whatever the combination of factors, whether "genetic" or "non-genetic," they interacted with one another. Some were "physical," such as the flu, and some "mental," such as life experiences. They affected his brain, and the activity and functions of his brain, known as his mind. His response to these changes was also affected by what we might call "Jim Himself," (although obviously most of Jim Himself is embodied in his brain/mind). Jim, at age 34, had already become a unique person, a composite of mental strengths and weaknesses, who lived in his own unique social world. Jim was fortunate, as many people noted, because it was basically a pleasant world. Jim was surrounded by people who liked or loved him and who wanted to help him as soon as he asked, and even before he asked. He also had adequate financial resources and access to health care. He himself was accustomed to making the best of things, and to making the best of himself, as his own natural adaptive response. So when his mind/brain became disturbed and depressed, his "case" was not as bad as it might have been, and treatment was ultimately effective.

The medications that he took created a cascading and interactive reversal of his depression. They created changes in the chemical functions of his brain that also changed the functions of his mind. His ability to concentrate improved, his depressed mood lifted, and his state of hyper-vigilance abated. As these effects were occurring, he also was influenced by both professional and personal psychosocial and supportive therapy. His mother, Mary, and Dr. Walker ensured that his suicidal thoughts did not lead down a dangerous path. Dr. Walker gave him brief supportive therapy even during his first visit. As anyone who has been a patient knows, a few encouraging (or discouraging) sentences can have a major impact on how we perceive ourselves and our situation. Jim did not choose (or apparently need) extensive psychotherapy in addition to med-ications. Had he not been surrounded by supportive friends and family, and had he not already learned a variety of adaptive ways to deal with problems, he might have needed more psychotherapy.

Jim is a unique person who had his own personal experience with a specific mental illness. Each person who suffers from a mental illness is equally unique, and each has a slightly different mixture of things going on and therefore a slightly different story. Some stories are much sadder.

The best hope that all of us have to help Jim, and others like him, is to keep reminding ourselves that "nothing is ever one thing." If scientists fail to recognize the complexity of causes, they will never find all of them, and perhaps none of them. If society assumes that mind and brain are sep-arate and that mental disorders are "different" or "bad," misunderstanding, mistreatment, and stigma will persist. If we see people as categories (e.g., schizophrenics, depressives), we will not see them as people.

PART II

MIND
MEETS
MOLECULE

CHAPTER 4

THE BRAIN
The Mind's
Dynamic Orchestra

My brain is my second-favorite organ.
—Woody Allen, in *Sleeper*

The brain forms the essence of what defines us as human beings. To understand its structure and its workings is to understand ourselves. The normal healthy brain is a complex, miraculous, and ingeniously created organ. It permits us to achieve the wonders of music, art, science, architecture, engineering, political organization, and economic structure. Each of us has been endowed with a unique brain with particular capacities that we can either enhance through learning and productive work or waste through intellectual inactivity and unhealthy living habits. We can use our brains to good ends, as when we build bridges or heal illnesses, or we can use our brains to harm or destroy one another in many innovative ways.

The brain can also become "broken" in many ways that lead to the disorders known as mental illnesses. Most of the factors that produce a "broken brain" are outside the control of the individual who develops the illness, although he or she does have "free will" in deciding how to cope with its consequences. In order to understand how disturbances in the brain lead to disturbances in the mind, we need a rudimentary understanding of how the brain is organized and how it works to produce thoughts, emotions, and personal identity.

Basics about the Brain

The human brain is an amazing piece of engineering that allows us to process billions of bits of information within a compact, powerful, continuously changing "living computer" that we carry around on our shoulders our entire lives. It weighs just over two pounds. We each get issued only one. We therefore need to understand its components, how it works, and how we can take good care of it—continually updating its software and keeping its system running smoothly with a minimum of glitches and incompatibilities.

The brain is composed of three types of tissue: gray matter, white matter, and cerebrospinal fluid (CSF). Figure 4–1 shows the differences between these three kinds of brain tissue, as visualized by a routine mag-

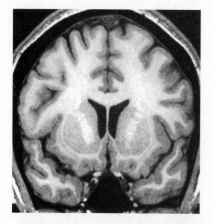

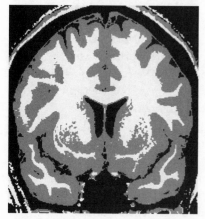

Figure 4–1: Types of Tissue in the Human Brain. The upper figure shows a conventional MR scan, while the lower figure shows the tissue composition of the brain after it has been classified as gray matter, white matter, and CSF.

netic resonance (MR) brain scan (left), and as visualized when a computer program has been used to classify the tissue into these three types (below, left).

Gray matter is called "gray" because it looks relatively dark in anatomical brain specimens (postmortem tissue). Its dark color is produced by densely packed cell bodies of nerve cells (neurons). Figure 4–2 shows a schematic drawing of a neuron.

The cell bodies of neurons perform the basic "command functions" within our brains. We all have approximately 10^{11} of these "command cells" in our "big brain" (the cerebrum), at least during our prime. We slowly lose them as part of the normal aging process, and people who develop neurodegenerative disorders such as Alzheimer's or Parkinson's disease lose them more rapidly. Nerve cell bodies contain the nucleus of the cell, which is composed of DNA (deoxyribonucleic acid, the source of the genetic code, described more fully in chapter 5). They also contain mitochondria, which are the "power packs" of the cell that give it the energy to do its metabolic work. And they contain other machinery that permits the cells to make enzymes, other proteins, and some neurotransmitters, which all perform communication and "housekeeping" functions in the brain—creating the chemical messengers that permit nerve cells to communicate with one another, maintaining the cell's energy levels, or eliminating waste products. The nerve cell bodies are surrounded by smaller, highly branched projections ("dendrites," or tree branches) that permit neurons to talk to one another over relatively short distances and receive many messages simultaneously.

Nerve cell bodies are highly concentrated on the surface of the brain, giving it the appearance of being covered with bark, and so this outer surface is called the cerebral cortex (bark of the brain). Small concentrated islands of nerve cells also occur deep inside the brain, and these are called subcortical regions (regions below the cortex).

White matter is called "white" because it looks much lighter than gray matter in both postmortem tissue and most MR images. Neurons are connected to one another by long "wires" that project out of them, called axons, which permit individual neurons to send messages back and forth to one another across relatively long distances. Although the neuronal cell body may perform the majority of the cell's executive functions, most of the cell's total volume is in the axons—around 90%. The axons are covered with a fatty insulation, called myelin, which makes them appear white. The thickness of the insulation varies depending on the function of the cell. Cells that have small amounts of insulation (e.g., cells that sense pain) send messages faster than those that have heavy insulation. While most brain illnesses strike at the command center (the cell body), a few work their destruction on the white matter. These illnesses, called demyelinating diseases, include multiple sclerosis (MS) and amyotrophic lateral sclerosis (ALS, or Lou Gehrig's disease). White matter diseases effectively "cut the wires" that permit nerve cells to talk to one another.

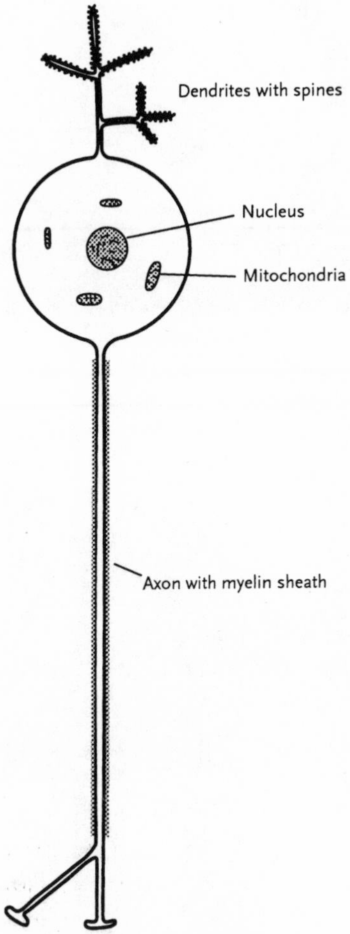

The brain is bathed inside and out by CSF, a fluid that contains nutrients and byproducts of brain activity. The regions inside the brain that contain CSF are called ventricles. Healthy young brains have relatively small amounts of CSF on the surface and in the ventricles. For many years monitoring the amount of CSF present inside or on the surface has been used to determine whether brain injury or degeneration has

Figure 4–2: The Structure of a Neuron

occurred. If the ventricles grow larger or the CSF increases on the surface of the brain, doctors have a clue that brain tissue has decreased and that something has probably gone wrong, since the increased CSF replaces the missing brain tissue.

The surface of the human brain looks very wrinkled—perhaps reflecting the phylogenetic maturity or "advanced age" of the human mind. A picture of the surface of the human brain, reconstructed from an MR scan, is shown in Figure 4–3. It is covered with ridges and furrows called gyri and sulci. In comparison with other earthly creatures, we human

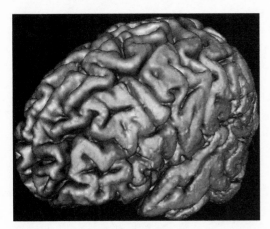

Figure 4–3: The Surface of a Human Brain

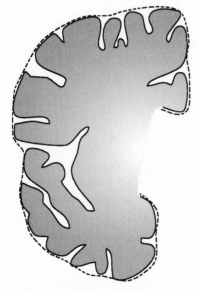

Figure 4–4: Measuring the Complexity of the Brain Surface: The Gyrification Index

beings have brains that are highly "gyrified." The level of gyrification has been compared across various species. Rabbits, for example, have brains that are nearly smooth. Human beings have brains that are crumpled and folded. Our wrinkled brains are nature's solution to an important problem in mechanical engineering. How can we pack enough neurons inside our skulls to survive in an increasingly complex world and still be able to carry our heads around on our shoulders? If we increased our gray matter without some compensatory mechanism, our heads would get so big that we would simply topple over. Instead, our heads stayed the same size, and our brains crumpled and folded, producing gyri and sulci. A German neuroscientist, Karl Zilles, has developed a method for measuring this crumpling that is known as the gyrification index or GI. Shown in Figure 4–4, the GI is simply the ratio of the distance around the entire surface of the brain (i.e., dipping deep into the sulci) to the distance around the outer surface. The normal adult human brain has a GI of around 2.6 (a GI of 1.0 means the brain is totally smooth).

Measuring the GI has been a powerful tool for understanding both brain evolution and neurodevelopment. Not only did the GI increase as human beings evolved and differentiated from old-world monkeys and chimpanzees, but (as the saying goes) "ontogeny recapitulates phylogeny." (The development of the individual reflects or repeats the development of the species.) Measurement of the GI has demonstrated that human brain development proceeds very slowly. It has scarcely begun at the time of birth, and it continues on through childhood and adolescence and appears to become complete some time in the early to mid-twenties. The human brain is essentially smooth until around the sixth month of fetal life, with a GI of around 1.06. At this time the major sulci and gyri begin to form, as the complexity of the connections between nerve cells in the cerebral cortex increases.

By the time of birth, the human brain is still relatively immature, with a GI of around 2.15. This first neuroanatomic fact is intuitively obvious to anyone who observes the nearly total helplessness of the newborn human infant, which would be unable to survive without adult help for at least four or five years after birth. The GI continues to change, as the human brain continues to mature, until it finally reaches normal adult levels in the early twenties. This second neuroanatomic fact, although less obvious, is also evident to anyone who contemplates the behavior of adolescents and the steady changes that occur in impulse control and social responsibility as adolescents mature into young adults in their early to mid-twenties.

The evolutionary facts send a revolutionary social message.

Brain development is an ongoing process in human beings until the early twenties. Both bad and good influences may affect whether the human brain grows poorly or well in our children and in future generations.

How the Brain Grows: A Miraculous Process

Brain growth (neurodevelopment) is a miraculous process. Somehow the mass of nerve cells and their associated axons and dendrites must connect themselves to one another so that the correct messages get received, sent to the right places, stored or thrown out, and reflected on or acted on. The stages of neurodevelopment are summarized in Table 4–1.

TABLE 4–1
Stages of Neurodevelopment

Neuron formation
Neuronal migration
Proliferation of dendrites and spines
Synaptogenesis
Myelination
Pruning
Apoptosis

The process begins a few months after the fetus is conceived. The programs coded in our DNA (which is the same in all cells in the body) begin to send instructions that cause some cells to differentiate into nerve cells (while others are differentiating into liver or heart cells in other parts of the little fetal body). Tiny germ cells of neurons form in the middle of the brain. After a sufficient number accumulate, they begin a journey more magical and mysterious than the seasonal flights of birds. The mechanism for this process, known as neuronal migration, was identified by Pasko Rakic of Yale University. The nerve cells begin to move outward, guided in their journey by pathfinder cells known as glia ("glue"), ultimately arriving in a new territory that they create by their colonization: They form what will become the cerebral cortex and the various subcortical gray matter regions.

After they migrate, they begin to form connections by sending out axons ("axis"), the long wires of the brain. For example, the brain quickly divides into two sides, the right and left cerebral hemispheres. These two halves of the brain send axons back and forth so that the two hemispheres can talk to one another, creating the corpus callosum (firm body). The cells line up and arrange themselves in orderly layered patterns according to the role that they will play in the brain's overall activities. In many regions they form as many as six layers. Each neuron sprouts branches, or dendrites, and the dendrites further expand their communication capacities by sprouting spines.

Initially, more cells and connections are formed than will ultimately be needed, much as a gardener plants an excess of seeds. As the brain matures throughout childhood and adolescence, the excess is steadily removed in order to create the correct balance of connections, a process known as programmed cell death (apoptosis, or turning away) and pruning (trimming back overgrowth of dendrites and spines). The communication points that permit many cells to talk to one another simultaneously—the synapses ("fasten together")—also form during fetal life and continue to mature and change, in a process known as synaptogenesis. Many different kinds of chemical messengers are created to travel across the synapses, such as dopamine, serotonin, norepinephrine, or glutamate. These must also interact with one another in a way that will create the right balance to achieve correct, rapid, and efficient communication.

Achieving the right connections on both the large scale of axonal connections and the smaller scales of the spines, synapses, and chemical messengers is an enormously complicated process. Further, it is an ongoing process that continues into young adulthood. How does the brain know how to grow correctly? How does it update its plan? What kinds of things interfere with this grand and magical process of human brain development?

The genetic blueprints coded in our DNA provide the basic instructions, in ways that we are still learning to understand. But we understand already that the genetic blueprints are shaped and modified by the experiences that the mind/brain encounters as each individual human being navigates his or her way through life. This is a very important concept, known as "brain plasticity."

How the Brain Teaches Itself to Learn: "Brain Plasticity"

The concept of brain plasticity has nothing to do with plastics and other novel chemical polymers that are pervasive in our contemporary world. Instead, this concept emphasizes that the brain is dynamic: It changes rapidly, from moment to moment, in response to challenges from the world around it. Many of these changes are permanently coded or stored for later use. The concept of brain plasticity is based on the recognition that brain development is shaped in each individual by both physical and psychological experiences—and, indeed, that the distinction between physical and psychological may be quite arbitrary.

The notion of plasticity was introduced by a Canadian psychologist, Donald Hebb, back in 1949. He argued that our ability to change our brains by learning new information occurs because of changes that occur at the level of the nerve cells. His notion was that the brain remodels itself by changing the connections at the level of the synapse. If several nerve cells receive a stimulus at the same time that causes them to "fire" (i.e., produce what neuroscientists call an "action potential"), they begin to share more and more synaptic connections. You might anthropomorphize this by thinking of the nerve cells as a band of friends who have shared experiences and gradually bond in a kind of neuronal buddy system.

This idea is often referred to as Hebbian plasticity, and it is expressed in the motto "neurons that fire together wire together." Sometimes the bands of neurons that are created through these shared experiences are called "neuronal assemblies." Hebbian plasticity was an interesting concept, but neuroscientists have only been able to explain its how's and why's during the last few years.

We now know that the new wiring is created through a mechanism called long-term potentiation (LTP). Our understanding of long-term potentiation, achieved primarily by studying nerve cells in the hippocampus, has given us the explanation as to how our brains change at the cellular and molecular level when learning occurs—exactly how it is that neurons that fire together wire together. LTP is the process by which the size of a neuronal response increases after stimulation. The increase in response ("potentiation") is relatively long lasting ("long-term"). This increase in neuronal response is one important mechanism by which long-term changes such as learning occur.

During recent years we have learned about several important properties of LTP. One is that the potentiation is usually relatively specific. That is, when Cell A talks to Cell B, "dendrite to dendrite," the potentiation occurs only on those particular dendrites, not the whole of both nerve cells. The specificity of LTP means that the transfer of messages between cells can be quite fine-tuned and detailed, rather than being a grossly generalized process. This explains why our brains are able to connect, record, and retain very specific tiny bits of information. Another important feature of LTP is that it occurs cooperatively. That is, if Cell A and Cell B both get a message from Cell C at the same time, the potentiation of both is increased, and it also becomes tied together or associated. We now think that this associativity forms the physiological basis for Hebbian plasticity. Finally, we also understand exactly how this happens at the level of molecules and neurotransmitters. Glutamate (an amino acid neurotransmitter) facilitates the development of LTP. Glutamate communicates with two different receptors, which are called the AMPA and the NMDA receptors. Many studies have examined long-term potentiation or LTP in the hippocampus, one of our major memory regions, and have observed that LTP is enhanced by glutamate activation of NMDA receptors. Importantly, they appear to show the principles of associativity when glutamate activates the NMDA receptor, thereby explaining Hebbian plasticity at the molecular level.

Early work in neuroscience by David Hubel and Torsten Wiesel added to our understanding of plasticity from another perspective. They examined how our experiences in the environment affect brain development—thereby linking the outer world to inner biology. Their work has introduced the important concepts of critical periods and activity-dependent learning. They were awarded a Nobel Prize for their discoveries. They studied the development of the vision center in the brain by determining what happened to brain growth when the developing nerve cells failed to achieve guidance from the environment as an adjunct to the instructions provided by the genetic plan. Studying the effects of experience on brain development in cats and monkeys, they found that when one eye was covered or removed in very young animals, normal cellular alignment did not occur in the visual centers of the brain that would normally receive input from that eye. Organized columns of cells (called ocular dominance columns) did not line up correctly, and the "good eye" adapted and took over the functions of the "bad eye." There is a prime time when the vision center forms, called a critical period by neuroscientists. If the eye is later unpatched and given input after this critical period has passed, it is too late: the ocular dominance columns can no longer form correctly.

Thus the concept of brain plasticity has two important components: critical periods and activity-dependent changes. The idea of critical periods teaches us that for some aspects of brain development, timing of environmental input is crucial, and that important abilities will be lost or diminished if stimulation does not occur at the right time. This applies not only for vision, but also for other functions such as auditory discrimination or the sense of touch. It may also apply for "higher" functions such as language abilities. For example, it is difficult for many people to learn to speak a foreign language fluently and with a "good accent" if they are not exposed to it until after childhood. Americans cannot say the French "r," and Japanese cannot say the English "r." Yet they probably could learn to discriminate and speak these sounds readily if they were exposed to them in infancy and childhood. Early exposure to other languages is no doubt one of the reasons that people in countries that begin language training early, such as Scandinavia or the Netherlands, speak so many languages so well. These observations have important implications for education that often pass unrecognized, both by parents and by public and private educational systems.

The idea of activity-dependent learning teaches us that exposure to either psychological or biological environmental influences causes changes in the brain. At the grossest level, cellular alignment may be affected. At finer levels, spines may expand on dendrites, synapses may form, or concentrations of chemical messengers may increase or decrease. Sometimes single powerful experiences affect our brains for life. A ten-year-old child who watches her father die from cancer will code that memory in her brain irreversibly, and no amount of conscious effort can erase it. But smaller and subtler psychological influences also change our brains. People who learn to hit a tennis ball well through practice change their brains. People who learn to listen keenly to musical pitches and patterns change their brains. People reading this book will have slightly different brains as they continue to read it and absorb its contents and implications. Each of us is a different person because we are composites of the

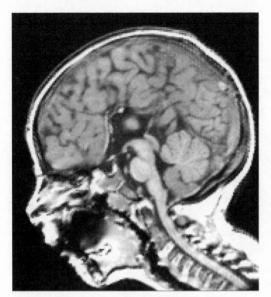

Figure 4–5: Agenesis of the Corpus Callosum in a Child with Fetal Alcohol Syndrome

effects of different arrays of psychological and physical experiences. Again, the implications are powerful. We *can* change who and what we are by what we see, hear, say, and do. It is important to choose the right activities for our brains to be well trained. This principle applies not only to childhood, but also to adulthood and even to the aging process.

These principles also have important implications for psychiatry and mental illness. For example, brain plasticity explains how and why psychiatric treatments that are not "biological," the various types of psychotherapy, can be effective for relieving the symptoms of illnesses such as depression or anxiety. These treatments, which we tend to think of in a false polarity between the physical and psychological (or brain and mind), help people reframe their emotional and cognitive responses and approaches. This reframing can only occur, however, as a consequence of biological processes in the brain—a form of activity-dependent learning.

These principles also explain some of the ways that brains can be injured and can produce a variety of mental illnesses. For example, exposure of the fetal brain to large amounts of alcohol during critical periods in brain development can produce a problem known as fetal alcohol syndrome (FAS). Children with this syndrome have impaired growth at the time of birth, an unusual facial configuration, and learning disabilities or mild mental retardation. Brain imaging studies have recently shown that many children with FAS also have neurodevelopmental brain anomalies caused by exposure to large amounts of alcohol. The most severe abnormality is failure of the nerve cells in the two hemispheres to send out their axons and connect the two halves of the brain—an abnormality known as agenesis of the corpus callosum. (See Figure 4–5.)

FAS is an extreme example of a severe physical injury to the brain during a critical period. It is also possible, however, that milder and more psychological injuries may occur during critical periods. For example, overexposure to television during early childhood may provide children with

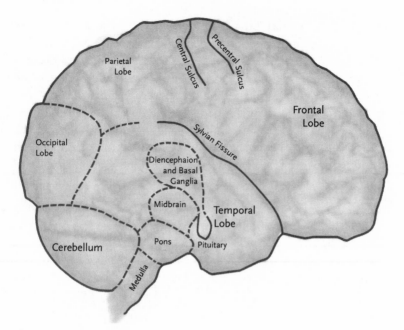

Figure 4–6:
Major Subdivisions of the
Human Brain

a very passive learning style and deprive them of the opportunity to learn a variety of important and more active learning skills—reading, using their bodies well, learning to find their way around a city or the surrounding countryside. Overexposure to scenes of violence on TV or in movies may desensitize them to pain and suffering and teach them to be indifferent or even actively violent themselves. The Freudian belief that all mental illnesses are due to early childhood experiences has largely been abandoned. However, most psychiatrists concur that life experiences during both childhood and adulthood (broadly defined to include many factors, such as nutrition, exposure to toxins, exercise, accidents, and relationships with peers) affect brain development and may either protect people against or predispose them to the development of mental illnesses. As discussed in subsequent chapters, the relative balance of influence probably varies among the various mental illnesses.

Neuroanatomy for Literati: Cortical and Subcortical Regions

The vocabulary of modern psychiatry is filled with terms from neuroscience and neurobiology. Twenty years ago people impressed one another at cocktail parties by discussing cathexis, counterphobic reactions, or libidinal drives. Today they get together informally and sip Evian laced with a lime slice while discussing the amygdala or the frontal lobes.

One can no longer carry on a conversation with many people without knowing at least a smattering of elementary brain anatomy.

On the large scale the nervous system is divided into several major territories: the cerebrum (composed of the left and right cerebral hemispheres, joined by the corpus callosum), the cerebellum (little brain), the diencephalon, the midbrain, the pons (bridge, so-called because it connects the cerebrum with the spinal cord and cerebellum), and the spinal cord. These are shown schematically in Figure 4–6. Inside the brain small islands of neurons are clustered together in a sea of white matter or CSF. Many of these territories or islands were given poetic names in Greek or Latin by early neuroanatomists. (See Table 4–2.)

TABLE 4–2
Neuroanatomy for Literati

Region	Latin/Greek	Function/Location
Cortex	Bark	Gray matter on brain surface
Neocortex	New bark	Newer (more highly evolved) cortical areas
Paleocortex	Old bark	Older, more primitive cortical areas
Subcortical	Below the bark	Any gray matter region below the cortex
Corpus callosum	Firm body	Axon tracts connecting the two hemispheres
Hippocampus	Seahorse	Memory
Amygdala	Almond	Emotional memory
Limbic system	Border system	Appetites, emotions, and memory
Caudate	Tail	Motor/emotional modulation
Putamen	Stone	Motor/emotional modulation
Globus pallidus	Pale globe	Motor/emotional modulation
Nucleus accumbens	Nucleus lying beside	Emotional modulation
Basal ganglia	Lower nerve knots	Combination of the four above
Lentiform nucleus	Lenslike nucleus	Combination of putamen and globus pallidus
Thalamus	Marriage bed	Filter or central switchboard
Hypothalamus	Under the bed	Modulation of appetites and drives
Diencephalon	Between brain	Combination of the two above
Cerebellum	Little brain	Coordination of movement, thinking, and emotion
Tentorium	Tent	Separation of cerebellum and cerebrum
Substantia nigra	Black substance	High concentration of dopamine cells
Locus ceruleus	Skyblue place	Center for norepinephrine cells

Over the years we have learned more and more about the functions that these regions perform. For example, regions that are involved in memory include the hippocampus, which looks like a seahorse, and the amygdala, an adjacent almond-shaped structure. Another cluster of gray matter structures is called the basal ganglia, which play an important role in regulating movement and emotions. This group of structures includes the caudate, a C-shaped structure with a long tail, which affects mood and emotions and helps to regulate movement; the putamen, or "stone," which is adjacent to it and also regulates movement and emotion; the globus pallidus, adjacent to the putamen and probably also involved in the regulation of movement; and the nucleus accumbens, a part of the basal ganglia that is known to have many connections with the limbic system (a more primitive part of the brain that regulates emotions). The diencephalon, or "between brain," includes the thalamus and the hypothalamus. The thalamus, or "marriage bed," is the central switchboard or filter in the center of the brain. Signals come in to it from all our sense organs and our bodies, and it screens out those that are unimportant and forwards on those that require action or response. The hypothalamus (below the thalamus) is an important regulator of appetite and other hormonal functions. These various small gray matter structures are shown in MR images in Figures 4–7, 4–8, and 4–9.

As shown in Figures 4–10 and 4–11, specific regions have also been identified on the surface of the brain. Looking at the outer convexities of the brain, we see the four major lobes: frontal, temporal, parietal, and occipital. These regions can also be seen when the brain is split down the middle, revealing the inner surface of the two hemispheres. The frontal lobe is separated from the rest of the brain via a large deep sulcus, the central sulcus. The gyrus in front of it is called the motor strip, because it contains the cells that control movement throughout our bodies. The gyrus right behind it, in the parietal lobe, is the somatosensory strip, which receives all the sensations that we experience. The close apposition of these two parts of the brain probably helps us respond more efficiently. If we place a hand on a hot surface, the sensory region in the brain notes this experience and sends a message rapidly to the adjacent motor region controlling the hand, letting it know that it needs to pull the hand away. The various regions of our bodies are mapped on these two gyri in a pattern that seems to be the same in all human beings. Neuroscientists have displayed this pattern for years as the comical "motor homunculus," shown in Figure 4–12.

The pattern was originally mapped by Wilder Penfield, a Canadian neurosurgeon, who placed tiny electrodes up and down the strip in anesthetized patients undergoing surgery and observed which body

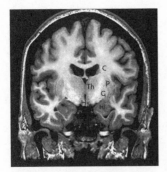

Figure 4–7: Subcortical Brain
Structures: Coronal View

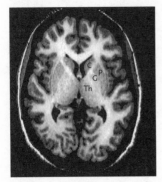

Figure 4–8: Subcortical Brain
Structures: Sagittal View

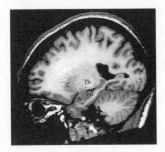

Figure 4–9: Subcortical Brain
Structures: Transaxial View

KEY FOR VIEWS OF BRAIN
A = Amygdala
C = Caudate
G = Globus pallidus
Hi = Hippocampus
Hy = Hypothalamus
P = Putamen
Th = Thalamus

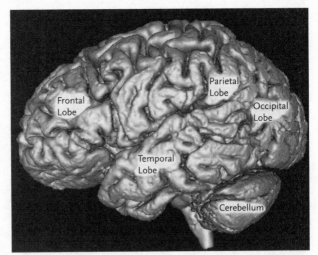

Figure 4–10: Lobes of the Brain Seen from the Side

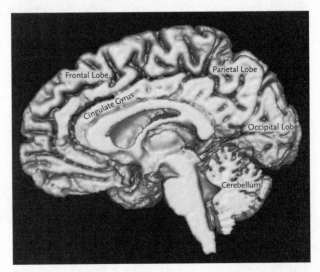

Figure 4–11: Lobes of the Brain Seen from the Middle

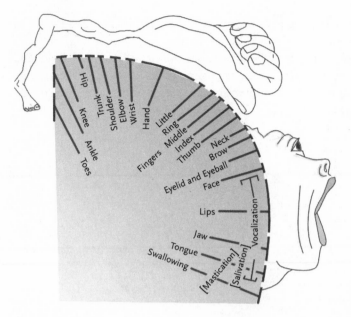

Figure 4–12: Penfield's Motor Homunculus

regions were stimulated by receiving a small electrical current. As the homunculus (little man) reveals, some parts of our bodies are given lots of space and other parts very little. Space allocation is determined by need rather than size. Our lips, tongue, and fingers, which we use for tasting, kissing, talking, and touching, get the highest amount of brain real estate.

The divisions into various lobes are somewhat arbitrary. They only help to get a very rough handle on how various regions of the brain work. The frontal lobes are often thought of as the home of "executive functions" such as planning, deciding, remembering, making moral judgments, or formulating abstract thoughts. The temporal lobes are used for language and some aspects of memory. The parietal lobes are used to make sensory and visuospatial associations, while the occipital lobes contain regions used for perceiving visually. On the inside surface of the brain, we see the large bundle of white matter fibers, the corpus callosum, that connects the two hemispheres, with the cingulate gyrus above it. The cingulate gyrus is used for many things, particularly to help us focus our attention. Frontal, parietal, and occipital lobes also wrap around to the inside of the brain.

The kinds of neurons and the way they line up and connect to one another is somewhat different in these different brain regions, reflecting the fact that they do somewhat different kinds of work. Early neuroscientist spent many years of painstaking work mapping these differences. The most famous was done by Korbinian Brodmann, who worked in Emil Kraepelin's department of psychiatry in Munich. Although developed nearly one hundred years ago, Brodmann's maps of "cytoarchitectonics" (the architecture of the alignment of nerve cells in the cerebral cortex) are still used today. They are shown in Figure 4–13.

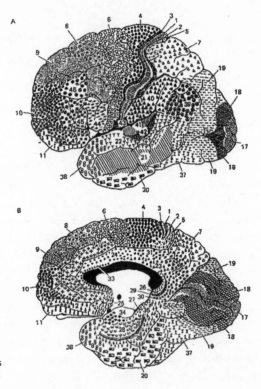

Figure 4–13: Brodmann's Maps of Cell Organization

The Map of the Mind: The Unity of Form and Function

In addition to these general anatomical divisions of the brain, neuroscientists also divide the brain into functional systems, or what we sometimes think of as "the mind." These include mental activities such as remembering, communicating with one another through language, or focusing our attention. Form and function—brain and mind—are a unity, however, and so we also think of functional systems in terms of specific brain regions, such as the frontal cortex or the limbic system, which generate mental activities such as making moral judgments or experiencing emotions.

The mapping of these functional systems has steadily progressed over the past one hundred years. The earliest methods for mapping functions exploited unfortunate accidents of nature, such as head injuries or strokes, which wiped out a particular part of the brain. Scientists inferred what the brain must have been doing in that region by observing what the victim was no longer able to do. This approach, known as the lesion method, provided pivotal information on how we produce creative thoughts, exercise self-discipline, or communicate using language.

More recently, functional imaging techniques such as positron emission tomography (PET) or functional magnetic resonance (fMR), which are described in detail in chapter 6, have complemented the lesion method. These newer imaging techniques permit us to directly visualize which parts of the healthy intact brain are used to perform specific mental activities. These neuroimaging techniques, which are more powerful and accurate than the lesion method, have shown us that most of the earlier divisions into functional regions or systems are oversimplifications. We perform most mental activities using multiple regions that are distributed throughout the brain. Nonetheless, the traditional divisions into functional systems are a useful starting place for understanding how the brain works to create a mind that thinks and feels.

The Language System: Our Link to One Another

As far as we know, the capacity to communicate in a highly developed and complex language is limited to human beings. Although a few higher primates have been able to acquire a small vocabulary using sign language (they have no voice box), and although dolphins and a few other sea animals apparently can send messages to one another, only human beings have a language that uses a specific grammar and word order. And only human beings record their present and past in written form in order to pass this information on to future generations. Our ability to record our history and to communicate with other human beings across the span of time has permitted us to create science and literature, to build complex civilizations and social systems, and (unfortunately) to design weapons to destroy one another.

The reason that only human beings have the capacity for language is that only the human brain has specialized regions dedicated to language. A simplified schematic diagram of the human brain circuitry traditionally considered to mediate language functions appears in Figure 4–14. Early lesion work originally suggested that the language system was located almost totally in the left hemisphere in most individuals.

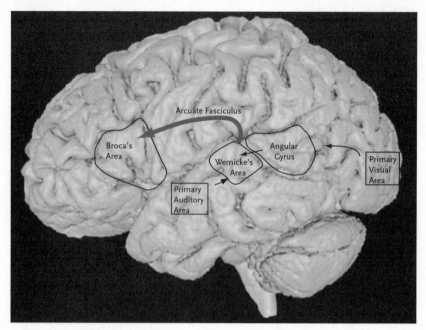

Figure 4–14: The Language Circuits

Two of the subregions in the language system have received their names from the nineteenth-century physicians who originally described them, based on observing people who had suffered from strokes.

Broca's area was first described by a French neurologist, Paul Broca, who had cared for a patient who suffered a stroke and experienced loss of speech and right-sided paralysis. He was only able to say the nonsense word "tan," and so he has become famous in the history of neuroscience as the patient "Tan." Broca concluded that the stroke must have affected regions in the left hemisphere, and declared at a meeting in Paris in 1864 that "nous parlons avec l'hemisphere gauche" (we speak with the left hemisphere). Broca's area *is* the region dedicated to the production of speech. It contains information about the syntactical structure of language, provides the "little words" such as prepositions that tie the fabric of language together, and is the generator for fluent speech.

Wernicke's area, described in 1876 by a German psychiatrist, Karl Wernicke, is often referred to as the "auditory association cortex." It encodes the information that permits us to "understand" or "interpret" information that is presented to us in auditory form. The perception of sound waves, which encode speech, occurs through transducers in the ear that convert the information to neural signals. The signals are received in the auditory cortex, but the meaning of the specific signals cannot be understood (i.e., perceived as constituting words with specific meanings—as opposed, for example, to the wordless music of a symphony) without being compared to associations and other information in Wernicke's area. An analogous process occurs when we understand written language. In this case the information is collected through our eyes, relayed via the optic tracts back to the primary visual cortex in the occipital lobe, and then forwarded on to the angular gyrus, a visual association cortex that contains the information that permits us to recognize language presented in visual form.

When strokes strike these regions, people show interesting differences in their problems with using language. For example, "Wernicke's aphasia" occurs as a result of damage to Wernicke's area and leaves the individual without the ability to understand what is said to him. This is a direct consequence of his loss of the auditory association cortex, which attributes meaning (aphasia = loss of language) to the sound waves that he hears. In addition, an individual with Wernicke's aphasia loses the ability to speak coherently because he has lost the "meaning" of language. Individuals with Wernicke's aphasia produce fluent, disorganized speech that is sometimes referred to as "word salad." Wernicke's aphasia is sharply distinguished from Broca's aphasia. In Broca's aphasia the individual can comprehend what is said to him, but cannot express himself, a situation that typically leads to great frustration. Damage to the angular gyrus leads to loss of the ability to read and write, the two forms of language that are visually mediated, with no loss of auditory comprehension or spontaneous speech.

Our understanding of the language system based on the study of stroke patients has been supplemented by studies using functional neuroimaging (i.e., PET and fMR). The direct study of the intact living brain that is made possible by functional imaging suggests that language functions are not quite as simple as suggested by lesion studies. Specifically, the strong left hemisphere dominance for language is being called into question. PET and fMR studies indicate that auditory language perception occurs in both hemispheres and that blood flow may also increase in both hemispheres during language generation. This probably explains why many people are able to recover language functions after experiencing a stroke. They can learn to draw primarily on the companion language regions in the right hemisphere. The traditional maps of brain language regions, as shown in Figure 4–14, are now being redrawn, as more *in vivo* data are added to our knowledge of the human brain map.

The Memory System: Remembering the Future

The memory system is also uniquely human. Our ability to store information about our past, to recall it, and to use it to plan for the future is as important as language in giving us the ability to perform a variety of complex tasks. In addition, memory is probably what gives each of us our highly individual and personal identities. Starting from early childhood, we begin to encode and store our various perceptions and experiences. Each of us has different ones. The formation of this repository of memories of personal experiences, referred to as episodic memories by Endel Tulving (episode = event, experience), continues throughout our lives and is one of the major forces guiding brain plasticity.

We are our memories. The personal identity, the sense of self that each of us has, is the composite of the episodic memories that we have retained and draw on each time we think a thought, experience a feeling, or make a decision. Episodic memory, a sequentially time-linked memory system, permits us to have a sense of the future as well as the past and present. Because we can place ourselves within the linear context of time, we can look forward as well as back. In a sense, when we draw on our episodic memory system to plan the next week's activities or wonder if our grandchildren will have a healthy earth to live on, we are "remembering the future." Our capacity to think within the context of time is the backbone of our psyche.

Laypeople speak of memory as if it is a single thing. However, as cognitive neuroscientists have studied memory over the past few decades, they have realized that there are many different kinds of memory. Several different classification schemes have been proposed, which are summarized in Table 4–3.

Tulving has distinguished episodic memory from semantic memory, which is a memory system for words, facts, or information that is objective rather than subjective. Alan Baddeley and Patricia Goldman-Rakic

TABLE 4–3
Systems for Classifying Types of Memory

Episodic vs. Semantic
Working vs. Associative
Declarative vs. Procedural
Explicit vs. Implicit

have identified "working memory" and distinguished it from "associative memory." Working memory is what we use for information that we may only want to keep for a short period of time, while we use it to perform some particular piece of work. The classic example is hearing or reading a telephone number. We keep it briefly, dial the number, and talk to the person we have called. By the time we hang up the phone, the memory for the number has been erased. Sometimes we keep things in working memory and refer to them as we do a task (i.e., use them to perform "work"). For example, performing a long string of calculations "in our head" is a working memory task (e.g., add five and seven, multiply by eight, and divide by three). Associative memory is contrasted with working memory because it is composed of more long-term stores of associations that we refer to when we remember (e.g., who discovered America, or the color of our father's hair when he was a young man). Yet another contrasting way of subdividing the various memory systems, proposed by Larry Squire, is the distinction between declarative and procedural memory. Declarative memory is about "what" and consists of things we know in the sense that we can state or declare them. Procedural memory is about "how." It consists of our knowledge about procedures—how to drive a car or ride a bicycle, make a cake, or navigate the road from Philadelphia to Pittsburgh. A final way of dividing memory, proposed by Daniel Schacter, is into implicit versus explicit. Implicit memory lies beneath the surface and has not been pulled directly into our conscious awareness; it consists of all the information that we are not aware of, at least at a given moment. Explicit memory, on the other hand, is precisely that: explicit. It consists of the memories that are immediately accessible and obvious to us.

Memory is also not a single thing in another sense. When we build our repository of memories in our brains, it is done in stages, each of which contributes to the overall process of using the memory system. The various stages of memory are summarized in Table 4–4.

TABLE 4–4
Stages of Memory

Encoding
Consolidation
Storage
Retrieval
Ecphory

Understanding memory requires understanding a process, as well as how the brain performs the process. Our brains are constantly besieged with information, most of which we ignore. (As I write this on my laptop on a train in Italy, I am ignoring the pressure of the seat, the conversation of the surrounding passengers, and the Tuscan countryside whizzing by the windows.) From all this information, we choose a few bits and pieces that we want to keep, at least briefly, by focusing our attention on them. That begins a process called encoding, which puts the memory in a short-term buffer, which may be similar to working memory. (The details of the train are now encoded for me, since I took notice of them.) After encoding occurs, the memory may stay with us, or it may be discarded.

Memories that are kept move on to the next stage, which is known as consolidation, and which is used to convert encoded memories from their short-term buffer to long-term storage so that they can be retained. Sometimes we make a conscious choice to consolidate a memory, and we all know strategies that can be used to help with this process—repeating a date or word over and over, turning it into a picture, converting a numeric code to a letter code, and so on. But often memories are retained without any subjective awareness that we have *chosen* to remember them. Why some memories become consolidated is somewhat mysterious. We do know that experiences or information are more likely to be consolidated if they have more personal meaning, a high emotional loading, a vivid sound or appearance, or other factors that make our processing "deep" rather than "shallow." Consolidated memories are then stored.

Storage of memories is also somewhat mysterious, since the amount of information that we keep on hand throughout our lives is clearly a massive challenge to our brains—beginning when we first start to recognize parents and family, learning words and then sentences, reading, writing, on and on—an endless sea of information ebbs and flows in our brains. Sometimes a fact or idea rises to the surface without our asking for it—again a mysterious process that is the stuff of creativity, impulse, and spontaneity.

When this happens, the process is called retrieval, or moving a memory out of storage so that it can be used. Sometimes we make a conscious decision to remember something and begin to make a systematic search through our memory stores—a process called conscious retrieval. Who is that familiar face across the room? What word is perfect to describe this feeling? What was the name of that person I met yesterday afternoon? The search hums along . . . and sometimes it fails. We just can't find the name or word. We may get another one that is obviously wrong, and we are able to tell that it is wrong—another mysterious business. But when the retrieval is successful, we have a subjective sense of gratification that we have found the right thing—a process called ecphory by Tulving (= brought out).

The mechanisms that our brains use to form short-term (encoded) or long-term (consolidated) memories are not completely clear, although a consensus is that short-term memories are coded through different mechanisms than long-term memories. Encoding probably occurs through activating synaptic circuits using neurochemical transmission; this type of neural activity is both rapidly implemented and rapidly reversed. On the other hand, consolidation has to use a more permanent process, most likely through the creation of a sequence of molecules that code the information in a form that can be stored. Eric Kandel of Columbia University has done seminal work on the mechanisms of consolidation and storage. Using the gill withdrawal reflex in the small seasnail *Aplysia* as a model, Kandel has suggested that long-term memory may depend on the synthesis of proteins and RNA in neurons that are synaptically connected during the time that short-term learning occurs. Kandel won the Nobel Prize in 2000 for his discovery of the different ways that long-term and short-term memories are preserved. Both kinds of memory are due to plasticity in the synapse, which changes in response to stimulation and experience. Long-term memory occurs when a stimulus is intense and long lasting, prompting the nerve cell to send a message to the cell nucleus that leads to the synthesis of proteins that in turn produce long lasting changes in the shape of the synapse. Through Kandel's work, we now understand the molecular mechanisms of consolidation and storage.

For many years, scientists conducted a search for the place where memories were stored, sometimes referred to as the "search for the engram." Some assumed that there must be a single place, while others believed that memories might be stored in many different places. For example, Karl Lashley of Harvard spent much of his career placing lesions in various parts of the brain in experimental animals, demonstrating that no specific lesion could produce memory deficits. However, in 1957 Scoville and Milner of the Montreal Neurological Institute changed the conceptualization of memory radically by reporting a single informative and very famous case: H.M. was treated surgically for intractable epilepsy by removal of the anterior temporal poles on both sides of his brain. Removing overactive tissue that was producing "fits" that were nearly continuous (and therefore very dangerous) was not unusual, but normally the overactive tissue was only on one side of the brain. In H.M.'s case both sides had to be removed. Afterward, the surgeons were dismayed to discover that he had totally lost the capacity to remember any new information. However, his memory for everything that he had learned or experienced *before* the surgery was completely normal. H.M. was frozen in a time warp. This famous case focused attention on the gray matter structures located in the anterior poles of the temporal lobes, the amygdala and hippocampus, and on the possible importance of bilateral as opposed to unilateral lesions.

We now know, based on the case of H.M., and a large quantity of other human and animal evidence, that injuries to the temporal lobes on only one side do not typically produce memory deficits, but that bilateral lesions in certain specific locations can completely destroy learning and memory. We now suspect that some aspects of learning and memory are mediated in two brain regions, the hippocampus and amygdala. Both these brain regions work together to store memories. The amygdala may work primarily to integrate memories learned from different modalities and memories with strong emotional valence.

These lesion studies have been supplemented by studies of memory using functional imaging (PET and fMR). These technologies permit the direct observation of the intact brain while it is in the process of forming or retrieving a memory. Lesion studies show what functions are lost when a portion of the brain is lost. In vivo imaging techniques show what parts of the brain we use when everything is working well. In that sense, they give a better picture of how the normal brain functions. Functional imaging studies suggest that we use many different parts of our brains when we remember, and in particular that many regions of the cortex are involved, in addition to subcortical or limbic regions such as the hippocampus and amygdala. Subcortical regions are often stimulated during PET or fMR studies in specific kinds of memory, such as those that involve intense encoding or intense emotion. PET studies indicate that memory processes are widely distributed in the brain and involve most parts of the cerebral cortex, including parts of the frontal, parietal, temporal, and occipital lobes (depending on the content of the memory). Further, it is now clear that the frontal cortex in particular plays an important role not only in working memory (as had been demonstrated in animal and lesion studies), but also in encoding and retrieval. Our current view of memory is now not unlike the early ideas of Lashley.

Functional imaging studies have illuminated many more specific aspects of memory as well. For example, how do we manage to remember so much so efficiently with a brain that is only around 1,200 cubic centimeters in volume? We subjectively recognize that as we learn, we initially have to concentrate on what we are learning, but as the process becomes smoother and more practiced, we seem to do it more efficiently and almost automatically. A great American psychologist, William James, described this phenomenon at the turn of the century.

> The great thing, then, in education, is to make our nervous system our ally instead of our enemy. It is to fund and capitalize our acquisitions, and live at ease upon the interest of the fund. For this we must make automatic and habitual, as early as possible, as many useful actions as we can. . . . The more of the details of our daily life we can hand over to the effortless custody of automatism, the more our higher powers of mind will be set free for their own proper work.
> *William James, Principles of Psychology (1890)*

In one of our PET studies at Iowa, we examined how the brain achieves economic memory retention. We measured brain blood flow when people remembered lists of words that they saw many times one week earlier and practiced remembering until they had learned them well (making the task very easy for them). Within the same people we compared brain blood flow when they remembered lists of words that they just saw 60 seconds before being imaged and only one time (which requires more effort). Two interesting observations emerged. First, many widely distributed regions were used with nearly the same pattern—even though the retention interval for the two conditions was enormously different by the standards of cognitive psychology—one week versus 60 seconds. Both practiced and novel memory use regions that would never have been predicted from the old lesion literature—including not only the frontal cortex but also the cerebellum. Second, when remembering is well practiced, the size of these regions gets much smaller—indicating that practice permits the brain to work more efficiently. The insight of William James, formulated nearly one hundred years ago, was prescient. Nature does work in wondrously economical ways.

The Attention System: The Spotlight in the Brain

Attention is the "spotlight" that our brains use to identify stimuli within the context of time and space, to select what is relevant, and to ignore what is irrelevant. We are bombarded continually with sensory information in multiple modalities, as well as with the information that our memory system has stored and that some stimulus causes to float up to the surface of our consciousness. A person driving a car on a busy highway is receiving information about other cars, the road, and the surrounding terrain from his visual system, as well as auditory input from the car motor or the rush of other vehicles as they pass, and tactile input from hands on the steering wheel and the foot on the gas pedal and the physical sensations experienced by the rest of the body as the car grips the road or bounces and sways. The person may also be talking on a cellular phone, listening to music, or thinking about a recent conversation. Attention is the cognitive process that permits this person to suppress irrelevant stimuli (e.g., to ignore most of the interior of the car), to notice important stimuli (e.g., that the car in front is putting on the brakes and slowing down suddenly), and to shift from one stimulus to another (e.g., from thoughts about the recent conversation to the traffic). If we lacked this capacity, we would be overwhelmed. Attention is the spotlight that we all use to highlight what is important for our survival.

Attention is a very "central" cognitive system, and therefore it is difficult to study in isolation. Like memory, it is also classified into different types. These are listed in Table 4–5. Sustained attention involves focusing for a prolonged period of time (as when studying for an examination). Directed attention involves consciously selecting a particular feature or stimulus from the large array available (as when Sister Wendy comments on the erotic male fury of the bull in Neolithic cave paintings). Selective attention involves focusing attention on a stimulus that may have importance for personal or practical reasons (as when a person at a party hears his name spoken in a nearby conversation). Divided attention involves focusing attention on several things at the same time, or in rapid shifting sequence (as when the person at the party tries to listen to two conversations at once). Focused attention involves directing attention to some particular stimulus or task (as when solving a mathematics problem or developing the outline for a paper to be written).

TABLE 4–5
Types of Attention

Sustained
Directed
Selective
Divided
Focused

Attention is created by the cooperation of multiple brain systems. Information flows through a region in the base of the brain, the reticular activating system. Midline circuitry passes this information through the thalamus, which acts as a "filter" or "gate." Other brain regions also appear to play a role in attention, including the cingulate gyrus, the hypothalamus, the hippocampus and amygdala, the prefrontal cortex, and the temporal, parietal, and occipital cortices. Functional imaging studies have shown that the cingulate gyrus has increases in cerebral blood flow during tasks that place heavy demands on the attention system, such as those that involve competition and interference between stimuli. They have also shown that blood flow can be shifted from one hemisphere to another as a consequence of directed attention; increased blood flow is seen in the right superior temporal gyrus as a consequence of instructions to listen to the left ear, and the increase shifts to the left superior temporal gyrus in response to instructions to attend to the right. (The wiring for hearing is set up so that most crosses to the opposite side.)

The Thalamic Filter: How the "Marriage Bed" Helps Us Set Priorities
The thalamus was given its Greek name, "marriage bed," long before we began to understand exactly what its functions were. The name must have been chosen based on the realtor's motto: location, location, location. The thalamus sits in the very center of the brain, and therefore it must perform some very central function. This rather small region (occupying only about 12 cubic centimeters out of the approximately 1,200 in our entire brain) has a very big role in our mental life.

We usually think that acquiring information is the most important thing that our brains do. We never stop to realize how much information we have to *ignore*. In fact, our brains continually deal with so much information that we would be overwhelmed if we did not have a mechanism to filter out most of it and select what we need to notice and keep. We have five senses, and all five are "on" all the time. Think about just one of them—for example, vision. Our eyes pick up hundreds of bits of information every minute as we walk down a street. But we cannot afford to actually "see" all of them, either consciously or unconsciously.

Imagine that you are in a new city, perhaps Chicago, looking for a specific store, perhaps Crate & Barrel. You leave your hotel, having been told how to find the store—five blocks north, on the left side of the street. You just have to get there without being run over or having your wallet stolen, and you need to buy some wine glasses and get back to the hotel in less than an hour. Otherwise you will be late for dinner, and your spouse will be furious. This is a pretty easy task, compared to most things that we have to do!

As you walk along, your eyes take in cars rushing by on the street, people thronging the sidewalk, shop windows filled with tempting potential purchases, buildings stretching up to the sky, traffic lights, street signs, the gray of the pavement, an occasional piece of litter or trash, the blue of the sky, the drift of a few small clouds . . . an endless panorama of visual sensations. Your eyes invite you to look at that new computer in the window, that new Audi TT parked illegally in front of a store, or that well-dressed curly-haired man attired totally in black leading a black poodle on a leash. And then there are the smells, sounds, and jostles of other people. But, given your mission, your brain has to ignore all of this . . . or anyway as much as it can, so that you can do what you have to do: count blocks, notice street names, walk briskly, stop and wait when traffic lights are red, and ultimately find the store. Once there you are confronted with another panorama of visual information that you must also suppress, as you make your way to the part of the store that sells glasses.

The little thalamus, a central switchboard in your brain, is the filter that helps you ignore and suppress information so that you can perform this simple task. All the input from all our senses flows into it. It takes selective note of what should be given a high priority and lets you throw the rest away. Zap. Gone. If I hadn't mentioned the Audi, the computer, or the man in black, you would not bother to waste any of your precious neurons noticing them, encoding them, and later remembering that you saw them. And you don't even realize that the process of filtering is happening. The thalamus has connections back and forth with almost every part of the brain—"higher" cortical regions, motor regions, sensory regions, emotion regions, memory regions. Through checking between them (a process that occurs at lightning speed), the thalamus manages to help our brains figure out what all the billion of bits of information around us at any given moment mean, whether they are important for what we are trying to accomplish at that moment and should be noticed, or whether they are irrelevant and should be ignored.

The Prefrontal System and "Executive Functions": Our Moral Monitor

The prefrontal system, or prefrontal cortex, is one of the largest cortical subregions in the human brain. Brodmann estimated that it constitutes 29% of the cortex in human beings, as compared to 17% in chimpanzees, 7% in dogs, and 3.5% in cats. Its high degree of development in human beings suggests that it too may mediate a variety of specifically human functions, often referred to as "executive functions," such as abstract thought, creative problem solving, and the temporal sequencing of behavior.

The lesion method provided an early landmark in our understanding of the prefrontal cortex through the case of Phineas Gage, a quarry worker who was injured by an explosion that drove an iron bar through his left frontal lobe. Gage survived the bizarre accident, but afterward he began to show serious personality changes that were described by Harlow, the physician who cared for him and saved his life. Prior to the accident Gage was conscientious, serious, and hardworking, but after he recovered he became immature, child-like, socially inappropriate, and irresponsible. Harlow's early description of frontal lobe functions has been supplemented by many subsequent studies of people with frontal tumors, injuries to the frontal lobes, and surgical treatments for epilepsy, psychosis, or obsessive-compulsive disorder. This work indicates that substantial damage to the prefrontal cortex produces a syndrome quite similar to that of Gage. Although general intelligence is not necessarily impaired by frontal lesions, individuals with substantial frontal injury lose other capacities such as volition, the ability to plan, and social judgment.

Two different subtypes of "frontal syndromes" have been observed. Lesions to the orbital region of the prefrontal cortex (a more "primitive" part of the frontal cortex on its lower surface, just above the eyes) make people euphoric, overactive, and inclined to inappropriate social behavior such as sexual overtures to unknown people. Lesions to the dorsolateral portion (the outer convexities of the frontal lobes on the sides of the brain) make people apathetic, physically inactive, and less able to perform complex cognitive tasks such as formulating an abstract concept. Within both of these syndromes, however, lies a common core: impairment in the capacity to pursue goal-directed behavior, based on the integration of environmental and internal cues. This is probably the basic function of the prefrontal cortex.

The intactness of the prefrontal cortex can be assessed by a variety of cognitive tasks, and it has been explored through neuroimaging as well. The Wisconsin Card Sorting Test, which assesses the capacity to think abstractly and to modify the assumptions that shape our responses, and the Tower of London and Porteus Mazes, which assess the capacity to plan ahead, are three standard "frontal lobe" tests in neuropsychology. The Continuous Performance Test (CPT) is a measure of attention that is also thought to tap prefrontal cortical functioning. Common to all of these is a need to draw on working memory, which is also an important frontal lobe function. Pioneering work in neuroimaging has been done with these tests by Monte Buchsbaum and Daniel Weinberger, who have used the techniques of functional neuroimaging to study the prefrontal cortex in schizophrenia and noted that patients have decreased flow in this region during rest and while doing "frontal" tasks such as the CPT and the Wisconsin Card Sorting Test.

The Limbic System: Where the Wild Things Are

If the prefrontal cortex is the superego of the brain that functions as a moral and social monitor and formulates abstract concepts, the limbic system is the id that feels and monitors emotions and basic survival drives. While the frontal system is the phylogenetically newest cortical region, the limbic system is the oldest and most primitive. Although it would be easy to say that the one is the seat of reason and the other of emotion, it is clear to most of us that this distinction is arbitrary, and the existence of anatomic connections between frontal and limbic regions make this point even clearer.

The word "limbic" means "border" in Latin. This term was first used by Broca to refer to the circular ring of tissue that appears to "hem" the prefrontal, parietal, and occipital neocortex when the brain is viewed from a midsagittal perspective. (He also called it "the great lobe of the hem.") Because the olfactory nerve is connected to this region, it was also known for a time as the rhinencephalon, or "nose brain." Cytoarchitectonic maps also revealed that the cellular structure in some of these regions was paleocortex rather than neocortex.

The function of this "primitive" central brain region was assumed to be related to olfaction until the 1930s, when James Papez proposed another alternative. He introduced the idea of the "Papez circuit," which is illustrated in Figure 4–15. He suggested that the major input to this circuit was not the olfactory portion of the brain, but rather a group of association cortices, which collected information from a variety of neocortical regions and then relayed this information on to the Papez circuit in

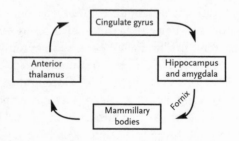

Figure 4–15: The Limbic System as Conceptualized by Papez

the limbic system. Papez suggested that the major function of this brain region was to experience and regulate emotion. Within the circuit, messages would flow from higher cortical regions to the cingulate gyrus, hippocampus, mammillary (breastlike) bodies, hypothalamus, and anterior thalamus. Papez suggested that emotions were concentrated in deeper structures such as the hippocampus, while awareness of them occurred in the cingulate gyrus.

Since the early formulation of Papez, the concept of the limbic system has been expanded and revised. Discussions of what constitutes the limbic system are sometimes inconsistent or confusing because our knowledge about its functions has been steadily growing during the past several decades, based on new contributions from neuroimaging and neuroscience. Modern formulations frequently add other regions to the original Papez circuit, including the parahippocampal gyrus, the nucleus accumbens, and inferior frontal regions such as the orbital gyri. Its many interconnections suggest that the limbic system may integrate body sensations, the experience of the external environment through multiple modalities (e.g., sights, sounds), emotions, and the memory system.

The limbic system is clearly pivotal for regulating basic body functions such as thirst, hunger, and sex drive. However, it is also important for helping us respond to our environment and in the association of emotional experiences and memory. In animal studies Joseph LeDoux has shown, for example, that the amygdala plays a key role in our response to threatening or dangerous stimuli. Exposure to an unpleasant or frightening stimulus lays down powerful memory traces that are difficult to extinguish and that are stored in a circuit in which the amygdala plays a central role, but that has thalamic, hippocampal, and cortical components. There appear to be two parallel tracks for processing emotional stimuli. A fast track engages subcortical regions (thalamus and amygdala) and permits rapid but relatively crude responses, while the slower track permits more refined identification of features and formulation of plans and includes the prefrontal cortex. The hippocampus also plays a role by assessing contextual cues. The importance of this aspect of the limbic system for the development of anxiety disorders is discussed in more detail in chapter 11.

The Cerebellum and Mental Coordination:
Walking and Chewing Gum at the Same Time
The cerebellum sits below the cerebrum and is separated from it by a large sheet of tough fibrous tissue, called the tentorium (tent). For many

years, doctors used this fact to speak to one another in a private code language. Since all "higher" mental capacities such as memory or personality were thought to reside in the cerebral cortex of the cerebrum, they would convey that they thought a patient's complaints were "all in her head" rather than "real" by saying, "Oh, her stomach problems are really supratentorial." In fact, supratentorial (above the tentorium) was a code term for "neurotic" or "due to mental illness." The cerebellum, sitting below the tentorium, was not thought to have anything to do with mental functions or with mental illness. Its only function was physical coordination—the ability to walk, catch or throw a ball, perform a ballet, or do push-ups.

Yet this simple view of the cerebellum has recently been revised, based on evidence from many different sources. Phylogenetically, the cerebellum has enlarged substantially in human beings. The human cerebellum, like the prefrontal cortex, is one-third larger than it is in chimpanzees. It also has more neurons than the cerebral cortex: 10^{13}. Further, studies have shown that the connections between the cerebral cortex and the cerebellum include many different cortical areas, including regions in the frontal and temporal lobes, in addition to the purely motor and sensory regions that are required for motor coordination. This suggests that in human beings the cerebellum must be designed to coordinate "higher cognitive functions" in addition to purely motor activities. Further, the design of the cerebellum suggests that it works as a finely tuned feedback system. The connections run from a specific region in the cerebral cortex down to the cerebellum (crossing in the midline as so often happens, for some mysterious reason), up to the thalamus, and then on up to the same specific region in the cerebral cortex. This pattern is shown in Figure 4–16.

Although early PET studies simply cut off the cerebellum, based on the assumption that it could not be doing anything of interest to students of cognition and emotion, more recent studies have indicated that it is used in many different kinds of mental activity: memory, language, emotional response, facial recognition, and consciously focusing attention. In the human brain the cerebellum appears to play an important role in cognitive and emotional functions. We now recognize that the cerebellum probably performs the very basic function of keeping track of information within the context of time. It serves as a "metron" that permits us to recognize both where our bodies are moving in space (its motor function) and where our thoughts are going in our minds (its mental function). Surprisingly, it may be particularly important in schizophrenia, as is discussed in chapter 8.

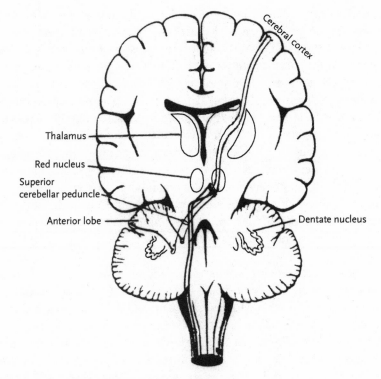

Thalamus

Red nucleus

Superior
cerebellar peduncle

Anterior lobe

Cerebral cortex

Dentate nucleus

Figure 4–16: Circuits Linking the Cortex to the Cerebellum

The Basal Ganglia: Getting Tied Up by "Lower Nerve Knots"
Until recently, many neuroscientists also thought that the primary function of the basal ganglia was to regulate and mediate motor activity.
Again, however, this view has been revised. The basal ganglia ("lower
nerve knots") are also important for the expression and regulation of
emotion and cognition.

The major structures of the basal ganglia include the caudate, putamen, globus pallidus, nucleus accumbens, and substantia nigra. Except for
the substantia nigra, these are well visualized in Figures 4–7 through 4–9.
The caudate is a C-shaped mass of gray matter tissue that has its head at
the lateral anterior borders of the frontal horns of the ventricles. It arches
back posteriorly in a circular fashion and then curls forward again, ending
in the amygdala bilaterally. Separated from it, and lateral to it, is the
lentiform nucleus, so called because it is shaped like a lens. The medial
portion of the lentiform nucleus, which is darker and more densely full of
gray matter, is the putamen, while the globus pallidus is lateral to it.

The basal ganglia are considered to play a role in cognition and emotion for several reasons. Abnormalities in these regions often lead to symptoms of mental illness. The first symptoms of Huntington's disease, caused by atrophy in the caudate nucleus, may be similar to those seen in schizophrenia or mood disorders. Patients with Huntington's disease often develop delusional thinking, depression, or inappropriate impulsive behavior. Parkinson's disease is another syndrome affecting the basal ganglia. It is due to neuronal loss in the substantia nigra, the midbrain region of the basal ganglia that sends projections to the caudate, using dopamine as its primary neurotransmitter. In Parkinson's disease, loss of pigmented neurons, and the associated loss of dopaminergic activity, produces a blunting of the ability to express emotion, impairment in the ability to remember, and a loss of volition. These symptoms are very similar to the negative symptoms of schizophrenia.

Consciousness: The Last Frontier

As philosophers such as John Searle have pointed out, consciousness is a word that means many different things to many different people. A surgeon says that "the patient is conscious" when she comes out of the anesthetized state and begins to respond to those around her and ask questions about what happened. A cognitive psychologist might say that consciousness refers to the capacity to recognize a boundary between the self and the external world. Another might disagree and say that consciousness is the ability to introspect and to be aware of internal states, such as anger or guilt. Another might reply that the ability to conceptualize experiences and actions within the context of time defines consciousness (i.e., to see oneself in terms of past, present, and future). Metaphors have been used to discuss consciousness—for example, it is the "mental theater" in which the self is the actor and in which perceptions, experiences, and actions converge. Others have argued that consciousness is the director or supervisor of actions on our "mental stage." Still others equate consciousness with the soul, self, or spirit. For them it is the essence of an individual human identity. For people who embrace religions that include belief in an afterlife, it is the part of an individual that survives after death.

Many gifted thinkers have probed the essence of consciousness as a property of the human brain. One of the most original is Gerald Edelman, a Nobel laureate in medicine who has been drawn to the siren's song of neuroscience. He has written a trilogy of books on the topic, beginning with *Neural Darwinism* and ending with *The Remembered Pre-*

sent. He postulates that consciousness occurs as a consequence of a process of "re-entry," by which functional mental components (e.g., categories, concepts, memories, experiences) are connected to one another to produce a sense of self within a context of nonself. He has proposed a neural circuit that could be the basis for consciousness, which has contributing components in cortex, basal ganglia, thalamus, cerebellum, hippocampus, and other key nodes. Francis Crick, codiscoverer of the double helix of DNA, is another Nobel laureate who has been drawn by the seductive song of neuroscience to the study of consciousness. He focuses on the thalamus as a key structure in permitting us to maintain a conscious state. Michael Gazzaniga, a pioneer in the study of hemispheric asymmetry, has used his studies of split-brain patients to explore other aspects of consciousness—how the narrator in the left hemisphere can discuss and interpret the voiceless life of the right and how these dual experiences can be integrated in a functioning and aware human being.

How Nerve Cells Talk to Each Other: The Chemical Messengers of the Brain

The daily conversation between neurons, which makes possible such complex functions as language or memory, is conducted via messages sent through chemical couriers: the neurotransmitter systems of the brain. These neurochemical systems provide the "fuel" that permits the functional and anatomic systems to run (or run poorly, when an abnormality occurs). The neurochemical systems do not map neatly on the anatomic and functional systems. We cannot say that dopamine is the "limbic neurotransmitter" or that norepinephrine is the "attention neurotransmitter." Rather, the neurochemical, anatomic, and functional systems are interwoven and interdependent. Any anatomic subsystem within the brain usually "runs" on multiple classes of neurotransmitters. This complex anatomic and neurochemical organization permits much greater "fine tuning" of the entire central nervous system.

Neurons, Synapses, Receptors, and Second Messengers

Neurons are designed to talk back and forth to one another as efficiently as possible. The structure of a neuron is shown in Figure 4–17.

They all consist of a cell body containing the nucleus and at least one axon of variable length. The axon carries an electrical message from the cell body to terminals that contain chemical messengers. The fatty insulation provided by the myelin sheath protects the axon, just as plastic does for a copper wire. The cell body is surrounded by dendrites that enlarge

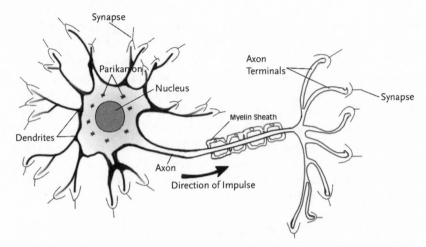

Figure 4–17: Nerve Cell Communication through Synapses

the capacity of the cell body to receive information through synaptic input from other neurons. Likewise, axons may branch as they terminate, and then produce multiple synaptic contacts.

The communication between nerve cells begins with electrical activity. The inside of a nerve is set at a specific voltage when it is "at rest" (-70 millivolts). If the various excitatory influences become predominant, causing the inside of the cell to rise as high as −35 millivolts, the cell

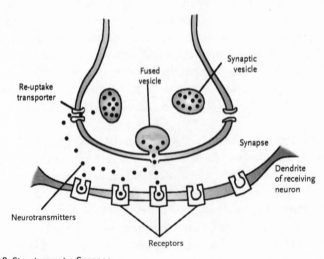

Figure 4–18: Structures at a Synapse

TABLE 4–6

Common Neurotransmitters Relevant to Mental Illnesses

Acetylcholine

Monoamines

Dopamine

Serotonin

Norepinephrine

Amino Acids

Gamma amino butyric acid (GABA)

Glutamate

"fires" and sends an excitatory message down the axon, ending at the terminal and causing the release of neurotransmitters. This firing is called an "action potential."

Although sending messages begins with an electrical process, communication at the synapse is chemical. Nerve cells talk to one another using chemical messengers, the "neurotransmitters." Some of the most common neurotransmitters that are relevant to understanding mental illnesses are listed in Table 4–6. They are all produced in the synapses of neurons, where they are stored in little "sacks" (vesicles), and released when the neuron fires.

A schematic representation of a synapse appears in Figure 4–18.

The neurotransmitter molecules must be sequestered in vesicles in order to prevent breakdown from enzymes that may be floating around inside the cell trying to keep it cleaned up. The concentration of neurotransmitters is constantly fine-tuned in order to maintain good communication. Basically, the neuron tries to keep an adequate (but not excessive) supply of neurotransmitters on hand. If it becomes very busy because of repeated firing, it sets to work to synthesize more. Maintaining an optimal concentration of neurotransmitters is regulated by proteins called enzymes, which typically end in "ase" and are named for the chemical reaction for which they act as helpers (catalysts).

After a neurotransmitter is released, it can experience a variety of fates. In the free-floating world of intersynaptic connections, enzymes hover and are ready to attack, leading to the neurotransmitter's metabolic inactivation or breakdown. If its concentrations in the intersynaptic space are high, the neurotransmitter may saturate presynaptic receptors, telling the transmitter neuron that it is time to "slow down." If it is lucky, the neurotransmitter may cross the synapse and occupy a postsynaptic receptor, thereby actually succeeding in sending a message to another neuron. Finally, because nature often loves efficiency and abhors waste, the chemical messenger may be returned to the original transmitter neuron and again be stored in the vesicles, eventually to be released again.

Receptors are large protein molecules that are embedded in the lipid sandwich of the neuronal membrane and that "recognize" specific neurotransmitters in a highly selective way, which is related to the chemical structure of the receptor. There are two main superfamilies of receptors: ion channel receptors and G-protein receptors. The receptors in the ion channel family are smaller and faster. GABA and glutamate receptors are members of this superfamily.

G-protein receptors are slightly larger and more complex. These are the receptors that catecholamine neurotransmitters such as dopamine occupy and use to send their messages. A picture of the dopamine type 2 (D2) receptor is shown in Figure 4–19. G-protein receptors wind in and out across the fatty sandwich of the neuronal membrane seven times. When the receptor is occupied by a neurotransmitter on the outer surface, it changes its shape slightly. This change in shape activates a substance known as a G-protein (the G stands for guanine, which is one of its components). This activation leads in turn to the release of a "second messenger" within the nerve cells whose receptor has been occupied.

The various types of G-protein receptors are named according to the neurotransmitter that they recognize. Many neurotransmitters are able to occupy several subtypes of receptors. For example, there are five subtypes of dopamine receptors, which are called D1, D2, D3, D4, and D5. Each differs in amino acid sequence, which determines its affinity (willingness to be occupied) for the neurotransmitter and also for specific drugs. Each receptor has a long tail on the extracellular side, which also determines drug affinity. Three loops of variable length are also present on the intracellular side, as well as an intracellular tail. These intracellular components provide the mechanism for passing the message on to the G protein.

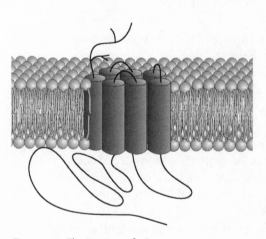

Figure 4–19: The Structure of a Receptor

Many neuroreceptors that are thought to be involved in mental illnesses and/or modulated by the drugs used to treat them are in this G-protein superfamily (e.g., serotonergic, dopaminergic, cholinergic). During recent years, the powerful tools of molecular biology (described in the next chapter) have been used to clone and sequence most of the G-protein receptors. Detailed information about their specific amino acid sequences opens up the exciting possibility that specific drugs can be designed to interact with receptors in a key-in-lock manner for various chemical systems or brain regions involved in specific illnesses.

The neurotransmitter is the "first messenger" that communicates with another neuron. The stimulation of the "second messenger" system is perhaps the most important part of the process of nerve cell communication, if the goal is to exert a long-term effect on the brain. Paul Greengard, of Rockefeller University in New York, has devoted much of his career to explaining how second messengers work, and he has been awarded a Nobel Prize for this pivotal contribution to neuroscience and psychiatry. The second messengers help create the proteins that regulate the expression of genes, the structural components of the cell, and the enzymes that aid in the synthesis of neurotransmitters. They do this by adding phosphate groups to a variety of different proteins, thereby setting up a highly skilled army of leaders, watchmen, and defenders of the integrity of the cell and its neighbors. The effects of activating G-proteins and stimulating second messengers occur relatively slowly and remain for a long time. This fact has important implications for understanding how the medications used to treat mental illnesses do their work. First, the slow pace of the G-protein receptor process may explain why so many psychoactive drugs require several weeks to exert their therapeutic effects. Second, it may also explain why the effects from many medications that work on the G-protein receptors also wear off very slowly—again often over the course of several weeks. Third, the effects may potentially be very long-term. Medications affect not only processes such as neurotransmitter synthesis, but also the transcription and expression of genes and the structure of the cell itself.

The Dopamine System

Dopamine, a catecholamine neurotransmitter, is the first product synthesized from tyrosine through the enzymatic activity of tyrosine hydroxylase. Because it has a close chemical relationship to two other important neurotransmitters, its synthetic pathway, as well as the subsequent ones of norepinephrine and epinephrine, is shown in Figure 4–20. Its importance

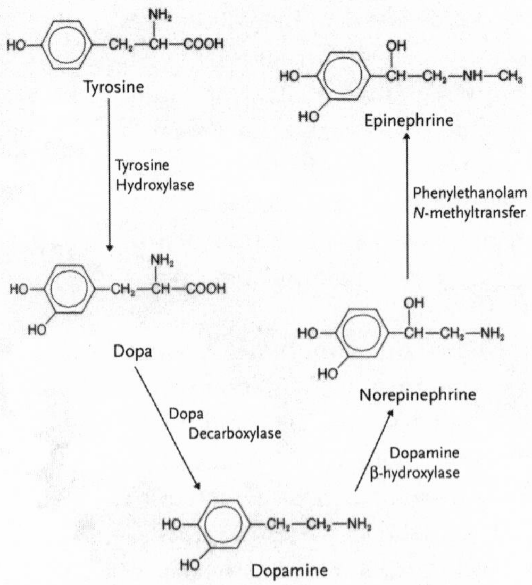

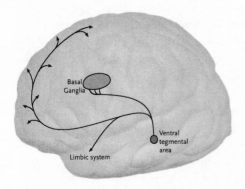

Figure 4–20: The Pathway for Synthesizing Dopamine, Norepinephrine, and Epinephrine

Figure 4–21: The Dopamine System

as a key neurotransmitter was discovered by Arvid Carlsson of Sweden, who earned a Nobel Prize for this discovery. We now know that dopamine plays an important role in many diseases, such as Parkinson's disease and schizophrenia.

There are three subsystems within the brain that use dopamine as their primary neurotransmitter. These all arise in the ventral tegmental area. One group, arising in the substantia nigra, projects to the caudate and putamen and is referred to as the nigrostriatal pathway. A second major tract, called the mesocortical or mesolimbic (or mesocorticolimbic), projects to the prefrontal cortex and temporolimbic regions such as the amygdala and hippocampus. The third component of the dopamine system originates in the hypothalamus and projects to the pituitary. These various dopamine subsystems are summarized in Figure 4–21. The dopamine system is fairly specifically localized in the human brain. Since its projections include only a limited part of the cortex and are distributed to brain regions important for cognition and emotion, it is considered to be one of the most important neurotransmitter systems for the understanding of these functions.

For many years, schizophrenia was explained by the "dopamine hypothesis." Its characteristic disturbances in cognition and emotion were thought to be due to overactivity in the dopamine system. Understanding the projections of the dopamine system also explains some of the side effects of neuroleptic drugs. The older "typical antipsychotics" are powerful D2 blockers and have potent side effects as a consequence of blocking dopamine receptors in the nigrostriatal pathway. The newer "atypical antipsychotics" have a weaker effect on D2 receptors and therefore have fewer side effects. The side effects are called "extrapyramidal" because they block transmission in basal ganglia regions, which are outside the pyramidal motor tracts (extrapyramidal). They are often referred to as extrapyramidal side effects, or EPS for short.

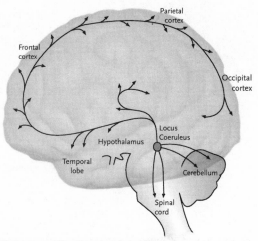

Figure 4-22: The Norepinephrine System

The Norepinephrine System

The norepinephrine system arises in the locus ceruleus and sends projections diffusely throughout the entire brain. These projections are summarized in Figure 4-22. As this figure illustrates, norepinephrine appears to have effects on almost every region in the human brain, including the entire cortex, the hypothalamus, the cerebellum, and the brain stem. This distribution suggests that it may have a more diffuse modulatory or regulatory effect.

Many of the medications used to treat mental illness, especially mood disorders, affect the norepinephrine system. Soon after they were developed, it was demonstrated that tricyclic antidepressants inhibit norepinephrine reuptake, thereby enhancing the amount of norepinephrine available to stimulate postsynaptic receptors. Julius Axelrod of the National Institute of Mental Health (NIMH) received a Nobel prize for discovering the mechanism of norepinephrine reuptake and antidepressant drug action. Monoamine oxidase (MAO) inhibitors also enhance noradrenergic transmission by inhibiting the ability of this enzyme to break down norepinephrine. These observations led to the proposal of a "norepinephrine hypothesis" for mood disorders. However, many antidepressants have mixed noradrenergic and serotonergic activities or purely serotonergic effects (e.g., the selective serotonin reuptake inhibitors, or SSRIs).

The Serotonin System

Serotonergic neurons have a distribution strikingly similar to norepinephrine neurons. This is shown in Figure 4-23. Serotonergic neurons arise in the raphe nuclei, localized around the aqueduct in the midbrain. They project to a similarly wide range of brain regions, including the

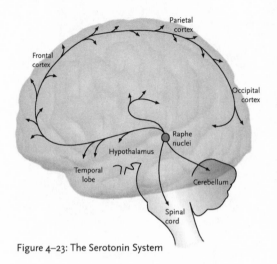

Figure 4–23: The Serotonin System

entire neocortex, the basal ganglia, temporolimbic regions, the hypothalamus, the cerebellum, and the brain stem. As is the case with the norepinephrine system, the serotonin system appears to be a general modulator.

A "serotonin hypothesis of depression" has also been proposed, largely because some antidepressant medications such as Prozac facilitate serotonergic transmission by blocking reuptake. However, the newer atypical antipsychotics, such as risperidone, also have a potent effect on the serotonin system, suggesting that it may be disregulated in schizophrenia as well. Thus, there are probably no simple "single neurotransmitter = single illness" relationships.

The Cholinergic System

Like dopamine, acetylcholine has a relatively specific localization in the human brain. This is shown schematically in Figure 4–24. The cell bodies of a major group of acetylcholine neurons are located in the nucleus basalis of Meynert, which is in the medial part of the globus pallidus. Neurons from the nucleus basalis of Meynert project throughout the cortex. A second group of acetylcholine neurons, originating in the diagonal band of Broca and the septal nucleus, projects to the hippocampus and cingulate gyrus. A third group of cholinergic neurons are local circuit neurons within the basal ganglia.

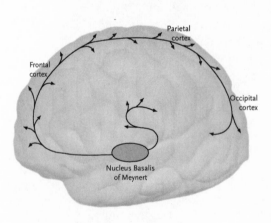

Figure 4–24: The Acetylcholine System

The acetylcholine system plays a major role in the encoding of memory, although the precise mechanisms are not understood as yet. Patients with Alzheimer's disease show losses of acetylcholine projections both to the cortex and to the hippocampus, and blockade of cholinergic receptors produces impairment in memory. Dopamine and acetylcholine share heavy concentrations of activity within the basal ganglia, and the drugs used to block the extrapyramidal side effects of neuroleptics are cholinergic antagonists, suggesting a possible reciprocal relationship between dopamine and acetylcholine in the modulation of motor activity and possibly of psychosis as well. Cholinergic antagonists ("anticholinergics") may also impair cognitive functions such as learning and memory in individuals for whom they are prescribed. Many commonly used drugs, such as certain decongestants, have anticholinergic effects.

The GABA System

GABA is an amino acid neurotransmitter, as is glutamate. These two major amino acid neurotransmitters appear to serve complementary functions, with GABA playing an inhibitory role, while glutamate plays an excitatory role.

GABAergic neurons are a mix of local-circuit and long-tract systems. Local-circuit neurons are those that stay within a given brain region. Within the cerebral cortex and the limbic system, GABAergic neurons are predominately local-circuit. The cell bodies of GABAergic neurons in the caudate and putamen project to the globus pallidus and substantia nigra, making them relatively long-tract, and long-tract GABA neurons also occur in the cerebellum. The distribution of GABA is shown in Figure 4–25.

The GABA system has substantial importance for the understanding of the neurochemistry of mental illness. Many of the anxiolytic drugs act as GABA agonists, thereby increasing the inhibitory tone within the central nervous system. Loss of the long-tract GABA neurons connecting the caudate to the globus pallidus releases the latter structure from inhibitory control, thereby permitting the globus pallidus to "run free" and produce the choreiform movements that characterize Huntington's disease.

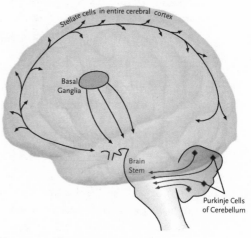

Figure 4–25: The GABA System

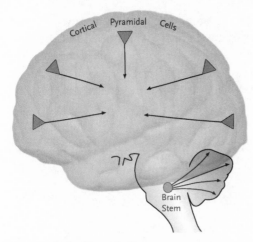

Figure 4–26: The Glutamate System

The Glutamate System

Glutamate, an excitatory amino acid neurotransmitter, is produced by large cells throughout the cerebral cortex and hippocampus. The distribution of the glutamate system is shown in Figure 4–26.

We have known for many years that glutamate, in addition to being a neurotransmitter, may also be a neurotoxin if present in amounts that produce excessive neuronal excitation. (This is one reason people avoid the use of monosodium glutamate, or MSG, as a salt substitute or flavoring agent . . . and probably why it gives some people headaches.) Recently, this knowledge was coupled with observations about the psychological and biochemical effects of phencyclidine (PCP) to suggest a possible role for glutamate in either psychosis or neurodegenerative diseases such as Huntington's disease. PCP blocks the effects of activating one subgroup of glutamate receptors, the NMDA receptors. PCP intoxication produces a psychosis characterized by withdrawal, stupor, disorganized thinking and speech, and hallucinations. The possible relationship between PCP, its characteristic psychosis, and its effects on the glutamate system suggest that glutamate may play some role in producing (or protecting against) the symptoms of psychosis. Some neurodegenerative diseases may be produced by excessive glutamatergic activity, which might cause neuronal degeneration through excessive excitation.

The Distributed Circuits of the Brain:
The Whole Is Greater Than the Sum of Its Parts

We all find it easier to think simply: Where is fear located in the brain? Where is memory?

The advances in neuroscience over the past several decades have taught us that such simple thinking is no longer possible. Localizing memory in the hippocampus or language in the left hemisphere served us well for many years. This way of thinking has more than a kernel of truth to it. But by the third millennium such simple localizations have become outdated neophrenology. They are only helpful if we recognize them for what they are: attempts to break the brain into component parts in order to make its complexity more tractable to human understanding. But after we complete this process, we have to put it all back together again.

A modern view of the brain sees it as composed of multiple distributed circuits. Some quasi-specialized regions exist, such as the motor strip, but no single region can or does perform any mental or physical function without coactivation and cooperation from multiple other regions. When we move, we usually use a circuit that includes motor cortex, somatosensory cortex, basal ganglia, cerebellum, and thalamus. We cannot say that movement is located in any one of those places. Movement occurs when all of them act together. As functional imaging techniques permit us to visualize how the brain performs more complex mental functions, such as memory or attention, we see that their functional circuitry is even more complicated.

Furthermore—amazingly—we almost never do only one "mental activity" at a time. Think about what happens in the brain when we tell someone about a past personal experience that means a great deal to us, such as when we attended the funeral of a close friend a few weeks ago. We use the "language system" to describe what happened. We use the "memory system" to recall the various events at the funeral and to describe our friend's life or appearance. We use the "limbic system" as the emotions of sadness and loss wash over us and a tear comes to our eyes. We use the motor system to move our lips, and perhaps to make a gesture or two with our hands. We use the "frontal executive system" as we think about our relationship with our friend and explain how generous and altruistic he was. In fact, I could probably go through all the "specialized brain systems" described in this chapter—and some that were not described—and show how they were used during this single activity of describing a past personal experience. Frequently, we use many systems to perform a single act . . . and all at the same time! That is what we mean when we say that the brain/mind is composed of distributed parallel circuits. The complexity of what the human mind can do is awe-inspiring.

The functioning human brain is like a large orchestra continuously playing a great symphony. We cannot point to any single part, or even combination of parts, and say that it constitutes either the orchestra or the symphony. Violins, violas, cellos, oboes, clarinets, horns, and other components all play together to create a rich texture of sound. At the right moment the trumpets join in, the cymbals are struck, or the cadence of the drums is added. Themes are introduced and re-echoed to produce a sense of coherence. The emotional coloring shifts and shimmers. The miraculous process of mental activity occurs, routinely, in all of us, all of the time, whether we are considered gifted or ordinary. Each of us—each individual mind/brain—not only plays a uniquely rich and complex symphony but also spontaneously composes its own score at the same time . . . and conducts it as well.

CHAPTER 5

MAPPING THE GENOME
The Blueprint of Life
... and Death

To see a World in a Grain of Sand
And a Heaven in a Wild Flower,
Hold Infinity in the palm of your hand
And Eternity in an hour.
—William Blake
Auguries of Innocence

Neatly packaged inside every cell in our bodies (except red blood cells) is a dense structure called the nucleus of the cell. We have been able to see the nucleus for as long as we have had microscopes—more than one hundred years. The discovery of the contents and purpose of the nucleus during the last century has been one of the major achievements in science—an achievement perhaps more important than discovering the law of gravity or formulating the theory of relativity. We now know that the nucleus contains DeoxyriboNucleic Acid (DNA) and the genetic code: the blueprint used to create and destroy life. In human beings, the nucleus contains between 30,000 and 40,000 genes, which are located on 23 pairs of chromosomes. (The exact number is still being debated.)

We tend to think that genes are destiny. One set of chromosomes is passed to us through our fathers and the other set through our mothers. The 23 pairs produced through this union give us hereditary traits that are sometimes viewed as creating genetic determinism. We fear that we are likely to inherit father's heart problems or alcoholism or mother's breast cancer. No one likes to feel deprived of autonomy or freedom. Because mental illnesses are among the most familial of human diseases, the possibility of genetic determinism is all the more frightening.

Fortunately, the story is not so simple—or so grim.

Here is the clue as to why. Every cell in our bodies has exactly the same DNA and the same genes. Yet these cells, built from the same genes, may be quite different from one another. We have brain cells, liver cells, kidney cells, heart cells, skin cells, stomach cells, eye cells, hair cells, and many others—all produced from the same set of instructions coded in our DNA and our genes. By some magical process, the basic blueprint of life laid down in the DNA gets modified and expressed so that cells differentiate to form hearts and brains and lungs and blood. Just as the organs in our bodies differentiate, so too we differentiate as people, from our parents, and from one another.

The metaphor of DNA as a blueprint is a good one. The blueprint coded in our DNA and genes is similar to the blueprint for a house that is going to be built. It lays out a footprint and a very general set of structural guidelines. A particular house may have three bedrooms, two baths, a pantry near the kitchen, a living/dining area and a two-car garage. But the blueprints do not go on to say whether the house will be made of wood or brick, whether the walls will be painted or papered, whether the floors will be wood or tile, whether the stove will be electric or gas. So too, genes set general rules: Each of us will have a brain, a heart, a pair of lungs, two arms and two legs, and so on. Genes create limits as to their size and specify a few "decorative details," such as whether our eyes will be blue or brown, or whether we will be male or female. But for many aspects of our lives and bodies, the genes interact with the environment around them inside the cell, inside the body, and in the complex world where the body must eat, sleep, breathe, and think.

Genes are not rigid autocrats that dictate our destiny. They are instead a responsive group of legislators that must listen to biological messages and respond. They decide to "turn on" and become active or to remain silent, depending on the circumstances that they must confront. Biologists call this process gene expression. The flexibility created by gene expression is our source of autonomy and our liberator from genetic determinism. Further, it is the means by which we can eventually use the new and powerful tools of molecular genetics and molecular biology to treat and prevent mental illnesses and other diseases.

Molecular biology and molecular genetics have begun a revolution in medical science that will have an impact on our lives far greater than that already created by the scientific revolution or the more recent electronic revolution. One major product of the molecular revolution has been the Human Genome Project, a major research initiative that has required the investment of billions of dollars and that is mapping all of the genes in the human body in increasingly fine detail. The next step is to identify the causes of the multiplicity of human illnesses at the level of the most basic mechanism, the genetic code. Armed with this information, we hope ultimately to be able to alter DNA's impact on suffering victims in order to either cure illnesses or prevent them from arising. The tools of molecular biology are so fundamental that they can be applied to illnesses as diverse as mental illnesses, cancer, and cardiovascular disease.

We have already witnessed some of the power (and the threat) of molecular biology. A feat that sounds like science fiction was achieved a few years ago: the cloning of Dolly the sheep. A duplicate model of an entire complex organism has been created. Although this sounds ominously like the human cloning envisioned in Huxley's *Brave New World*, scientists have immediately recognized the ethical dangers, and firm sanctions have already been put in place forbidding this in our current brave new world of the Human Genome Project. A few years after her creation, we observed that recently cloned Dolly was aging prematurely, since she was created through "older" DNA. This consequence was unexpected, and it has given a further early warning of the risks inherent in exploring uncharted territories of the human genome.

Many other powers and premonitions will unfold before our eyes over the next decade or two. If we want to understand how the molecular revolution can both improve and impair our personal destiny, that of our children, and that of future generations, we have to understand its fundamentals. What is DNA? What are genes? What are chromosomes? How are traits and diseases transmitted? How do genes work? How can they be modified? What affects their expression? A whole new confusing language bombards our brains: alleles, phenotypes, clones, recombinant DNA, mutations. If we want to exert control over our destiny, we have to learn and understand the language and concepts of molecular biology and molecular genetics.

Beginning with the Basics: Birds, Bees, and Peas

Human beings have observed for many years that both normal and abnormal traits are transmitted within families—e.g., eye color, hair color, a tendency to bleed excessively or to develop mental illnesses. The process by which much of this transmission occurs was formalized through the painstaking observations of Gregor Mendel, an Austro-Hungarian monk who was also a superb amateur botanist. He studied science in Vienna and then returned to his Augustinian monastery in Brno, located in what is now the Czech Republic. In the mid-1860s he conducted meticulous experiments in which he manipulated the sex life of peas in order to explore how traits were transmitted. He observed that the seeds for peas could be either wrinkled or smooth, that the flowers could be white or purple, and that their stems could be tall or short. He also recognized that each pea plant was essentially a hermaphrodite. That is, it contained both male and female sex cells. Sexual reproduction occurred when a pea plant transmitted male sex cells (contained in the pollen) to the female ovules.

This ingenious monk, who eventually became the abbot of his monastery, experimented with his peas by castrating the plants in order to do "genetic engineering." He removed the male portion known as the stamen, which contains the pollen, so that plants could not self-pollinate. Instead, he took pollen from other plants with different characteristics and brushed it on to the stigma (the female sex organ). In this manner, he could experiment with normal sexual reproduction and observe how traits were transmitted when they were "unnaturally" manipulated. For example, he could mate two plants with purple flowers or two plants with white flowers, or a white and a purple. Likewise, he could focus on the seeds and mate smooth with smooth, wrinkled with wrinkled, and smooth with wrinkled.

Mendel's observations, which we now refer to as "classic Mendelian patterns of transmission," created the framework within which genes and genetic transmission are currently understood. Mendel did not know that he was studying the effects of genes, because the word "gene" had not yet been coined. He was observing "traits" or "factors."

Back in the 1860s, the terms and concepts that we use in modern genetics did not exist. Many of them were proposed later because of Mendel's work. This Austrian monk was able to simply observe what happened to his peas based on experimental manipulation, and to deduce the principles of genetics . . . when he did not know that genes existed! All he could see were plants that varied on multiple traits, such as smooth or wrinkled seeds, or yellow or green pods. We now know that in his experiments he was observing what we call a phenotype (pheno = apparent, type = kind), or the form that is expressed from the genetic blueprint. What he was in fact manipulating was the genotype: the genes that produced the phenotypes, which existed in different patterns and were transmitted in different ways. His observations of patterns of transmission permitted him to develop fundamental genetic concepts, such as dominant versus recessive inheritance.

Mendel was struggling to identify patterns by manipulating a total of seven different traits in his peas. To simplify things in order to understand his observations more easily, we will focus on only one, smooth versus wrinkled seeds. When he crossed plants from a smooth father and a wrinkled mother, he noticed that all the offspring were smooth. These offspring are called the F_1 generation (the first filial generation). Next generations are called F_2, F_3, and so on. He then went on to cross the F_1 generation. Three quarters of the plants produced smooth seeds, while one quarter were wrinkled. In order to guarantee a strain of wrinkled seeds, the only certain method was to cross two parents that had wrinkled seeds. He also noticed that crossing plants that appear to come from two different pure strains did not produce a mixture, as one might expect as a first guess. Plants with white flowers crossed with purple flowers did not create lavender flowers in the second generation, nor did smooth and wrinkled seeds produce a mild mixture of wrinkled. Mendel concluded that the traits he was studying are "segregated." That is, they are passed on as an "either-or." He also concluded that the predominance of smooth seeds or purple flowers, in a rather consistent 3:1 ratio when two types were crossed, indicated that some traits must be dominant while others were recessive. For example, smooth seemed to be dominant, as did purple flowers.

We can summarize Mendel's observations and conclusions by using modern terminology. Mendel was observing the phenotype, but behind the phenotype was a genotype. When he crossed two plants that represented different phenotypes by placing pollen from the father on the stigma of the mother, he was transmitting their genes. Through sexual reproduction, genes are always inherited in pairs.

When he crossed his various strains, he was creating pea plants that were either homozygous (carrying identical genes from both "father" and "mother") or heterozygous (carrying different genes from each parent) (homo = same, hetero = different, zygote = germ cell). Each member of the pair is called an allele. The two alleles can be the same or different. If the alleles are the same (e.g., both code for smooth seeds), then the genotype is homozygous, and if they code for two different traits (one for smooth and one for wrinkled), then the genotype is heterozygous. Mendel could not see the genotype—only the phenotype. Nonetheless, he inferred the presence of something that must represent a genotype, since this was the most plausible explanation of the patterns of transmission that he was observing.

Figure 5–1 illustrates Mendel's experiments. In the figure, a large "A" stands for a dominant trait and a small "a" for a recessive trait. He started out by matching two plants that appeared

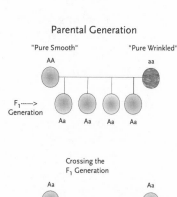

Parental Generation

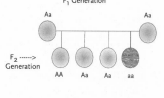

Crossing the
F₁ Generation

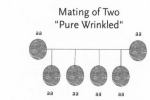

Mating of Two
"Pure Wrinkled"

Figure 5–1: Mendel's Experiments

to be "pure smooth" and "pure wrinkled." Since he did not in fact know the genotype, he could only guess as to the pureness. Their offspring, the F_I generation, would all be a mixture of "Aa," if indeed each of the parents was a pure "AA" or a pure "aa." At the level of the phenotype, the offspring could not be differentiated from one another. They all had smooth seeds. When he went on to produce the next generation, the F_2 generation, by crossing parents from F_I, he obtained three plants with smooth seeds and one plant with wrinkled seeds. The assortment of the alleles via sexual reproduction created a pattern that Mendel could not see directly, but could only infer: AA, Aa, Aa, aa. If he took his newborn plant with wrinkled seeds and crossed it with another plant with wrinkled seeds, then he could be assured of obtaining offspring that would all have wrinkled seeds. The logical inference from these multiple experiments is that the difference in seed type is controlled by a gene that codes for the surface coating of seeds. The gene has two alleles, one of which is dominant and one of which is recessive. The dominant allele causes smooth seeds, while the recessive allele causes wrinkled seeds.

Applying Mendel's Observations to Understanding Human Traits and Diseases

We are all fascinated with how traits get passed on from parents to children. Astute men, who never quite know what to say when confronted with a red-faced newborn, learn to make comments such as, "Oh, he looks so much like his mother." We enjoy looking at family photos and seeing ski-jump noses passed down through multiple generations. When we wander through the shopping mall, we may smile at the sight of a blue-eyed blond mom and dad with their three blue-eyed blond kids in tow. We all know that some traits, such as blue eyes, are recessive. We are especially interested, however, in whether diseases are being transmitted in our families—or more correctly, *which* diseases are being transmitted in our families. Mendel's observation that inheritance can be dominant or recessive has permitted us to understand the transmission of many diseases and to predict with some certainty the likelihood that these diseases will be transmitted from parent to child.

Human genetics are more complicated than the genetics of peas. We now know that our genetic information is located on 23 pairs of chromosomes. One pair, which determines whether we are male or female, are the sex chromosomes. Two X's make us female, while an X plus a Y makes us male. The other 22 are called "autosomal," and they govern the remainder of our body traits. These 23 pairs of chromosomes are shown in Figure 5–2.

We know what these 23 human chromosomes look like. Like human beings, they have individual personalities, as well as some common traits. Each one looks slightly different from the others. The area where they all taper in at the waist is called the centromere. They all have four arms. For some, such as Chromosome 1, the arms are approximately equal in length, while others have a pair of relatively short arms on top, as in the case of Chromosomes 13–22. The banding seen on the chromosomes in Figure 2 is produced when they have reacted with a stain called Giemsa. As was the case with Mendel's peas and all other things in nature that undergo sexual reproduction, we obtain one set of our chromosomes from our mothers and the other set from our fathers. Thus we are (theoretically) a 50/50 mixture of the genes from our two parents. The actual extent to which paternal or maternal genes influence us depends, however, on whether the particular allele that we have picked up for a given trait is dominant or recessive.

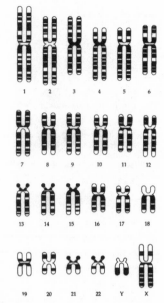

Figure 5–2: Appearance of the 23 Pairs of Human Chromosomes

Looking at pedigrees that summarize patterns of inheritance is a common tool in modern genetics. In order to understand the language of pedigrees, one needs to

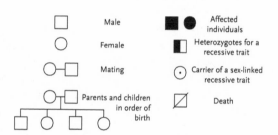

Figure 5–3: Common Symbols Used in Pedigrees

learn a few symbols. The symbols are summarized in Figure 5–3.

Supplementing Mendel's observations with our understanding of the multiple chromosomes in human beings and the specific role of sex chromosomes, modern medical genetics now recognizes four classical patterns of Mendelian transmission: autosomal dominant, autosomal recessive, sex-linked dominant, and sex-linked recessive. The four major types of classic Mendelian transmission are shown in Figure 5–4. The dominant and recessive alleles are noted symbolically, with "A" indicating dominant and "a" indicating recessive. The pedigrees are schematic, in that they all show families of four offspring. The sex and order of the offspring, and the distri-

1. Autosomal Recessive
(E.g., Tay Sachs, PKU, Cystic Fibrosis)

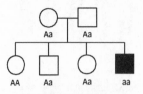

2. Sex-Linked Recessive
(E.g., Hemophilia, Duchenne Muscular Dystrophy, Fragile X Syndrome, Lesch-Nyhan Syndrome)

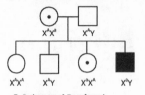

3. Autosomal Dominant
(E.g., Huntington's Disease, Male Pattern Baldness)

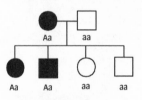

4. Sex-Linked Dominant

(Female Affected)

(Male Affected)

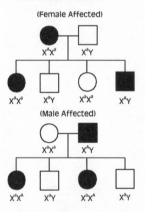

Figure 5–4: Four Types of Mendelian Transmission

bution of the alleles, is done in a standard way that probably would not occur in real life. In each figure the order of offspring is female, male, female, male. The distribution of the alleles starts with the first in the mother, which is paired with the first in the father, followed by the second in the father. Then the second allele in the mother is paired with the first and second in the father. The dominant allele is always shown before the recessive allele. This schematic notation creates a pattern where the disease is always expressed in the youngest offspring, who is always male. In real life, however, the patterning of the alleles could occur in any random order, so that a recessive disease could easily affect a first-born, or an autosomal-dominant disease could affect a last-born.

We have now identified a number of different diseases that are transmitted through these classical Mendelian patterns.

Recessive Diseases

The recessive illnesses are perhaps the most intriguing, since they sneak up on us "out of the blue." Recessive illnesses often strike families that are not expecting them. As Figure 5–4 shows, neither of the two parents has the disease. The only way they might know that they were carrying the disease gene (i.e., were "carriers") is if one of their antecedent relatives had the illness, such as parents, grandparents, aunts, or uncles. Examples of autosomal-recessive diseases include Tay-Sach's disease, phenylketonuria, and cystic fibrosis.

Tay-Sach's disease affects Jews of Ashkenazi descent and French Canadians. Children with Tay-Sach's disease are normal at birth, but subsequently begin to develop signs of nervous system degeneration. Nearly all die within three to four years after birth. Tay-Sachs can be due to one of several different mutations. Although all of these mutations are relatively rare, the risk of producing a child with Tay-Sach's is increased in communities or situations where the pool of people available for intermarriage is relatively restricted, as has happened when Jews were segregated into isolated villages or ghettos within cities.

Phenylketonuria (PKU) is also inherited as an autosomal-recessive Mendelian phenotype. In this instance, the illness appears to be due to two different mutations, each of which affects the enzymes that break down the amino acid phenylalanine, which is present in all the proteins that we eat. As in Tay-Sach's disease, children with PKU are normal at birth, but the accumulating quantities of phenylketones that occur as a consequence of the defective enzyme gradually damage the central nervous system, and the affected child begins to show signs of intellectual impairment and personality change. The urine also develops a peculiar odor. PKU was one of the earliest genetic diseases to be identified. Children are now routinely screened at birth for PKU and, if the test is abnormal, are placed on a special diet that is free of phenylalanine until they reach adulthood and the brain has matured. In this instance, identifying the abnormal physical process (inability to digest phenylalanine) and a preventive treatment (changing the diet) occurred before the genetic mechanisms were fully understood. The example of PKU illustrates some of the paradoxes of genetically transmitted diseases. One can make major strides in diagnosis and treatment without knowing anything about the genetics. Further, the case of PKU illustrates that "environmental treatments" (i.e., a change in diet) can be used to treat a genetic disease. PKU demonstrates that our genes do not govern our destiny.

Cystic fibrosis is another autosomal-recessive disease. Children with cystic fibrosis secrete abnormally large amounts of mucus in their lungs, which affects their ability to breathe and often leads to death from respiratory infection some time in the childhood or through young adult years. Although cystic fibrosis does not produce any brain injury, the children and their parents endure considerable mental suffering because of recurrent episodes of infection and the recognition that the disease will ultimately be fatal unless (or until) new treatments are developed. One allele that causes cystic fibrosis was identified in 1989, and its DNA was sequenced and found to involve a deletion of three bases that correspond to a single amino acid. (There is more information about the DNA code in the next section.) This mutation is seen in about 70% of cystic fibrosis carriers. The remaining 30% have turned out to be quite heterogeneous, however. It now appears that many different alleles can produce the cystic fibrosis phenotype, illustrating another facet of the complexity of modern molecular biology.

Albinism is another autosomal-recessive trait that occurs in human beings. It has no effect on mental function, but it is interesting because we know the biological mechanism by which the condition occurs. Albinos are "white" because their bodies cannot make melanin (melanin=black), the pigment which produces a dark coloration in skin, hair, and other parts of the body. Their DNA lacks the code that gives the "make melanin" command.

Sex-linked recessive diseases and traits are perhaps even more fascinating, and also better known than the autosomal-recessive disorders. Two of them, Fragile X syndrome and Lesch-Nyhan syndrome, affect the central nervous system and produce mental retardation and abnormal behavior. Hemophilia has been widely recognized because it affects European royal families, and muscular dystrophy because it has been the topic of Jerry Lewis telethons.

These various sex-linked diseases and traits come about because the Y chromosome, which makes men male, is embarrassingly small and conveys very little information. Traits carried on the sex chromosomes are subject to "genetic hen-pecking." For sex-linked traits, the information carried on the female X chromosome gets to dominate and have the final say most of the time. The upshot for X-linked genetic diseases is that women do pretty well. Even if they are carriers of a disease gene, their second healthy X chromosome kicks in and takes care of the problem. However, the guys frequently get stuck. X-linked diseases occur primarily in men and boys, because their weak Y chromosome is unable to stand up to the nasty commands in the genetic code on the corresponding X chromosome that they received from their mothers.

As shown in Figure 5–4, and as is the typical pattern for recessive disorders, both parents are usually phenotypically normal. They have no signs of disease at all. One of them, however, is a "carrier." Further, in sex-linked recessive syndromes, women are "carriers," while the disease phenotype is expressed only in men and boys. As shown in the sex-linked recessive pedigree, the mother carries the recessive allele on her X chromosome (X^a). When the abnormal recessive allele is linked to a normal dominant allele (X^A) in a girl or woman, the disease is not expressed because of the influence of the dominant gene. The disease arises when the abnormal X^a links with a Y chromosome—i.e., when the person is male. Since the Y chromosome carries very little information, the recessive gene from the mother dominates and controls the specific trait that the gene codes for. Some simple rules that describe sex-linked recessive inheritance include the following:

1. Most people who develop the disease are male.

2. There is no father-to-son transmission because sons become male by inheriting their father's Y chromosome, and cannot inherit the father's abnormal X chromosome.

3. Although none of the children of a man with the disease is affected, the daughters are all carriers, since they all receive his X chromosome.

Hemophilia is perhaps the most famous of the sex-linked recessive conditions, largely because Queen Victoria was a carrier who passed the disease on through her many children to many other royal families of Europe. Perhaps the most famous was Alexis, the great-grandson of Queen Victoria. He was the last of five children and the only son of Alexandra and Nicholas II, the last czar of Russia. Through the luck of the draw, the English royal family has escaped this recessive illness. Hemophilia is a disorder in which one of the proteins necessary for blood clotting is missing, and so the individual is predisposed to bleed excessively if cut or bumped. It is a very unpleasant and painful disorder that is treated with transfusions of the missing clotting factor. A blessing of modern medicine, blood transfusion has not been without its downside. Many hemophilia victims were accidentally transfused with the human immunodeficiency virus (HIV) before its existence was known, and therefore many have died from AIDS.

Fragile X syndrome, perhaps the most common cause of mental retardation, is also due to a gene on the X chromosome. In this case, the genetic abnormality has been well described. The term "Fragile X" refers to the fact that the information carried on the X chromosome is unstable. Boys with Fragile X have an abnormally large number of "trinucleotide repeats." These are a large number of copies of a DNA sequence (CGG), which interferes with the transfer of genetic information. This abnormality comes about when genes are copied. Fragile X is interesting because it varies in severity depending on the number of repeats. Normal people have somewhere between 5 and 50 copies of CGG on the X chromosome. People who have Fragile X usually have somewhere between 300 and 1,500 copies, and the severity of retardation is related to the number of the repeats. Since the level of abnormality is related to the number of repeats, female "carriers" may also be affected, while men are always affected.

Lesch-Nyhan syndrome is another example of X-linked recessive inheritance. In this case the genetic mutation has also been identified. Lesch-Nyhan syndrome is caused by an absence of the enzyme that metabolizes purines, which are essential chemicals in the body. The name of the enzyme is a real mouthful: hypoxanthine guanine phosphoribosyl-transferase. This syndrome was popularized by Richard Preston's book *The Cobra Event,* in which a "mad scientist" developed a plan for chemical warfare by designing a highly infective virus that transmitted a disease similar to Lesch-Nyhan syndrome. In circumstances other than *The Cobra Event,* only males can be affected, and both parents are normal. Although we do not know the exact mechanism, the enzymatic abnormality affects the brain and causes the child to have a variety of striking behavioral abnormalities. Children with Lesch-Nyhan syndrome mutilate themselves, chewing on their hands and hitting at themselves, and also strike out at others. This behavior usually requires that they remain restricted and confined. Some children with Lesch-Nyhan syndrome seem to have some insight into their condition but are unable to control it.

Some "normal" traits are also X-linked and recessive. Red-green color blindness, or inability to see the difference between red and green, is one of the most common. In this instance, the abnormal genes fail to code the information that permits the cone cells in the retina to recognize the wavelengths of red or green.

Dominant Diseases

X-linked dominant conditions, shown at the bottom of Figure 5–4 for the sake of completeness, are in fact extremely rare. One example is an unusual type of vitamin D–resistant rickets known as hypophosphatemia. As shown in Figure 5–4, women who are affected (i.e., carry a dominant allele on their X chromosome (X^A) pass the disease on to half their sons and half their daughters. Men who are affected (i.e. carry X^A) can only pass the disease on to their daughters.

The final major type of Mendelian transmission is autosomal dominant. In this form of transmission, a single dose of the gene is enough for the disease to be expressed. Figure 5–4 shows an example of this type of transmission. When a disease is autosomal, a father or a mother who has the illness can transmit it on to either a son or a daughter. The fact that the illness affects both sexes normally rules out sex-linked inheritance. Further, the phenotype appears in every generation, and each child has a 50/50 chance of developing the illness.

Huntington's disease is perhaps the most famous example of an autosomal-dominant disorder that is also a mental illness. Huntington's disease is discussed in more detail in chapter 10, but is also described briefly here. Unlike Lesch-Nyhan syndrome, Fragile X syndrome, Tay-Sach's disease, or PKU, Huntington's disease is not evident early in life. In fact, quite the opposite. It is a late-onset disease, which illustrates the fact that purely genetic diseases can be caused by mechanisms that do not take effect until relatively late in life. People who develop Huntington's disease usually do not begin to have symptoms until some time between age 30 and 60. Unlike people with Lesch-Nyhan syndrome or Tay-Sachs, they live long enough to have children and to pass the disease along to their offspring, who also have a 50/50 chance of developing the illness, which is characterized by personality change, abnormal movements, and the eventual onset of dementia and premature death.

Huntington's disease is an interesting example of the fact that scientists can succeed in locating the gene and understanding the nature of the mutation, but then can get terribly stuck on figuring out how the gene actually works to cause the manifestations of the disease. We know that Huntington's disease is due to a single genetic mutation on the short arm of Chromosome 4. Like Fragile X, the mutation causes multiple trinucleotide repeats. The good news for Huntington's disease is that an accurate genetic test is now available, so that individuals who are at risk for possessing the gene can determine whether they have it. If the test is positive, they can choose to avoid having children so as not to risk passing the gene on.

People can also suffer from a pattern of inheritance that is not sex-linked, which is referred to as sex-influenced inheritance. In this instance the mode of transmission is autosomal and can be transmitted by both men and women. However, the trait is more common in one sex than the other because sex hormones are involved in its expression.

Male pattern baldness is perhaps the best-known example of sex-influenced inheritance. Baldness is an autosomal dominant trait that affects only men, because the gene causing baldness can only be expressed when the person carrying it is producing high levels of testosterone. For this reason, boys and girls who possess the autosomal-dominant baldness gene have full heads of hair. As they grow up, women with the gene continue to have plenty of hair, but men start to lose their hair after their bodies have been under steady testosterone influence, usually in the early to mid-twenties.

Sex-influenced and sex-linked inheritance can appear indistinguishable if we look at disorders only from the point of view of the sex ratios of diseases and do not trace them back to their specific genetic mechanisms. Males have higher rates of Fragile X, Lesch-Nyhan syndrome, hemophilia, baldness, schizophrenia, and hyperactivity, to mention only a few conditions. We do not know why schizophrenia or hyperactivity are more common in males, but it seems likely that both are due to some type of problem in the process of brain development. Schizophrenia is intriguing because, like male pattern baldness, the symptoms become manifest at around the same time that testosterone levels begin to rise. Consequently, sex hormones could be playing some role in its manifestation or expression. Alternatively, however, the higher incidence of disorders such as hyperactivity and schizophrenia in males could indicate that they are partially caused by a gene that is sex-linked and recessive: an abnormality on the X chromosome that affects brain development or brain chemistry. These two possibilities both must be evaluated for disorders that affect men or boys more frequently or more severely.

One Hundred Years after Mendel:
The Double Helix and the Central Dogma

Although Mendel's work eventually provided pivotal insights for understanding the molecular and metabolic basis of inherited diseases and inherited traits, it passed unnoticed for many years, in part because Mendel himself did not publicize it. People speculate that he may have been concerned about disapproval from the Roman Catholic Church.

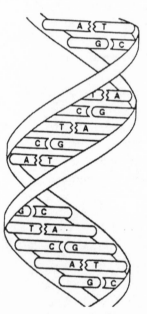

Figure 5–5: The Structure of the Double Helix

The next major breakthrough occurred nearly one hundred years later. Two brash young men, the American James Watson and the Briton Francis Crick, published one of the great head-turners in the history of science in *Nature* in 1953. Titled "Molecular Structure of Nucleic Acids: A Structure for Deoxyribose Nucleic Acid," and occupying only one page, this article presented a model that explained the structure of DNA: the famous double helix. The structure of the double helix is shown schematically in Figure 5–5. It consists of two spirals, each with a sugar-phosphate backbone, which are joined together by two pairs of bases, guanine-cytosine (or GC) and adenosine-thymine (or AT), which are like rungs on a twisting ladder. DNA contains the genetic code that dictates the structure and development not only of human beings, but of all living things. The amount and complexity varies from one species to another. DNA is also the mechanism by which life is preserved and conserved. It contains the germ lines that permit all living cells to reproduce and create a new generation of themselves. Watson and Crick ended their short article with a charming understatement: "It has not escaped our notice that the specific pairing we have postulated immediately suggests a copying mechanism for the genetic material."

The story of how DNA works has unfolded rapidly during subsequent years.

What Exactly Is a Gene?

One of the first great mysteries was how so much information could be summarized by something as simple as four base pairs. The number of ways that A and T and G and C can be combined is relatively limited. Most scientists at the time, including Watson and Crick, recognized that the function of DNA must be to code information that would control chemical functions—and specifically to create the protein substances that regulate chemistry and metabolism. Amino acids, of which there are only 20, join together to create proteins, and so the most plausible hypothesis would be that DNA somehow contained a code for the sequence of amino acids needed to create particular proteins.

Crick began to work with a South African molecular biologist, Sydney Brenner, and together they demonstrated that the "code" was based on a series of three bases, called triplets. Brenner named these triplets "codons." In 1961 two scientists at the National Institute of Health (NIH), Marshall Nirenberg and Johan Matthaei, identified the first letter in the genetic alphabet. They reported that the amino acid phenylalanine had the code UUU. Subsequently, the rest of the code was steadily broken. The code, and the 20 amino acids that it codes for, is summarized in Table 5–1. This code is sometimes referred to as "degenerate," because some amino acids have multiple codons. For example, leucine has six. Perhaps because codons come in threes, molecular biologists prefer to talk in three-letter words, and so the 20 amino acids are conventionally referred to by three letters.

Note that, in addition to the 61 codons for amino acids, there are three codons for a "STOP" command. The STOP codons indicate the end of an amino acid sequence coding for a given protein. Most proteins are comprised of 100 or more amino acids, and so a long sequence of codons is needed for any given protein. The complete sequence of codons for a particular protein is known as an open reading frame. Oddly, most of our DNA consists of long, useless sequences of nucleotides, interspersed between relatively short bursts of open reading

TABLE 5–1

The Genetic Code

Amino Acid	Abbreviation	Codons
Alanine	Ala	GCA GCC GCG GCU
Cysteine	Cys	UGC UGU
Aspartic acid	Asp	GAC GAU
Glutamic acid	Glu	GAA GAG
Phenylalanine	Phe	UUC UUU
Glycine	Gly	GGA GGC GGG GGU
Histidine	His	CAC CAU
Isoleucine	Ile	AUA AUC AUU
Lysine	Lys	AAA AAG
Leucine	Leu	UUA UUG CUA CUC CUG CUU
Methionine	Met	AUG
Asparagine	Asn	AAC AAU
Proline	Pro	CCA CCC CCG CCU
Glutamine	Gln	CAA CAG
Arginine	Arg	AGA AGG CGA CGC CGG CGU
Serine	Ser	AGC AGU UCA UCC UCG UCU
Threonine	Thr	ACA ACC ACG ACU
Valine	Val	GUA GUC GUG GUU
Tryptophan	Trp	UGG
Tyrosine	Tyr	UAC UAU
	STOP	UAA UAG UGA

frames. These large meaningless stretches are called "junk DNA." Only about 2% of the human genome contains the useful information that directs protein synthesis.

So we now have a definition of a gene. It is a sequence of codons that codes for the production of a particular protein. For example, the gene for monoamine oxidase (MAO) creates the enzyme that is used for breaking norepinephrine down into VMA (vanilmandelic acid). This particular enzyme regulates the tone of one of our key neurotransmitters, norepinephrine. Its activity is blocked by a group of antidepressant drugs that were widely used for many years, the MAO inhibitors. Other proteins, such as brain-derived neurotrophic factor (BDNF), regulate brain growth. Still other proteins form the neuroreceptors that nerve cells use to communicate with one another, such as the various dopamine and serotonin receptors. Clearly, the modestly complex story of how these proteins are created through our DNA is worth learning, since it will ultimately explain how mental illnesses are caused, treated, and prevented.

How Do Genes Work?

The role of DNA was summarized by Francis Crick in 1956, in a formulation often referred to as the Central Dogma. The code stored in DNA is a database. The purpose of DNA is to provide the instructions for the synthesis of proteins, which are the building blocks of cells and the regulators of metabolism and chemistry in our bodies. This concept is summarized by the following simple diagram:

The Central Dogma

	Transcription		Translation	
Duplication of DNA	→	RNA	→	Protein

The essence of the Central Dogma is that it specifies the pathway for the flow of genetic information, which must begin with the transcription of DNA within the cell nucleus, passing out to RNA (ribonucleic acid) located in the cytoplasm of the cell, which then translates the information into the synthesis of proteins. The flow of information, according to the Central Dogma, can pass only in one direction, with DNA acting as a template for RNA, and RNA acting as a template for proteins. Although Crick's Central Dogma has subsequently been found to have two exceptions, it still serves as the fundamental law about how the DNA database functions. The two exceptions are retroviruses, such as the human immunodeficiency virus (HIV) that causes AIDS, and the prions that cause Creutzfeldt-Jakob disease ("mad cow disease").

The double strands of DNA separate and act as templates for two fundamental purposes. The first function of DNA is duplication of the entire pair of strands of the double helix, in order to transfer the complete database on to new cells or new offspring. One type of duplication occurs during cell division (called mitosis). This process ensures that each of the two new cells will have precise copies of their single parent's DNA. In this instance, the two strands of the helix (comprising an entire chromosome) unwind. Each unwound strand serves as the template for the formation of a new strand that is complementary to it. The two pairs of complementary strands thus form the chromosomes within the nuclei of the two new cells that are created. Meiotic division, which forms the basis for sexual reproduction, is a variant of this process. In this instance, the

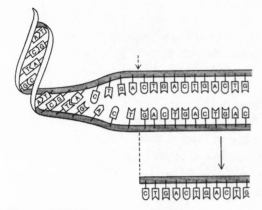

Figure 5–6: DNA Transcription and Translation

strands remain. One for each of the 23 pairs of chromosomes that human beings possess is assigned to each gamete (sperm or egg). When gametes from a man and a woman join one another, a new cell is formed, again with 23 pairs of chromosomes. This cell will form the basis of life for an entirely new organism, created from the mixture of paternal and maternal DNA.

The second major function of DNA, as formulated through the Central Dogma, is the production of proteins. In this instance, only a small component of an individual chromosome, the part containing a specific gene, is turned on or "expressed." In this instance, only a portion of the double-stranded DNA unzips in order to create a template on which transcription can occur. The first step in this process is done within the cell nucleus through the formation of messenger RNA, or mRNA. RNA is similar to DNA except for having only one helix (the sugar-phosphate backbone) and replacing uracil (U) for thymine (T). Figure 5–6 illustrates the process of DNA transcription and translation.

The process of transcription is initiated by a group of regulatory proteins called transcription factors. These factors cause the double-stranded DNA to unzip and open up, beginning at a point where a single gene starts. Through complementary base pairing, G's line up with C's and A's with U's to create a chunk of mRNA. Since the original formulation of the Central Dogma, we have learned much more about "junk DNA." We now know that a gene is not a compact series of codons that concisely summarizes commands to create proteins. Instead, units called "exons" are interspersed with units called "introns." The exons contain the codes for the sequence of amino acids that will be used to create the final protein. The introns, or intervening sequences, may contain regulatory information and serve as a critical basis for alternative splicing. This permits a single gene to have the potential to make several different proteins and multiplies the information conveyed. (Our 30,000 to 40,000 genes may be used to make more than 500,000 proteins.) Introns are removed during the transcriptional process that produces mRNA, and therefore they play no role in translation.

As the mRNA is reading the amino acid sequences from the original DNA template, it recognizes the conclusion of the sequence through the STOP codons. After the entire DNA sequence, including both exons and introns, is transcribed into a long RNA strand, the introns are then spliced out to yield a more compact and functional RNA that can direct protein synthesis. The process of protein synthesis occurs outside the nucleus in the cellular cytoplasm on small structures within the cell known as ribosomes. This process is mediated through smaller units of RNA known as transfer RNA (tRNA) composed of a three-nucleotide sequence known as an anticodon. The units of tRNA are used to attach the appropriate amino acids to a growing protein chain.

The Heart of the Mystery: Regulation of Gene Expression

This chapter began by pondering one of the great mysteries about our bodies: the fact that every cell contains the same DNA, and yet our cells have somehow managed to differentiate into many different types—brain cells, blood cells, liver cells, and so on. As cell differentiation proceeds in human beings, different organs are produced, such as hearts and lungs. Our brains further differentiate, creating their neurotransmitter systems and their functional systems such as memory and language, as well as the different personalities and abilities that make each of us an individual human being never to be duplicated in the history of mankind.

How does something so miraculously complicated arise from a single event—the unification of an egg carrying a single set of 23 chromosomes with a sperm containing a complementary set of 23 chromosomes. The two join to produce a single cell, which uses the directions coded in the DNA to build up a human life, first within the warm and watery world of the uterus for a brief nine-month period, and subsequently for another 70–80 years in an outside world that may be intermittently cold and cruel or warm and loving. All the physical growth or degeneration and all the physical and mental responses of that single human being are determined by DNA in the cell nucleus—as it interacts with external "events," from nearby changes in cell temperature to very distant ones such as the mental stress that occurs if a person is raped or mugged.

Regulation of gene expression is the secret behind differentiation and adaptability. Genes are not active and busy all of the time. They get turned on and off in response to some stimulus. The stimulus tells them to start producing proteins that can be used to build structural components of the body, such as cell walls, or to create the enzymes that will trigger chemical reactions that in turn regulate hormones, levels of neurotransmitters, and the like. Each gene is preceded by an "on-and-off switch," which either remains in the off position or somehow gets turned on and tells the gene to get busy and replicate.

The "off-on switch" was discovered by François Jacob and Jacques Monod, who earned a Nobel Prize for their discovery in 1965. They did their experiments in *E.coli*, a bacterium abundantly available in the intestines. They examined how its ability to use lactose, the food that it lives on, was affected by changes in its cellular environment. Their experiments led them to conclude that DNA must have a control system that would tell the gene when to start working. The system is a "signal box" located just before the part of the gene that contains the code for a particular protein. In bacteria it consists of two sections, called the promoter and the operator. Jacob and Monod described how the activity of these regions was determined by a complex group of factors. These included the amount of lactose (a simple sugar) in the cell's environment, the "*lac* off-or-on," and a regulatory substance that they called the "*lac* repressor." The *lac* repressor sat at the off-or-on site and kept the switch in the off position. Later experiments demonstrated that the *lac* repressor was a protein that changed its shape in response to the environment. The change in shape loosened its control over the operator and permitted the gene to be turned on. The gene could then start producing the enzyme needed to break down lactose and feed the bacterium. In other words, the previously silent and inactive gene was now being "expressed." The bacterium could begin to digest the food that it needed to continue living.

Subsequent experiments in molecular biology have expanded the story of the regulation of gene expression, particularly through the study of more complex organisms called eukaryotes. We know that eukaryotic genes contain multiple promotor and enhancer elements, which are short modular sequences that bind transcription factors. Each gene is regulated by a large number of such factors, working combinatorially.

We now know that some genes that need to be used frequently (sometimes called "housekeeping genes" because they perform so many key functions) tend to remain set in the "on" position, while the majority are set at "off" and only get turned on when some stimulus indicates that they are needed. We also know that regulatory substances can be either repressors or activators and that the control of gene expression can be either positive or negative. We know that the primary mechanism for turning genes on and off occurs because of the ability of the proteins to change their shape in response to the surrounding chemical environment, turning switches on and off as needed by fitting like a key into a lock. We know that some of the regulatory sites on genes have a shape that resembles fingers. Several famous examples are the "helix-turn-helix" pattern of proteins, which resembles a finger and thumb that can grab hold of the operator. "Zinc fingers," which are composed partly of zinc and also resemble a pair of grasping fingers, are present in more than 200 DNA transcription factors and are another important type of regulator.

Understanding the regulation of gene expression is likely to give us important clues as to how diseases are caused and how they can be treated. For example, many of the regulatory processes are mediated by the occurrence of stress, hormonal influences, and other factors that are not far removed from plausible mechanisms of mental illness. We know already, for example, that our bodies respond to stress by an outpouring of hormones from the adrenals—steroid hormones called glucocorticoids, such as cortisol. These steroid hormones bind to cytoplasmic receptors, which then move to the cell nucleus and bind to regulatory sites, thus initiating the transcription of proteins that we need to produce an adaptive response. Understanding this complex process gives us clues that may eventually be used to treat the many types of mental illness for which cortisol may play some causative role. We know that sex hormones such as testosterone and estrogen also activate receptor sites and initiate changes in the sex organs when their levels increase at puberty. Since at least one major mental illness, schizophrenia, has its typical age of onset after puberty and primarily affects males, understanding how testosterone regulates the expression of genes involved in postpubertal brain development may eventually lead to clues concerning one possible causative factor in schizophrenia.

Such insights have already been useful in developing treatments for several types of hormone-sensitive cancers, such as breast cancer, which is at least partially caused by estrogen turning on proteins that cause excessive cell division and proliferation. One breast cancer drug, Tamoxifen, works directly at the genetic level by blocking the site on the transcription factor that is normally turned on by estrogen. Contrariwise, estrogen may be a protective factor for the development of Alzheimer's disease. Understanding how estrogen might affect the mechanisms by which proteins such as amyloid or presenilin are overproduced in Alzheimer's disease (see chapter 10 for more detail) may ultimately provide a clue for treatment or prevention of this devastating illness.

"Wild Types" and Mutations

The replication of DNA for cell division, sexual reproduction, and protein synthesis is obviously an enormously complicated process. There are many opportunities for things to go wrong, and occasionally they do.

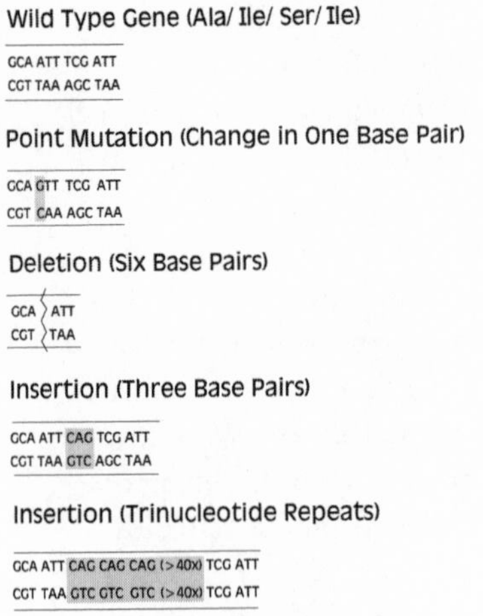

Wild Type Gene (Ala/ Ile/ Ser/ Ile)

GCA ATT TCG ATT
CGT TAA AGC TAA

Point Mutation (Change in One Base Pair)

GCA GTT TCG ATT
CGT CAA AGC TAA

Deletion (Six Base Pairs)

GCA ⟩ ATT
CGT ⟩ TAA

Insertion (Three Base Pairs)

GCA ATT CAG TCG ATT
CGT TAA GTC AGC TAA

Insertion (Trinucleotide Repeats)

GCA ATT CAG CAG CAG (> 40x) TCG ATT
CGT TAA GTC GTC GTC (> 40x) TCG ATT

Figure 5–7: Various Types of Mutations

Many of the diseases from which human beings suffer occur as a consequence of some disruption in the complex but orderly process of DNA replication. A variety of changes can occur in the genetic code, which are known as mutations. Some mutations may be completely harmless. Some may be beneficial. In fact, the creation of the rich array of animals and plants that surround us in the world today is at least in part a consequence of the occurrence of mutations that were desirable or useful, causing new forms of life to develop and then to survive. Some mutations, however, have undesirable consequences. These are the ones of interest for the study of disease. One of the goals in searching for disease genes is to identify the type of mutation that has occurred.

Figure 5–7 summarizes various types of mutations, using concise and oversimplified examples of short segments of DNA.

"Wild Types" Are Pretty Dull

The first example in the figure is a "wild type" gene.

When people who are not molecular biologists first hear the term "wild type," they sometimes find it puzzling. The term can be especially confusing to people who are interested in the genetics of mental illness, since it sounds as if it is referring to something "wild" or abnormal. But the term means exactly the opposite. It comes originally from the study of fruit flies, or Drosophila, which supplanted Mendel's peas as resources for studying genetic transmission in the laboratory, since their reproduction cycle only lasts 14 days. One of the traits often studied in Drosophila is eye color, which is red in the fly's natural state—that is, the genes controlling eye color produce red in the flies found in the natural world of orchards and fruit markets. Fruit flies can have a variety of mutations in the genes that regulate eye color. White eyes, for example, are a common mutation. When fruit flies were first studied in the laboratory, the form found in the natural world became the laboratory standard, the reference against which mutations could be compared. Thus the term "wild type" refers to the allele found most frequently in natural populations or the allele that is used in standard laboratory stocks. It represents the form of the gene that exists either prior to some genetic accident or before manipulation by scientific investigation.

In the example shown in the table, the gene contains four codons: GCA, ATT, TCG, and ATT, which code for alanine, isoleucine, serine, and isoleucine. Historically, fruit flies were very useful in studying the mechanisms by which mutations could be produced. For example, one method was to expose the flies to radiation, which caused a marked increase in new alleles. So, simply speaking, a wild-type allele is simply a fixed reference point, or standard, against which change can be identified. Wild type alleles and mutant alleles can be either dominant or recessive.

What Are Mutations?

Mutations are probably occurring in our bodies all the time, as our cells divide and our DNA is replicated. Most of these mutations probably have no effect on our health or well being or that of our children, because they are quickly repaired. So they pass unnoticed. Scientists are in the process of figuring out how and why mutations occur. Some reasons are already familiar to all of us. We know, for example, that exposure to radiation increases mutation rates, based on all the early fruit fly experiments. We also know that radiation exposure increases the risk for cancer, based on increased rates of cancer in some populations that have had greater exposure to radiation, such as survivors of the atomic bomb explosions in Hiroshima and Nagasaki, or health workers who used X rays in the early days before careful shielding was done. Mutations can lead to the uncontrolled overgrowth of certain types of cells, as in cancers such as leukemias, lymphomas, and myelomas. Some mutations arise for reasons that we do not fully understand as yet. For example, Queen Victoria, who passed hemophilia on to her descendants, had no known family history of hemophilia and may herself have experienced some type of mutation. We also know that "old DNA," or the DNA in the cells of older individuals, is more unstable and more likely to produce nondysjunctions, a phenomenon similar to mutation. This is the reason, for example, that Down's syndrome is more common in the children of women who conceive and give birth in their late thirties or forties.

When thinking about mutations in relation to diseases, it is important to be aware of the importance of mutation rate. In general, diseases that run in families tend to have low mutation rates. Huntington's disease, for example, is almost totally hereditary. Although due to a mutant allele, the mutation occurred quite a long time ago in the germ cells of a progenitor who then passed the abnormal allele on to subsequent generations. For example, most cases of Huntington's disease in the United States can be traced to only two immigrant families. Another pocket of Huntington's disease, which may have arisen from the same mutation in a different population, occurs in the people living around Lake Maricaibo in Venezuela. Obtaining DNA from this population permitted Nancy Wexler to lead the search that eventually permitted the identification of the gene for Huntington's disease.

At the other extreme, some disorders are characterized by a high mutation rate. In this instance, the disorders arise spontaneously in individuals who have no family history for the illness. Neurofibromatosis, or "Elephant Man Disease," is an example. This disorder is characterized by small tumors, which are called neurofibromas, that occur both on the skin and in internal organs of the body. People with this disorder also have abnormal skin pigmentation, known as café au lait spots. This disorder is autosomal-dominant, once it arises.

Cancer and many mental illnesses are probably also examples of disorders that have high mutation rates, since they sometimes occur in individuals with no known family history and then re-occur in subsequent generations. When the disease is autosomal-dominant, the pattern is more easily tracked, but the picture can be quite confusing if the disorder is recessive. Figuring out what causes mutations, in conjunction with understanding their effects and mode of transmission, is one of the several lines of attack for understanding the causes of diseases such as cancer and mental illness.

Table 5–2 shows the mutation rates of some well-recognized diseases. If a genetic mutation arises when there is no family history, it is referred to as "sporadic." Some sporadic mutations are not hereditary, since the bearers of the mutation are unable to have children. Other sporadic

TABLE 5–2
Mutation Rates of Sample Diseases

Autosomal-Dominant	Mutation Rate (frequency per gamete)
Huntington's disease	$.1 \times 10^{-5}$
Neurofibromatosis	$3^{-25} \times 10^{-5}$
Multiple polyposis (large intestine disease)	$1^{-3} \times 10^{-5}$
X-Linked Recessive	
Hemophilia	$2^{-4} \times 10^{-5}$
Duchenne muscular dystrophy	$4^{-10} \times 10^{-5}$

mutations, which do not interfere with fertility, become visibly familial (and are therefore often called "genetic"), since they are passed on to subsequent generations, sometimes in recognizable patterns. This discussion of mutations highlights the fact that the gene versus environment distinction is truly arbitrary, since mutations that arise from environmental causes, such as exposure to radiation, can later be passed on in genes.

Mutations can occur at many levels. Broadly speaking, there are two types: Gene mutation and chromosome mutation. In the case of gene mutation, an allele of the gene changes. The change in the allele can itself be of many different types. In chromosome mutation entire sets of chromosomes, whole chromosomes, or segments of chromosomes undergo change. An example of this type of mutation is trisomy 21, or Down's syndrome, in which three copies of Chromosome 21 are present.

Figure 5–7 on page 108 illustrates several types of mutations.

The first small sequence in Figure 5–7 is the "wild type" reference standard. The second example shows a point mutation, where a change in only one base pair has occurred. Through some unfortunate accident, such as an error in transcription or a change produced by ionizing radiation, the fourth base in the sequence, G, has been substituted for A. Because of this very tiny change, the entire sequence of amino acids will be altered. Whatever protein would have been produced can now no longer occur.

In a deletion mutation, a segment of DNA is permanently lost in the duplication process. Since a piece of the genetic code is missing, the gene can no longer command the production of the correct protein. The example shown in Figure 5–7 portrays the deletion of six base pairs, so that the coding information for two amino acids has been lost (first isoleucine, followed by serine).

Insertion is another type of copying error. In this instance, the insertion may be short or long. The first example of insertion in Figure 5–7 shows an insertion of three base pairs (CAG). The second type of insertion found in the figure is known as a trinucleotide repeat. This mutation is of particular interest to psychiatrists, since it is the mechanism for two well-known mental illnesses, Huntington's disease and Fragile X syndrome. In Huntington's disease, the base pair sequence CAG is repeated multiple times. A small number of repeats is normal, and the actual disease phenotype only becomes manifest when a person has 40 or more CAG repeats. Not only do trinucleotide repeats occur in these two important mental illnesses, but they also produce patterns of transmission that are sometimes seen in other disorders.

Trinucleotide repeats (and Huntington's disease) are associated with a phenomenon known as "anticipation." This means that the age of onset for an illness tends to become increasingly younger in each subsequent generation after the disease becomes manifest, and the symptoms also become more severe in successive generations. Thus, the children of a parent with Huntington's disease may become ill at an earlier age, have more severe symptoms, and die sooner. Since we now understand Huntington's disease at the molecular level, we can observe that the biological mechanism for anticipation is an increasing length of the trinucleotide repeats in subsequent generations. For example, individuals from early generations may have an onset at age 50 or 60, with only 40 or 50 repeats. In the next generation, however, the symptoms may develop at age 30 or 40, and the victims may have 100 repeats. There is some suggestion that other mental illnesses, such as schizophrenia, may also manifest anticipation, providing a clue that trinucleotide repeats may play a role there as well.

Searching for Disease Genes

By now readers are well aware that the human genome is not a rigid structure that immutably dictates our destiny. Genes are to the cell as our brains are to our bodies. They are created through dynamic processes, and they respond with flexibility and plasticity. They change in response to influences in the cell environment and in the body as a whole. They also govern the biology of cells by regulating the production of the building blocks of life. Learning how these important processes and functions occur abnormally, producing a vast array of human diseases, is the long-term goal of biomedical research using the powerful new tools of molecular biology and the mass of information produced by the mapping of the genome.

If we know so much, why has it been so hard? Why is it taking so long? These are the questions on the lips of many people who have a loved one with mental illness—or who have a mental illness themselves.

The Genetics of Complex Illnesses

"Keep it simple" has been the rule so far. Although novices to molecular biology may find autosomal versus recessive inheritance or codons and amino acids somewhat daunting, the story is in fact even more complicated. This story is about "complex illnesses" that sometimes disobey the simple Mendelian rules that we have just taken so much time to learn! In fact, most common genetic diseases are complex, while Mendelian diseases are much more rare.

As the tools and methods of molecular genetics and molecular biology became steadily more refined during the late 1970s and early 1980s, investigators had high hopes that they could be used rapidly to identify disease mechanisms for a variety of major illnesses. Achieving some breakthroughs increased expectations further. The gene for Huntington's disease was identified quickly, causing many to believe that other mental illnesses would be equally easy. Crucial substances such as oncogenes, genes responsible for cancer mutations, were also discovered, suggesting that new ways to treat or prevent cancer might also be identified. Much of this research occurred within the context of the Mendelian framework, which seemed valid at the time. Huntington's disease, after all, had well known Mendelian-dominant transmission with a low mutation rate.

Although many of us in psychiatry who had taken family histories from our patients for many years recognized that most mental illnesses did not follow classic Mendelian patterns, basic scientists who came from a background of molecular genetics nonetheless set out on an enthusiastic search for disease genes for mental illnesses. All the early studies were done within the Mendelian framework, since the techniques of linkage analysis (described below) only worked well if a particular mode of transmission was hypothesized. A few successes were announced, such as the discovery of the "bipolar gene," but most such claims proved ill founded, and ultimately led to disappointment and frustration.

Molecular geneticists were bumping their heads against a fundamental problem: Most human diseases are "complex illnesses." This pair of words, which is now heard over and over at scientific meetings, refers to the fact that most common diseases are probably caused by multiple genes (i.e., are polygenic), not all of which occur in every individual with the illness. They are also caused by a variety of nongenetic factors that influence gene behavior and expression (i.e., are multifactorial). Most of the common medical conditions that plague human beings fall into the category of complex illnesses that are not caused by single genes, do not follow classic Mendelian patterns of transmission, and are also caused by nongenetic factors. Diabetes, heart disease, and cancer are complex illnesses, as are nearly all the mental illnesses apart from Huntington's disease.

Consequently, scientists have begun to change their strategy. The search for the genes for cancer or Alzheimer's disease has been modified. Investigators are now assuming that most diseases will be caused by multiple genes, each of which may have a relatively small effect, and some of which must accumulate in an additive or interactive manner in order to produce the disease. Further, we also have begun to explore the role of nongenetic factors. The risk of developing Alzheimer's disease, for example, is increased by a variety of nongenetic factors, ranging from lower educational levels through experiencing a head injury or having general anesthesia.

Two More Complications: Penetrance and Expressivity

Two other concepts may also be helpful in understanding why classic Mendelian patterns are not followed in the "real life" of human diseases. Penetrance is the first. Penetrance refers to the fact that not all people who have a particular genotype actually show the phenotype associated with it. Some genes, such as the one for Huntington's disease, are "fully penetrant." This means that everyone who carries the genetic abnormality will eventually manifest the illness. Other disorders, however, have "incomplete penetrance." That is, the person carries the gene for the disease but never actually exhibits it. There are several reasons why a gene does not become penetrant, such as protective environmental influences or modulators of gene expression. Expressivity is a related concept. The distinction between penetrance and expressivity is summarized in Figure 5–8.

While penetrance is an "all or none" phenomenon, expressivity refers to a range of expression. When the genotype is present, it is expressed to a degree that may range from very mild to very severe, and different individuals may express different features of the same genetic syndrome. Neurofibromatosis is an example of a disease with variable expressivity.

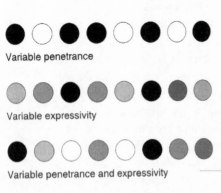

Variable penetrance

Variable expressivity

Variable penetrance and expressivity

Figure 5–8: Variable Penetrance and Expressivity

People who carry this gene may in fact have no physical signs of it at all, since their neurofibromas occur only on one or two internal organs. They may have a tiny brown spot and a fibroma on one toe or somewhere on the abdomen, which have passed completely unnoticed. At the far extreme is the severe neurofibromatosis manifested by "the Elephant Man," who was movingly portrayed a few years ago in a play and later a film. He had a deformed and ugly appearance that masked an intelligent and sensitive mind. Like penetrance, expressivity is modified both by nongenetic environmental factors and by the influences from the remainder of the genome.

These two factors, variable penetrance and expressivity, may make it difficult to identify genetic transmission even when diseases follow classic Mendelian patterns. In fact, for many years the existence of variable penetrance and expressivity led investigators to assume that these two problems were the primary "fly in the ointment" that was preventing easy identification of major illnesses. We now recognize others as well, such as the fact that there are complex illnesses or that we may have problems in defining the phenotype (discussed later).

The concepts of penetrance and expressivity also make it clear how hard it is to make absolute predictions from genetic tests. When a disease is absolutely known to be fully penetrant and fully expressed, there may not be an issue. Most diseases are neither, however. Therefore, a person may carry a gene and yet never manifest the disease phenotype. If genetic testing were used to make decisions about health insurance, job hiring, or other such things, such an individual could be treated very unfairly. On the other hand, a healthy person carrying a disease gene that is neither penetrant nor expressed carries an important risk for passing that gene on to an offspring, and therefore may have vital information to gain from genetic testing.

The Fivefold Path to Understanding Genes

Announcements that "the gene has been found for . . ." seem to occur nearly every week. Those uninitiated in molecular genetics and molecular biology tend to assume from such claims that some major public health problem is now solved. In the instance of complex illnesses—which constitute the majority of human diseases—it would be more accurate to say "*a* gene has been found," for *the* gene implies that there is only one. Since most human illnesses are a consequence of many genes of small effect, the initial announcement about *the* gene, presented with trumpet fanfares at press conferences, is followed by later reports that other investigators have been unable to replicate the results. At least some of the time the original report may have been correct, but replication will not occur consistently if the gene is one of many or if it makes a relatively small contribution to causology. If either of these possibilities is true, and they often are, then the genetic linkage or association will not necessarily be present at a statistically significant level when another population is studied. Further, even if *the* gene is found, as was the case with the original report of Huntington's disease, finding the gene is only the first step in the long process that is required for understanding the relationship between genes and human diseases.

The process of understanding the effects of genes in human illnesses involves five different steps. These five steps are summarized in Table 5–3. People who are really interested in keeping track of what is going on in "finding the gene" (more correctly, usually, "the genes") should become familiar with these five steps. The first one is the easiest. The job is not done until scientists reach the fifth.

Locating or finding the gene is the first step. Initially, this often means locating it on a particular chromosome. Methods for doing this are described in the next few paragraphs. A variety of methods are applied to determine on which of the 23 chromosomes the gene might be located. Thereafter, investigators focus in with finer and finer detail to a more specific location—e.g., the short arm, the long arm, and where the location is in relation to other known genes.

Cloning a gene refers to the process by which the particular gene is copied. Once the gene has been located, laboratory tools are used to isolate it in the chromosome and then transfer a copy of it to a vector, which is usually a virus. The hardworking vector then makes multiple copies of the gene. This technique, known as making recombinant DNA, was developed by Paul Berg, who earned a Nobel Prize for his efforts. When recombinant DNA techniques were initially developed, there was widespread concern that they could be abused or could cause unfortunate biological accidents. Consequently, a review panel was created to evaluate recombinant DNA research, and all projects were carefully scrutinized. Although anxieties about misuse of recombinant DNA in the laboratory are no longer high, since the risks have proven themselves to be minimal to absent, the technologies for gene cloning and reinsertion still have potential for abuse. Richard Preston's novel *The*

TABLE 5–3
The Five Steps in Understanding Genes

Finding or locating
Cloning
Sequencing
Identifying the product
Identifying the function

Cobra Event portrays a frightening scenario in which this technology is used for "germ terrorism."

Sequencing the gene is the task of identifying the order of the base pairs within it. This step was once laborious but is now very mechanized and straightforward. When this step has been completed, we can say that we really do "know" what the gene actually is at some meaningful level. Sequencing is the step that has permitted us to identify abnormalities such as the unusually large number of trinucleotide repeats in the Huntington's gene. The cloning and sequencing steps have been abridged by technological advances that permit scientists to identify genes in a certain region by computer search and use the sequence data to design "primers" to amplify the region they want to study more closely.

Identifying the product is the fourth step. Simply knowing the sequence of base pairs and amino acids does not tell us what product that particular gene creates. The "product" is predicted by computers, but it must be verified by biological techniques, such as showing that it reacts with the antiserum of a known protein. This step can be challenging. For example, we do not yet know what enzyme or structural protein the Huntington's gene produces. This is a crucial step in understanding the mischief caused by abnormal genes, but it can be extremely difficult to achieve.

Identifying the function of the abnormal allele is the final step. This is sometimes referred to as "functional genomics." Even if one knows the protein, one does not necessarily know its purpose. Initially, this fact may seem counterintuitive. If one knows that a gene creates a protein that works as an enzyme in a particular chemical process, such as the breakdown of norepinephrine to an inactive product, then wouldn't we know how and why the gene causes disease? Unfortunately, the process is not that simple. Many enzymes are widely distributed in the brain, and their chemical functions could lead to many different consequences, depending on location and timing in brain development or brain aging.

Methods for Locating and Understanding Genes

Our achievements in identifying disease genes have been supported by a steadily growing set of methods and technologies. While the average layperson, physician, or scientist from other fields does not need to know these technologies in detail, learning a modest bit about what they are can be helpful for interpreting the results reported in the popular press or even in the scientific literature. Table 5–4 summarizes some of the common methods that have been used or are currently being used for locating genes.

TABLE 5–4
Methods for Locating Genes
Linkage
Candidate genes and association studies
Genome scans
Snips and chips
Animal models

Linkage Studies

Linkage, the oldest of these methods, has been used in genetic studies for many years. Some of the very early psychiatric studies applied this technique to finding gene locations for mental illness by using knowledge of Mendelian patterns of transmission. For example, George Winokur, who was chairman of the department of psychiatry at Iowa for many years and an eminent investigator of bipolar disorder, made the observation that manic-depressive illness and red–green color blindness co-occurred in some families. He also noted that father-to-son transmission rarely occurred in bipolar disorder. This led him to propose that bipolar illness might be linked to the X chromosome. Although this observation has not been consistently replicated, perhaps because the gene is one of small effect in a polygenic multifactorial disorder, it may point to one of the genes involved in bipolar illness.

Linkage studies require collecting DNA from "multiplex families" (i.e., multiply affected families or families where two or more members have a particular disease). Locating multiply affected families, collecting samples to extract DNA, and making careful diagnoses is obviously quite labor-intensive. To extract the maximum power from linkage studies, scientists need families running through several generations with many ill members, and with the illness arising from only one side of the family. Because of the labor-intensiveness of this work, investigators usually band together in collaborative groups and pool their data. Much of the early work that reported "locating genes" for various mental illnesses has depended on these methods. At least until the late 1990s, when you read that "the gene for schizophrenia has been found on Chromosome 6," the data usually came from a linkage analysis. An alternate method is known as the the the "affected sib pair study." In this variant, DNA is obtained from two or more siblings who have the disease, as well as their parents.

Linkage studies began with simple studies using traits such as color blindness or blood groups, for which approximate locations (frequently referred to as "markers") were known. Later, newer methods were developed that use the tools of molecular genetics. As we began to understand the content and structure of DNA in increasing detail, and as we acquired the tools to cut and splice DNA, we developed the capacity to do more sophisticated linkage studies. Molecular biologists developed a set of markers, which are referred to as restriction fragment-length polymorphisms (RFLPs), which could be used as guidelines to locate genes. RFLPs are markers, not genes, and they can cover vast expanses of DNA. Further, they have no biological functions. Their utility has been that they simply set up signposts among the vast expanses of DNA in the human genome. They tell us which chromosome to look at for a particular gene, and where to look on that particular chromosome. RFLP technology has now been made obsolete by newer methods for linkage analysis, such as simple sequence repeat markers that can be amplified by primers, and most recently by single nucleotide polymorphisms (SNIPS), described later.

In a linkage analysis, multiple pedigrees are collected, and then statistical tests are used to predict the probability that a DNA marker is located close ("linked") to a particular disease. The most commonly used statistic is called the lod score. This is the logarithm of the odds ratio that linkage is occurring (i.e., log-odds = "lod"). Conventions have slowly evolved to decide how large a lod score needs to be in order to be clinically meaningful. Current guidelines are that a lod score of 3 is a reasonable cut off for Mendelian diseases, since at that point the odds favoring linkage are 1000:1. Likewise a lod score of −3 is considered to be strongly nonsignificant. Complex illnesses require a more stringent level, for a variety of reasons related to issues such as expressivity, penetrance, and the interactions of multiple genes.

Candidate Genes and Association Studies
Linkage methods are sometimes called "reverse genetics." They use information about known marker locations and familial patterns of disease transmission to identify the chromosome on which a gene may occur or the location on a given specific chromosome. They do not actually identify the gene itself, at least until further work has been done. Their contribution is to narrow down the place to search—from many millions of base pairs to a smaller and more workable number. Linkage studies are not usually driven by a theory as to what causes a given disease. The results of a linkage study are a bit like knowing that a terrorist group has set up its headquarters in Atlanta rather than Rome. They do not tell us about the mission of the terrorists, who they are, or who they work for.

Candidate gene studies start from the opposite direction. They begin with the theory that a particular gene might be involved in a given illness. Candidate genes are usually chosen based on theories about "what has gone wrong" to produce an illness such as schizophrenia, Alzheimer's disease, or bipolar disorder. Because many disease theories assume that the problem is with the function of neurotransmitters, the majority of the early candidate gene studies looked at genes for dopamine receptors or transporters, serotonin receptors or transporters, and so on. In candidate gene studies, DNA is collected from large groups of healthy volunteers and people who suffer from a specific disorder such as bipolar illness. These studies require that the candidate gene has already been cloned and that multiple alleles have been identified. The two groups are then compared for the frequency of the alleles. If the frequency is significantly increased in individuals suffering from the disease, as compared to the control group, then the likelihood is increased that a candidate gene has been found.

While this strategy sounds very promising, it has turned out to be difficult to apply in real life. Like linkage studies, this method has been plagued by initially optimistic reports, followed by multiple nonreplications.

Association studies can be done using either candidate genes or very closely spaced markers. The major limitation on the second possibility is the availability of closely spaced markers, but this problem will be reduced as the Human Genome Project produces finer and finer maps.

The candidate gene strategy has had some success stories. The most significant is the work originally done by Alan Roses, who successfully identified the ε4 allele of apolipoprotein E (or APOε4 for short) as a candidate gene for Alzheimer's disease. The original work was done by comparing a group of people with AD to a group of healthy normal controls (a case–control design).

The ease with which case-control comparisons of candidate genes can be done is a strength, but they also have the unfortunate weakness that confusing or erroneous results can be obtained. "Red herrings" such as ethnic imbalances can occur in the two samples, leading a scientist to think he has a finding about disease, when instead he only has a finding about ethnic differences. Association studies have therefore been strengthened by the introduction of an additional twist, family-based association studies. This approach is relatively easy to use, since scientists need only to identify one person suffering from the illness plus both parents and to obtain DNA from them. Alleles for a specific candidate gene are tracked through the two generations, using statistical tests to determine whether the parents are transmitting the allele for the disease to their children.

The identification of the apolipoprotein-ε (APOε)4 allele has been the great success story for this method. APOε has three alleles, with the ε4 allele being more highly associated with the disease in patients than controls in the original work of Roses and subsequent confirmation in numerous replications. The finding has intuitive appeal because APOε is a protein that binds to the senile plaques seen in patients with Alzheimer's disease. Although Roses' original findings were from case-control association studies, they have been confirmed in family-based association studies.

The case of the APOε gene also illustrates some of the problems of candidate genes/association strategies and molecular genetics in general. First, existing data indicate that approximately 60% of people with Alzheimer's disease carry the ε4 allele, while 40% do not. Further, a substantial percentage of people who have reached an age of reasonable risk (over 80) carry the allele but do not have Alzheimer's disease. Not only is the presence of an ε4 allele not a definitive or sole "cause" of Alzheimer's disease, but testing for ε4 in the population would be a risky strategy for identifying predisposed people for prediction or insurance purposes. Alzheimer's disease is almost certainly caused by a variety of nongenetic factors acting in concert with the ε4 allele when it is present, and people who carry the allele may not develop Alzheimer's because of a variety of protective nongenetic factors. Further, the ε4 allele is not the only "gene for Alzheimer's disease." Research to date has localized Alzheimer's disease to genes on many different chromosomes: 1, 14, 19, and 21. Even this illness, one of the current "stars" in the molecular biology firmament, has a long way to go in order for us to achieve a full understanding of its genetic mechanisms. It is important to realize, however, that this is not because of the quality of the science, which has been excellent. Rather, it reflects the inherent difficulty of the task.

An obvious strategy that is now often used is to study candidate genes in regions identified by linkage methods, combining the advantages of both methods. This will become an increasingly important strategy in psychiatry genetics, as findings are beginning to coalesce into a workable number of regions.

Genome Scans

The amount of information coded in our 23 pairs of chromosomes—the human genome—is obviously vast. The first step was to identify major reference points and locators, an accomplishment announced in 2000 as the "completion" of the Human Genome Project, conducted jointly from two different perspectives by Francis Collins and Craig Venter. This was only the first step in a large and long process. When mapping is totally complete, we will know the location and nucleotide sequence of all human genes. This does not, of course, necessarily mean that we will know either their products or their functions, but even the complete sequencing is an enormous achievement.

The long-term biomedical utility of the Human Genome Project will depend on linking its vast amount of information with carefully designed studies of human illnesses, using carefully collected samples of patients and affected families. Whole human genome scans have already been done for several mental illnesses, including schizophrenia and bipolar disorder. These studies have required the cooperation of large numbers of scientists working at multiple sites. To date these studies have used only linkage methods and have come up with a minimum of suggestive findings. These studies, like all genetic studies, must be interpreted cautiously, since the large number of statistical tests required leads to a risk for false positives, while variability in the disease phenotype or its genetic underpinnings may lead to false negatives. Nonetheless, scanning the entire genome, using either linkage/association studies or "snips" and "chips," will almost certainly substantially increase our knowledge of the genetics and molecular biology of mental illness during the next several decades, paving the way for improvements in treatment and prevention.

"Snips" and "Chips"

Like physicians, computer scientists, and government officials, molecular biologists seem to take a particular pleasure in speaking in code language that can sometimes be difficult for the average person to appreciate or understand. Snips and chips are among the buzz words that are currently in the air. "Snips" are an abbreviation for SNPs, which refers to single nucleotide polymorphisms. That mouthful refers in turn to variations (polymorphisms) in single bases (nucleotides) in the DNA sequence. The point mutation shown in Figure 5–7 is an example of the type of variation that can be detected through SNP testing. This type of change is relatively common, and SNP testing may be useful for whole genome association studies in which patients and controls can be compared to determine whether the patients have an increased rate of a particular allele that may cause or predispose them to an illness. The SNP strategy requires running through the genome and identifying the presence of these polymorphisms—a relatively daunting task, but one that is now underway.

Chips are a technological tool that permits scientists to rapidly process DNA samples using a "chip reader." This equipment is very expensive and therefore is only currently available in major laboratories. It permits scientists to scan all 23 chromosomes from a given individual or group of individuals and to determine their allelic variations. As this technology matures and suitable samples from affected individuals or families are obtained, chip technology will permit scientists to do "DNA crunching" much more quickly than has been possible in the past. Chip technology will expand enormously during the first several decades of the twenty-first century. It will be used to identify different alleles in patient and control groups in the search for disease genes. This will undoubtedly lead to many false-positive reports, because the risk of false positive findings is 5 out of 100 by standard scientific statistical conventions, and chips will compare thousands of alleles simultaneously and certainly find something. But eventually it will also lead to genuinely true findings, as reports are repeated and replicated.

Of Mice and Men

Animal models are yet another tool that is used to search for the genes that cause diseases in human beings. Animal models lend themselves very well to the study of diseases that affect the soma, such as cancer or diabetes, but they are more difficult to apply to diseases that affect the psyche. Nonetheless, they can and will be used to understand the genetics of mental illnesses. Animal models are potentially useful in two different ways.

First, animal models are sometimes developed in order to work out the basic neural mechanisms of disorders. The study of the fear-conditioning response, described in chapter 11, has been extremely helpful in delineating the neural circuitry and basic physiology of anxiety disorders. Since new medications for diseases as diverse as cancer and panic disorder are initially screened in laboratory animals, animal models are useful for drug development. As the example of fear-conditioning indicates, however, they may also lead back to more basic neurochemical mechanisms that "cause" the disorder and perhaps cause genes to turn on or be expressed.

Genetically engineered animals, particularly mice, are another important tool. (Mice are the preferred animal for genetic study of diseases that may affect human beings because they are relatively small and well behaved, unlike their well-named relatives the rats.) The "knock-out mouse" is the current darling of laboratory investigators interested in the genetics of human diseases. A knock-out mouse has been bred to have both copies of a gene removed from the mouse's genome, usually by replacing the wild type with a copy inactivated by genetic engineering. Genes are selected for "knock-out" because they produce a specific protein that may be interesting for understanding disease processes. For example, Marc Caron has developed a knock-out mouse that lacks the dopamine transporter gene. Since the dopamine transporter is no longer available to remove dopamine from synapses, dopaminergic tone is increased, and the mouse becomes hyperactive. A "mouse model" of this type obviously has interesting implications for the study of a disease like schizophrenia, in which dopamine activity appears to be abnormal, or other diseases that manifest behavior abnormalities such as attention-deficit hyperactivity disorder (ADHD). Some real surprises have already come from "knock-out" experiments. For example, mice that lack nitric oxide synthesase are also hyperactive. (Nitric oxide is a relatively recently discovered neurotransmitter and is the same substance that accounts for the efficacy of Viagra.)

A "knock-in mouse" has had a gene added rather than removed. In this instance, the gene carries a known mutation that predisposes to a given disease. The mouse is created by injecting the DNA mutation directly into a fertilized mouse egg. Knock-in (transgenic) mice containing oncogenes, or genes that predispose to the development of cancer, are currently one of the most widely used models. The transgenic mouse strategy requires that a known disease-producing gene has been identified. Once identified, the transgenic mouse can be studied in order to figure out how the disease-producing gene exerts its effects. Transgenic mice will be useful for the difficult area of molecular genetics that we call functional genomics.

The Problem of Phenotypes

The process of hunting for disease genes assumes that we know how to identify people who have the disease (the phenotype) so that we can then study them in order to find out what genes they are carrying (the genotype).

For some illnesses, recognizing the phenotype is easy. Some diseases are easily diagnosed, such as cystic fibrosis. In many cases, however, defining the phenotype and finding people who represent it is much more difficult. Several factors complicate our ability to define the phenotype.

One factor is variable expressivity, already described above. Even for an illness that has a clear and simple definition, such as neurofibromatosis, individuals who have it could pass unnoticed because they possess only one or two neurofibromas, which are present in their internal organs. They will be false negatives (erroneously diagnosed as negative for having the phenotype). In this instance, studies using the methods previously described, such as linkage studies or association studies, would be messed up because these people would not be identified as carrying the gene when they in fact did.

Another factor that complicates finding definite examples of the phenotype is variable age of onset. Genetic studies for disorders that arise early in childhood, such as Tay-Sachs disease or Fragile X syndrome, are easier to do because of this factor. All the people who are going to have the disease will have developed it, and one will not miss any cases who will become ill with the syndrome for the first time in another five or ten years. On the other hand, diseases with a late age of onset, such as Alzheimer's disease or Huntington's disease, can potentially be missed when one is seeking examples of the phenotype for genetic studies. Some people will not yet have manifested the illness, and therefore will not have the disease phenotype, even though they in fact possess the gene. These people will also have a false-negative phenotype when included in samples for study.

A third problem is variability of the genotype. This problem arises when we are unsure whether diseases that have the same clinical presentation and medical definition are actually the same illness at the genetic level. Diabetes is an obvious example. The disease phenotype is defined by inability to produce adequate quantities of insulin, and the resultant effects of insulin insufficiency, such as sugar in the urine, elevated levels of blood sugar, and weight loss. This clinical phenotype can be defined by simple laboratory tests that measure blood sugar levels under various conditions. Laboratory tests have traditionally been the means for identifying specific diseases and differentiating them from one another.

Yet people who develop diabetes vary in both severity and age of onset. Those who develop diabetes when relatively young are more difficult to manage clinically and have a worse outcome. They are said to have juvenile-onset diabetes. Other people develop diabetes in their thirties, forties, or fifties. They have a different level of severity and a milder clinical course. They are diagnosed with adult-onset diabetes. Do people from each of these "types" of diabetes have the same phenotype—the same disease? Or should geneticists consider these two different diseases? The current consensus is that these two types should in fact be considered two different diseases with different genetics and different pathophysiologies.

We have similar problems with mental illnesses such Alzheimer's disease, although in this case the "starting gate" for the onset of the disorder is set at a later age. We talk about early-onset Alzheimer's versus late-onset Alzheimer's. Here too, the severity and course tends to be different. Are these two different types of Alzheimer's disease, which should be separated in genetic studies, or are they in fact the same phenotype? As in diabetes, we have a definitive laboratory test that is used to define the disease: the presence of plaques and tangles in the brain, seen in post-mortem visualization after death. Both early- and late-onset Alzheimer's disease possess plaques and tangles. Further, the requirement that the definitive diagnosis can only be made after death makes defining the phenotype in genetic studies of Alzheimer's disease even more complicated. Out of convenience and necessity, most genetic studies continue to make the diagnosis based on clinical presentation before death rather than the definitive postmortem laboratory tests. In these samples, other types of dementia may be included accidentally. Such cases are called false positives.

Defining phenotypes for other mental illnesses is also plagued by many of these problems, since most have variable expressivity, variable age of onset, and questions about heterogeneity of the phenotype.

Mood disorders are a good example. We group them together under a single term, disorders of mood, but we are uncertain as to whether they all represent the same illness or whether they are many different illnesses. One test as to whether illnesses should be treated as "the same" in genetic analyses is whether they "breed true" within families. "Breeding true" refers to the fact that some disorders occur in only one form or type when they are traced through multiple generations of families. For example, early-onset Alzheimer's disease would be said to breed true if everyone who developed the illness in a given family manifested the symptoms prior to age 50. (Some families with Alzheimer's disease manifest this pattern, but most have either a late or a variable age of onset.) A classic question about mood disorders has been whether bipolar and unipolar forms should be considered the same, or whether they are two different illnesses. When pedigrees from families with bipolar illness are examined, it is clear that they contain mixtures of both people with bipolar illness and people who have only depression. Is this because people who are unipolar might become bipolar if they lived long enough? (This is not likely, since most people with bipolar illness develop their first mania before age forty, but the possibility of an onset of mania in the eighties cannot be excluded.)

As discussed in chapter 9, there is some evidence for breeding true for bipolar disorder. A person with bipolar illness has a higher rate of bipolar illness in family members than does a person who has unipolar illness (10% versus 5%). However, the rates for unipolar illness turn out to be about the same in the families of bipolar versus unipolar patients. So there also appears to be some overlap between these two potential subtypes. Some genetic studies of bipolar disorder treat severe depressions as if they are a mild form of bipolar disorder. This problem is not limited to the genetic study of mental illnesses. Family members of people with juvenile-onset diabetes also may develop adult-onset diabetes, making genetic analysis of these disorders challenging as well.

A major difficulty in conducting genetic studies of most mental illnesses arises because they lack laboratory tests that can be administered during the person's lifetime and used to define the presence of the illness. Among the major mental illnesses, only Alzheimer's disease has such a test. As pointed out in the case of diabetes, laboratory tests do not solve all the problems, but they do remove at least one aspect of the difficulty. With the exception of Huntington's disease, for which a single dominant and fully penetrant gene has been found, we have no definitive markers or diagnostic tests for any mental illnesses that can be administered to living people. The diagnosis of mental illness is usually based on clinical presentation—a clustering of signs and symptoms that co-occur in a familiar pattern.

As described in chapter 7, the development of standardized criteria for all mental illnesses has led to definitions that have excellent reliability. We can be reasonably sure that clinicians in Boston or San Diego, or in Tokyo or Ankara, will diagnose the same people as having schizophrenia or bipolar illness or panic disorder if they use standardized assessment procedures. For genetic studies, the concern is not about reliability (agreement between clinicians as to whether a person "has the disorder" using standard definitions). The concern is about validity (whether the standard definition refers back to a condition that leads to some kind of accurate prediction about causes, response to treatment, or outcome). During the next several decades, scientists seeking the genes for various specific mental illnesses will continue to struggle with finding the best way to define the phenotype. In the process, the existing clinical definitions will probably be refined.

Some investigators are currently exploring the possibility that the search for genes might work more efficiently if the definition of a given illness were pegged to some type of objective measure, rather than a clinical analysis. Such objective measures are sometimes called endophenotypes or intermediate phenotypes, since they measure "internal" aspects of the illness (endo = internal) rather than the symptoms seen on the clinical surface. Using this approach, an "intermediate" diagnosis can also be made in relatives who do not meet the full clinical definition, but who could possess the gene and manifest it only as an endophenotype. (These people would otherwise produce false negatives.)

For example, Phillip Holzman of Harvard University has shown that patients with schizophrenia have a variety of neurophysiological abnormalities, such as problems with eye tracking (smoothly following a moving target with their eyes). This eye-tracking abnormality is also relatively common in first-degree relatives of people with schizophrenia. This objective measure might be an indicator of a basic "misconnection syndrome" in schizophrenia or a basic "information-processing deficit," caused by a gene or group of genes that influence neurodevelopment. First-degree relatives might also share this gene, be clinically "normal," and yet have a subthreshold form of the illness that is not fully expressed. Breaking illnesses down into endophenotypes or component parts that may more directly reflect the underlying physiologic or biological processes is a useful strategy in the search for disease genes.

As the search for disease continues, investigators must also worry about another problem: environmental phenocopies. Environmental phenocopies are sporadic cases, but are caused by a "mutation" in the world around them rather than in their DNA. They are false positives from the genetic perspective. The concept of endophenotype implies that people may carry the gene, but not express the illness (false negatives). The concept of environmental phenocopies implies that people may express the illness, but not have the gene (false positives). Environmental phenocopies are particularly a problem if a disease is very common or has a large environmental component in its causology. For example, people may become depressed and meet the full clinical definition for major depression for a variety of reasons. At one extreme, they may come from a family that has a strong genetic diathesis for depression and develop the illness because they have a high genetic loading. At the other extreme, however, people may become depressed and manifest the full syndrome of major depression because of some early or recent life experience that predisposes them to become ill. Genetically, a person with "purely environmental depression" might be quite different from someone who has a strong familial loading. Pooling the two of them together in genetic studies could sabotage the results. For this reason, geneticists prefer to study multiplex families whenever they can, since this reduces the risk for environmental phenocopies.

Learning to Use the Blueprint of Life . . . and Death

As we complete the mapping of the human genome, we will have the tools in our hands to identify the molecular mechanisms of disease, to learn how specific genes interact with their environments, and to understand how we can manipulate these processes to treat a variety of diseases. The problems are complex, and progress will be slower than most of us would like. Nonetheless, the time is certain to come when we will be able to modify the course of illnesses such as schizophrenia or Alzheimer's disease through medications that intervene early in the disease and act directly on causative molecular mechanisms, rather than simply treating superficial symptoms. Very likely, these treatments will arise from the identification of aberrant gene expression that speeds up aging processes in the brain, modifies the orderly process of brain development, or permits hormonal regulators to fall asleep with the controls running "full speed ahead." Scientists will create medications that will correct the abnormal expression of disease-producing genes. Molecular biology will someday permit us to perform psychosurgery at the level of the gene.

CHAPTER 6

MAPPING THE MIND
*Using Neuroimaging
to Observe
How the Brain Thinks*

The important thing is not to stop questioning. Curiosity has its own reason for existing. One cannot help but be in awe when he contemplates the mysteries of eternity, of life, of the marvelous structure of reality. It is enough if one tries merely to comprehend a little of this mystery every day. Never lose a holy curiosity.

Nature reveals her secrets because she is sublime, not because she is a trickster.
—Albert Einstein

I decided to become a psychiatrist in the early 1970s, motivated by a strong desire to do research on major mental illnesses such as schizophrenia, depression, and dementia, each of which was fascinating in a different way. Cardiology (another specialty that I liked), appealing in its precision, was too easy by comparison. As a medical student I found mental illnesses to be the most interesting and challenging diseases that I had encountered. What could explain how some people experienced the loss of autonomy over their minds that characterized schizophrenia, leading to the intrusion of alien voices or the theft of their emotional vitality? What caused people to fall into a deep depression, depriving them of all confidence and self-esteem, just when things seemed to be going very well for them? Why did some older people, previously bright and alert, begin to lose their mental capacities, and ultimately their whole personalities, ending in a wordless fetal-like helplessness? Not only were these questions fascinating, but the diseases were very common. Getting a handle on any one of them would help millions of people.

Psychiatry and neurology were closely tied to each other in our medical school at the time, and no one in either department doubted that the three illnesses that interested me were brain diseases. Although the prevailing emphasis in American psychiatry at that time was psychodynamic, the Iowa department in which I trained had a broader view of the responsibilities of psychiatrists. Not only should we be skilled in individual and family therapy, like our colleagues at Harvard or Stanford, but we should be good general physicians who could handle most common medical illnesses if called upon, as any doctor might well be in a rural setting.

We were taught that even though we were specialists in psychiatry, we should be capable of taking a summer to work as a missionary doctor if we chose. Albert Schweitzer was a much-admired role model for doctors in that era. In fact, the former chairman of psychiatry whose family endowed the chair of psychiatry that I hold, Andrew H. Woods, was a missionary in China, and founded the second Chinese department of psychiatry in Shanghai. Trainees in psychiatry at Iowa did physical exams on all their patients, ordered routine laboratory tests, made routine medical diagnoses, and handled relatively simple medical problems on their own. Our identity was that we were doctors, who happened to specialize in diseases of the brain and the mind.

Research in the department of psychiatry in the 1970s was definitely brain oriented. The main tool was the electroencephalogram (EEG), as well as related measures of brain electrical activity occurring in response to stimuli, known as evoked potentials. Genetics was another major interest, particularly tracking the degree of genetic influence through the study of adopted offspring, which was applied to schizophrenia, alcoholism, mood disorders, and antisocial personality. A large neurochemistry division was busy studying the breakdown products of neurotransmitters (called peripheral metabolites), including serotonin, dopamine, and norepinephrine. A strong psychology division provided training in assessing cognitive functions and developing new tests that were more sensitive to early cognitive changes in dementia. This was a wonderful environment for a budding research scientist interested in psychiatry.

My problem was a nagging doubt about the power of most of these techniques to get to the questions *I* wanted to answer—those questions about how and why such fascinating symptoms could be produced by the human brain. Adoption studies could show that a disease had a genetic component, but they couldn't explain how that genetic component affected the brain and led to various symptoms and diseases (the field now referred to as functional genomics). Peripheral metabolites were messy, because they contained products from the rest of the body as well, and they couldn't really tell us anything all that specific about brain activity. The EEG techniques were the closest thing we had to measuring the brain, but only dementia seemed to produce EEG abnormalities, and these were pretty nonspecific. Further, the EEG techniques were very hard to use because they were prone to all kinds of artifacts that could blindside you (such as the effects of eye movements), and they produced a very weak signal. I reached the conclusion that none of these techniques was going to help with my questions.

My first decade of research in psychiatry was a lesson in humility. I could not do what I really wanted to do—literally "get inside the heads" of people who had illnesses such as schizophrenia or manic-depressive illness and figure out how their brains were producing such strange symptoms. The tools simply were not available. It was very frustrating. However, I tried to put my time and skills to use by doing what came to be considered solid descriptive psychopathology—designing instruments to define and measure the severity of symptoms in a reliable way, and applying the methods of experimental cognitive psychology.

Then, in the mid-1970s, something happened that changed my scientific life—and eventually that of many other research psychiatrists—forever. Iowa became one of the first medical schools in the United States to get a CT scanner. This, I said, would permit us to visualize and ultimately measure the brain in schizophrenia, and I was sure that we would find something interesting. It took a lot of convincing of my colleagues in radiology and internal medicine, as well as the government authorities and review committees to whom we applied for funding, but eventually we were able to do some of the pioneering work on brain abnormalities in schizophrenia. CT scanning launched a new era for psychiatry and was the first of an extraordinary array of tools available for visualizing—and above all for measuring—the brain in mental illnesses. After a decade or more of frustration, I suddenly was like a kid in a candy shop. Where other people saw "pictures of the brain," I saw a quantitative probe that could at last be used to *get inside the heads* of living people and measure what I saw. All that premed physics suddenly became very useful, since it formed the basis for most of the imaging technologies.

As time has passed, we psychiatrists have acquired more tools for studying brain structure and function than any single human being can become expert in. We now have the opportunity to study the living, working, thinking, feeling brain—a relatively new development in the history of mankind. A huge array of powerful technologies has launched a veritable voyage of exploration, in which investigators are mapping the human mind and brain in ways that were previously impossible.

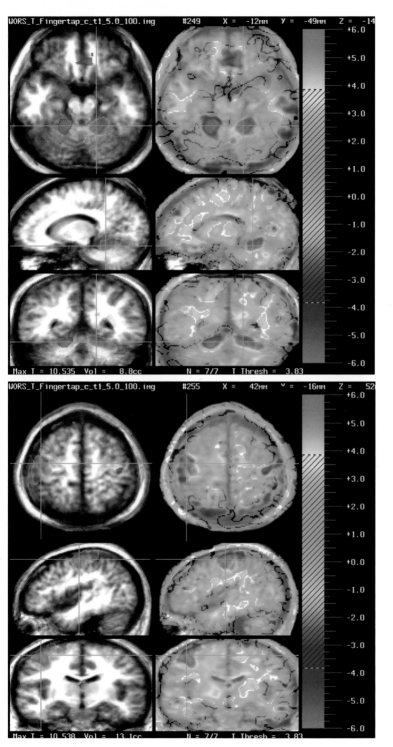

Figure 6–8: An fMR Study of Brain Regions Used in Finger-tapping Images are Shown Using Radiological Convention. (You are looking at the person face-to-face or from the foot of the bed, so her left is on the right side.) Figure 6–8A (top) shows activations in the motor region, with a large area of activation on the right (red), reflecting finger-tapping with the left hand, and a smaller area on the left (blue), reflecting finger-tapping with the right hand. Finger-tapping with the left hand requires more effort, and hence more blood flow and a larger activation. Figure 6–8B (bottom) shows the comparable activations in the cerebellum. The right motor region is linked to the left cerebellum (red), while the left motor region is linked to the right cerebellum (blue). This reflects a well-recognized anatomical circuit and validates the accuracy of fMR as a measure of blood flow.

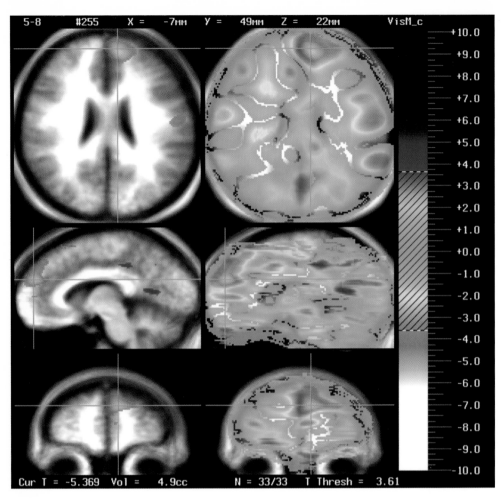

6-9A   Figure 6–9: PET Studies of Remembering New Faces versus Recognizing a Person as Male or Female. Figure 6–9A (above) shows increased blood flow in a "memory circuit" that includes right frontal, anterior cingulate, right parietal, and left cerebellar regions. Figure 6–9B (opposite) shows a very different circuit used for sex recognition, which activates more primitive regions of the brain, such as the straight gyrus and the inferior medial temporal area.

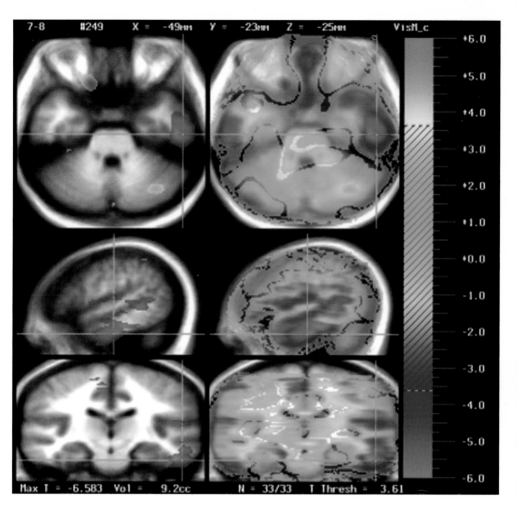

6-9B

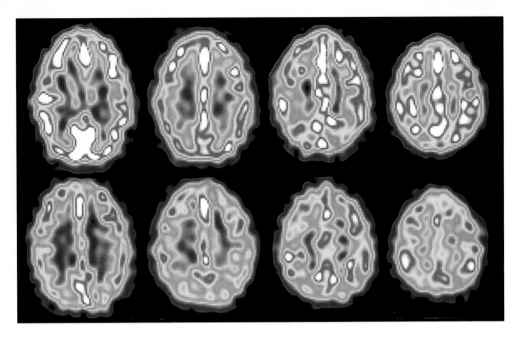

Figure 6–10: PET Studies Showing Differences in Blood Flow in a Pair of Identical Twins During a Memory Task.

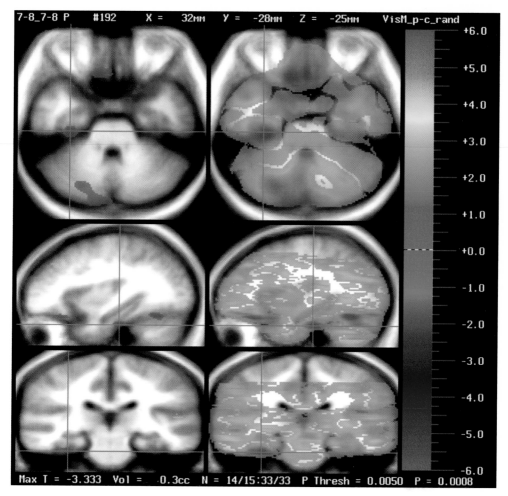

Figure 6–11: PET Study Showing Differences between Patients with Schizophrenia and Healthy Volunteers: Remembering New Faces. Areas of decreased flow in patients, shown in blue, include the straight gyrus, the parahippocampal gyrus, and the visual cortex.

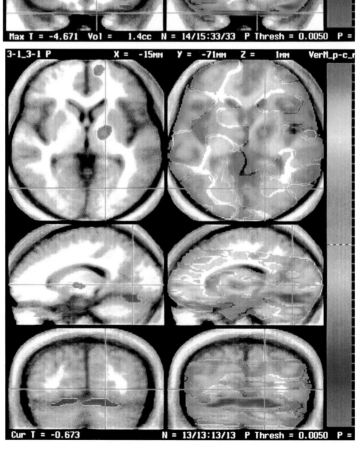

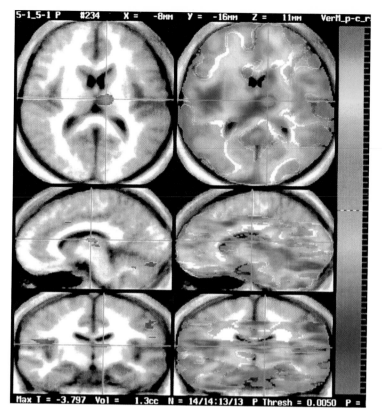

Figure 8–3: PET Study Showing Abnormalities in Schizophrenia during a Variety of Mental Tasks. The image is a difference map, showing regions in which patients have decreased flow (in blue) or increased flow (in red), when compared with healthy volunteers performing the same task. Figure 8–3A (opposite page, top) shows a memory task testing recognition memory of a group of words. The patients have lower flow in inferior frontal regions, cingulate gyrus, and the cerebellum. Flow is increased in anterior temporal regions. Figure 8–3B (opposite page, bottom) shows differences during a task requiring the subjects to remember a story that they have previously heard; flow is decreased in patients in anterior frontal regions, thalamus, and cerebellum. Figure 8–3C (above) shows differences during recall memory for a group of words; patients show decreased flow in the cingulate gyrus, frontal regions, temporal regions, thalamus, cerebellum. Although cortical regions change depending on the task, patients consistently have decreased flow in the cerebellum and often have decreases in frontal regions and the thalamus. This pattern supports the likelihood that abnormalities in brain circuits explain many of the symptoms and cognitive problems in schizophrenia.

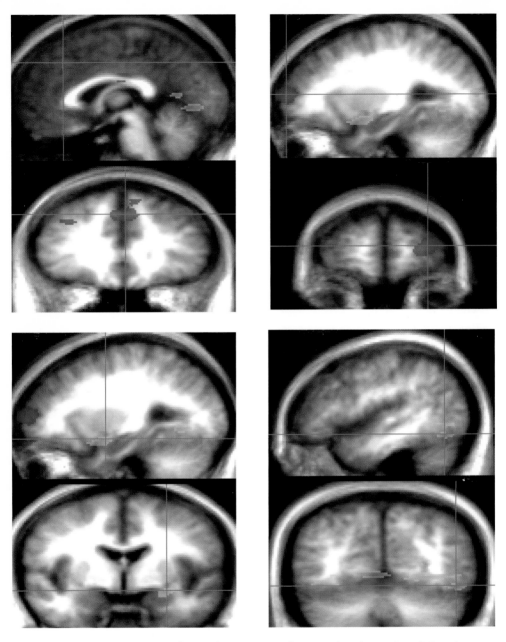

Figure 11–5: PET Images Showing the Response to Pleasant and Unpleasant Stimuli. Areas in yellow indicate increased blood flow when people look at unpleasant images. (See especially the bottom four, where blood flow is greater in the amygdala and visual association regions.) Areas in blue indicate increased flow in response to pleasant images. (See top two and middle left.) Pleasant images increase flow in cortical regions, especially in the frontal cortex.

TABLE 6–1
In Vivo Neuroimaging Techniques

Structural (Anatomical) Techniques
Computerized tomography (CT)
Magnetic resonance imaging (MR)

Functional (Physiological/Neurochemical) Techniques
Single photon emission computed tomgraphy (SPECT)
Positron emission tomography (PET)
Functional magnetic resonance (fMR)
Magnetic resonance spectroscopy (MRS)

Tools to See the Living Brain

The techniques used in this voyage of exploration are called *in vivo* neuroimaging. They are summarized in Table 6–1. In vivo neuroimaging permits us to visualize and study the brain during life (i.e., in vivo).

Mapping and Measuring Brain Anatomy: Structural Techniques

Computerized tomography (CT), the oldest of the new imaging technologies, was developed in the 1960s. Its inventors, Sir Godfrey Hounsfield and Alan Cormack, were awarded a Nobel Prize for this achievement in 1979. It has been available since the early 1970s and was the first in vivo brain imaging technique to become widely used in psychiatric research and in the clinical assessment of patients. Prior to CT, brain structure could be visualized only through the use of crude and invasive techniques such as pneumoencephalography. "Pneumos" were done by draining out the CSF from the ventricles, injecting air in its place, and then doing an X ray to visualize

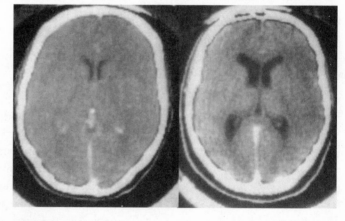

Figure 6–1: CT Scans from a Normal Control (left) and a Patient with Schizophrenia (right)

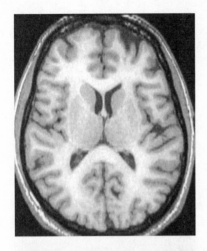

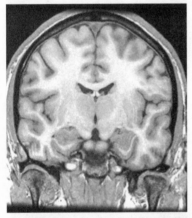

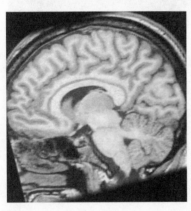

Figure 6–2: MR Scans in Three Planes: Transaxial, Coronal, and Sagittal

the size and location of the air-filled ventricles— a rather heroic procedure that was usually done only when a brain tumor was suspected. Figure 6–1 shows a pair of typical CT scans.

The slice runs through the brain in the transaxial plane, more or less parallel with the ground. It has been selected to pass through the ventricular system (the large cavities in the brain that are filled with CSF). In a CT scan CSF is black, while brain tissue is gray. The scan on the left, from a healthy normal control, has small ventricles, while the scan on the right, from a person with schizophrenia, has large ventricles. In early CT research, investigators measured the size of the ventricles and the size of the brain and produced a measure called the VBR (ventricular:brain ratio). In more than 50 studies this measure was found to be larger in people suffering from schizophrenia. These studies laid the early foundation for the application of neuroimaging tools to the study of mental illnesses by firmly documenting that in at least one major mental illness, schizophrenia, brain abnormalities could be consistently found when groups of patients were compared with controls.

Magnetic resonance (MR) came to the forefront in the 1980s. Edward Purcell and Felix Bloch, who discovered the phenomenon of nuclear magnetic resonance, were also awarded a Nobel Prize in 1952 for their work. MR permits us to see the anatomical structure of the brain in fine detail and has now largely supplanted CT as a scientific tool.

MR permits reconstruction and visualization of images from the brain in all planes. Because of the complex three-dimensional structure of the brain, multiple perspectives are very useful. Coronal cuts (i.e., made perpendicular to the ground, running from front to back) are especially valuable for visualizing small subcortical structures of great interest to psychiatry, such as the caudate, the putamen, the amygdala, and the hippocampus. Resolution is superb with MR, producing "slices" of brain that look as if they were obtained in a pathology lab at postmortem. For the sake of comparison with CT, Figure 6–2 shows slices through the brain in three planes: transaxial, coronal, and sagittal (down the middle, moving from left to right). The transaxial plane is in approximately the same place as the CT scan shown in Figure 6–1.

Mapping and Measuring the Mind: Functional Techniques

The second major group of neuroimaging techniques are known as the functional techniques. These include single photon emission computed tomography (SPECT), functional magnetic resonance (fMR), and positron emission tomography (PET). These methods permit us to observe the brain while it is actually functioning as a mind—thinking, remembering, seeing, hearing, imagining, experiencing pleasure or displeasure, or sending messages across synapses using chemical messengers such as dopamine. Structural and functional techniques complement one another very nicely. Structural techniques permit us to see the brain in fine detail and to perceive how various parts are related to one another spatially. The functional techniques permit us to examine how the brain creates thoughts and responds to challenges by changing its metabolism and blood flow, how nerve cells converse with one another via their chemical messengers, or how various medications change brain function. We can place a person's functional image on top of his structural image (a process referred to as image registration) and determine the exact anatomic location of the functional change. Using the combined methods, we can map virtually unexplored territories in human brain development and human cognition.

Single photon emission computer tomography (SPECT) is the oldest of the functional imaging techniques. It uses tracers that tag blood flow or neuroreceptors with a variety of isotopes that emit single photons, which gives the technique its name. Like CT, SPECT is an older technique that helped to launch the imaging revolution, but is rarely used now except for the study of neuroreceptors. The study of brain blood flow in mental illness was pioneered by Seymour Kety in the 1940s, but the technology at that time was too undeveloped to do precise mapping. David Ingvar was the first to demonstrate brain abnormalities in schizophrenia with SPECT, observing that patients had a pattern that he referred to as hypofrontality, or decreased perfusion in the frontal lobes.

This finding has now been replicated many times by others, using both SPECT and PET.

Positron emission tomography (PET) was the workhorse of functional imaging for many years. It uses tracers that emit positrons (positively charged antimatter electrons). The methods for using it to measure glucose metabolism, an indicator of which parts of the brain are using the most energy, were developed by Kety and Louis Sokoloff at the National Institute of Mental Health. Just as MR has better resolution and more power than CT, so too PET has more precision than SPECT. The ability to measure glucose metabolism with high resolution permitted neuroscientists to undertake the process of mapping the neural basis of thought and emotion in the living brain and to compare the results with the findings of the older lesion methods, as well as to examine differences between the healthy brain and the brain during various states of illness.

The addition of a method to measure blood flow with PET has further expanded its capacity; because the tracer for measuring blood flow, H_2O^{15}, has a very short half-life (about two minutes), it permits scientists to take brief serial "snapshots" of the brain during mental activity, much as was done in the early photographic time-motion studies of the movement of the human body. PET is also used to label a variety of neuroreceptors in order to examine their density, distribution, and response to psychoactive medications. PET is extremely flexible because the isotopes that it uses (commonly C^{11}, O^{15}, and F^{18}) are widely present in biological substances, and therefore they are relatively easy to attach to informative molecules such as medications. PET can therefore be used to study a wide variety of metabolic, neurochemical, and physiological processes.

Functional magnetic resonance (fMR) and magnetic resonance spectroscopy (MRS) are two techniques that expand the capacity of MR imaging to the study of brain function. Both of these techniques exploit the ability of MR to vary how the signal is emitted from brain tissue (known as pulse sequences) in order to get different types of information about brain physiology and chemistry. Using fMR, investigators can visualize changes in cerebral blood flow in response to various types of perceptual or cognitive challenges, much as had previously been done with PET. Unlike PET, MR has the advantage of not requiring any exposure to ionizing radiation. Likewise, MR spectroscopy is being applied to measure phosphorus, hydrogen, and fluorine spectra, which may provide information about the integrity of membranes or concentrations of drugs in the brain.

Using MR Technology: What Are We Learning?

How do these new technologies produce "pictures of the brain" that can be used to answer scientific questions about the healthy brain and about mental illnesses?

The development of neuroimaging techniques only became possible after the invention of efficient, high-speed computers. All of the neuroimaging techniques require the acquisition of mind-numbing quantities of numerical data. They all use similar principles.

The brain is divided into a series of slices. Each of the slices is then divided into a series of tiny cubes (called "voxels," which stands for volume elements). In early MR studies, slices were a centimeter thick, and voxels were therefore relatively large. As the technology has become more sophisticated, voxels have gotten smaller and smaller, and the pictures are clearer and can be used for new applications, such as the visualization of the surface of the brain. As of this writing, my MR research uses a voxel size of 0.5 cubic millimeters, and that will probably get smaller in another few years. With MR you can never get too thin. These tiny voxels permit us to see the brain in very fine detail.

A single brain contains thousands of these small voxels. The voxels are turned into picture elements (called pixels) by perturbing the brain tissue in some manner, which varies with the particular imaging method. That perturbation produces a number (called the signal intensity) that characterizes the tissue in each voxel, and the size of the number is then turned into a color or a shade of black/gray/white that reflects the nature or state of the tiny chunk of tissue. You can think of these images as a huge three-dimensional dot matrix of pixels, which are then usually portrayed two-dimensionally as individual slices.

The signal intensity in each voxel is produced by exploiting the fact that our bodies are full of hydrogen protons. Because they are protons (i.e., positively charged ions), they are like tiny bar magnets (referred to as paramagnetic), which possess a magnetic moment (a positive and negative end, just like a magnet). Therefore, they are sensitive to being placed in a magnetic field. When an MR scan is obtained, a person's brain is put inside a large donut-shaped magnet. Normally the little bar magnets/hydrogen protons in our bodies are lying around in random disarray, with their magnetic moments pointing every which way. Being placed inside the large donut-shaped magnet causes the hydrogen protons to be lined up neatly like a little column of well-behaved soldiers. This tidy formation concentrates the force of the magnetic moment produced by the hydrogen protons so that it is large enough to be measurable. Then the protons are stimulated with a radio frequency signal that gives them extra energy. The effect of the energizing jolt is brief, however, and they quickly begin to relax and return to their original energy level. As they return, they give off a signal, which is picked up by computers as a "signal intensity" and used to create MR images. By convention, MR uses a scale of black/gray/white to portray the size of the signal intensity, which is a number that ranges between zero and somewhere around 200.

MR is an amazing technology. All of us who were the first subjects for MR studies of course wondered if something in our bodies might dislike having our hydrogen protons subjected to such military discipline, but by now millions and millions of people have been scanned, and no adverse effects have been found. Unlike all other imaging procedures, this one requires no exposure to ionizing radiation and can therefore be used safely in large samples of healthy people. The simple description of "how MR works" in the paragraph above does not do justice to all the ways that collecting MR signals can be tweaked to produce information. The use of MR to obtain measurements of function is described in more detail in the next section. However, the application of MR to study brain structure is a fascinating and steadily developing discipline in and of itself.

For example, prior to MR all of our estimates of brain characteristics were based on postmortem tissue, which is necessarily a limited resource. It is relatively scarce. It is expensive to store. Once you slice it to examine it, you can't retrench and put things back again. And most people who die are older and have developed some kind of illness that led to their death, so even "normal" postmortem tissue is not exactly normal. On the other hand, using MR we can study large numbers of people of all ages, so that we can begin to examine fascinating questions such as how the brain changes during childhood and adolescence, or how it changes as we age, or how genes and environment affect the brain to produce individual similarities and differences.

We can use the beautiful three-dimensional capacity of MR and the wizardry of computers to visualize what brains look like more clearly. We can slice, reslice, and reslice again—looking in all three of the standard planes (coronal, transaxial, and sagittal) all at once. We can make our slices very thin. If we did this to postmortem tissue, it would be a bunch of cubes or would fall apart. But with an MR scan all we have to do is close the file and store it, and we can look at it again and again, changing how we examine it each time, depending on the question we want to ask. And we can also obtain very elegant quantitative measurements using our special locally developed software, some of which simply cannot be obtained from postmortem tissue.

For example, we might want to know how much of the brain consists of wires (white matter, WM), how much of it consists of neuronal cell bodies (gray matter, GM), and how much is cerebrospinal fluid (CSF). It is impossible to dissect those three things apart with postmortem tissue. In fact, the CSF all drains away when the brain is removed. But tissue classification programs have been developed that can provide this measurement for every individual person whom we scan. Figure 4–1 of chapter 4 provided an illustration of this technique,

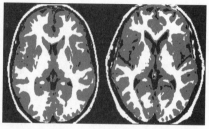

Figure 6–3: MR Scans of Tissue-Classified Images: A computer program has classified the tissue as gray matter, white matter, and CSF. The younger person is shown on the left and the older on the right.

which exploits the fact that the various tissues vary in their signal intensities. Figure 6–3 shows a pair of brains from a healthy twenty-year-old and a healthy seventy-year-old. The difference is obvious. The older person has an increase in CSF both in the ventricles and on the surface of the brain, as well as differences in the appearance of gray and white matter. We can also tell you (or anyone who obtains a scan from us and asks) how much of each brain consists of GM, WM, and CSF. For the youngster the measurements in cubic centimeters are GM 671, WM 401, and CSF 67, while they are GM 678, WM 346, and CSF 279 for the oldster. The CSF is markedly increased in the oldster, while the white matter is decreased. These changes reflect overall brain shrinkage and indicate that the primary loss appears to be in the connections between nerves. The gray matter is about the same in volume, but it is making fewer connections with friends and neighbors.

We might wonder how much individual variation there is in the sulci and gyri of the brain. Prior to MR the definitive book on this topic was based on a sample of 20 postmortem brains. With MR we can create highly accurate reconstructions of many different individual brains,

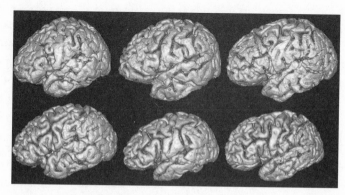

Figure 6–4: MR Scans Reconstructed to Show Variability in Brain Surface Anatomy in Six Different People

compare their variations, and measure surface characteristics as well. Figure 6–4 shows the brains of six different individuals. In general, their main sulci are similar (i.e., the sylvian fissure and central sulcus), but there is also a great deal of individual variability in gyrification. The next step, currently underway in our lab, is to determine whether individual variability in gyrification is related to psychological and mental characteristics. For example, we might expect that people with a higher intelligence quotient (IQ) might have a higher degree of gyrification, indicating that their brains were more intricately folded.

Our ability to produce quantitative measurements with MR has already allowed us to make one surprising discovery about brain characteristics and mental abilities. Lee Willerman was the first to report a relationship between brain size and IQ, using extreme groups—a group with very high IQ and another group with lower IQ. Using a larger sample drawn from the broad population range of IQs, we had been examining the same topic and came up with essentially identical findings—having a larger brain was correlated with a higher IQ, even after adjusting for overall body size (i.e., taller people also have larger brains, but being taller is not significantly correlated with being smarter). Several other groups have subsequently confirmed these results, and we have as well, in a second replication study, and so the finding seems fairly solid. The correlation is relatively small—around .3. A correlation like this can be squared and used to indicate how much of the variance (the variability in the population being studied) is explained by the correlation. A correlation of .3 indicates that only 9% of the variability in IQ is explained by brain size.

IQ, commonly thought of as a measurement of intelligence, is really just a measure of what IQ tests test for—they are probably more closely related to school performance than intelligence in its broadest sense (e.g., problem solving ability, street smarts, creativity). Nonetheless, until these studies were done, the standard teaching was that there was no relationship between mental ability and brain size. Why the relationship exists is another matter. We do not know the answer to that one. For example, did the people with bigger brains have better nutrition? A more stimulating environment during childhood? Bigger, healthier mothers?

What have we learned about the relative influences of genes and environment on human brain development?

This is not an easy thing to study in the human brain. Fortunately, identical twins provide a "natural laboratory" for studying this question. Identical twins have nearly identical DNA, and to the casual observer they often appear to be physically identical. But they also differ in some ways, and as they grow and mature, their different life experiences lead them to look and behave and think differently. The first sign of nongenetic influences is that one twin is invariably larger at birth, presumably because of greater placental nourishment. One twin is usually more outgoing and more of a leader. One often does slightly better in school. Identical twins can enhance their differences using nongenetic methods, such as wearing different clothing, eating different diets and gaining different amounts of weight, or changing their hair color. How similar or different are their brains? To the extent that their brains differ, we can infer that nongenetic environmental factors have affected brain development.

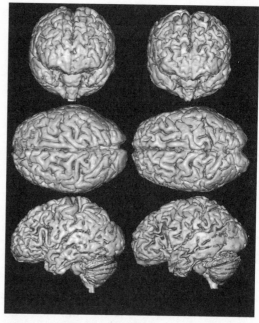

Figure 6–5: MR Scans Showing Differences in Surface Anatomy in a Pair of Identical Twins

In a sample of 12 identical twins that we have studied intensively, we have found them to have consistent and substantial differences in brain structure and function. The between-twin correlation of IQ scores in this sample is only around .6, suggesting that some of the variability in intellectual function is explained by environmental factors rather than genetic programming. (Some other studies have found higher correlations, perhaps because their samples were larger.) Using MR, we can visualize and measure the sulcal and gyral patterns of their brains as well. We can then use sulcal and gyral patterns as in vivo indicators of neurodevelopmental variance. As shown in Figure 6–5, which shows the brains from a pair of twins in this sample, the surface anatomy is variable. The twins are similar in the pattern and location of the large major sulci and gyri, but they have notable differences in the smaller ones.

What about aging? Can we measure the inexorable deterioration of the brain as it ages? Or does it deteriorate? At what point in time does the process of brain growth and development, characterized by proliferation and pruning back and programmed cell death, turn on itself and become unwanted loss of cells, spines, dendrites, and synapses?

Although MR cannot answer these questions in terms of the basic biological processes that the cell performs, it can look at the whole brain

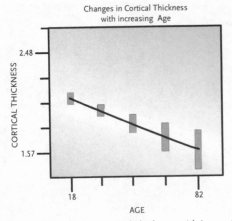

Figure 6–6: Decreasing Cortical Thickness with Increasing Age as Measured by Using MR Scans

and at least provide some clues about the answers. One of the great strengths of MR is its capability to examine large numbers of living human beings, to "remove" their brains and to perform measurements. So, for example, we used MR to collect brains from a group of 148 healthy normal volunteers who ranged in age from 18 to 82. We then applied some computer wizardry to measure various aspects of their brains in order to determine how the brain changes as people age. (The "wizard" program that we have developed at Iowa is called BRAINS, which stands for "Brain Research: Analysis of Images, Networks, and Systems.") Not surprisingly, although not very comfortably, we found that the brain does change with age, looking steadily more shrunken and atrophic in appearance over time. Our quantitative measurements of the cortical surface confirmed this. Figure 6–6 shows the results for one of the measurements, cortical thickness.

We noted that the thickness of the cortex shrinks as the surface becomes more atrophic. The thickness declines from a mean value of approximately 4.0 mm in the eighteen-year olds to 3.12 in the oldest individual in the sample (an eighty-two-year-old). Unfortunately for both men and women, the cortical thinning occurs significantly more rapidly in men than women. But Figure 6–6 also shows another interesting finding. In this graph the straight line shows the overall downward slope of the change, while the bars that cross the slope show the amount of variability at various points in time. The bars indicate that the young people are all very similar to one another. Apparently while the brain is developing or is in its early mature phase, there is very little variation in the thickness of the cortex from one person to another. However, as we begin to age and our brains begin to lose their nerve cells and dendrites and synapses, we start to diverge from one another quite considerably. Some of us apparently lose less, while others lose quite a lot. Some of us age gracefully or barely age at all, while others decline considerably. And remember, these are all healthy normal volunteers. Now, why are some of these older people drinking at the fountain of youth? What is the secret of keeping one's brain young? Is it mental activity? Good general health? A healthy life style? That is what we need to figure out next!

As fascinating as it is to use the technology of MR to examine gene/environment interactions, individual variability, or the aging process, the study of mental illness is the driving purpose of our research program and most other psychiatric teams that use imaging technology. What have we learned?

Schizophrenia has been studied with structural imaging for many years and therefore provides a good illustrative example of what it can tell us. CT provided the first major contribution. The first study demonstrating that abnormalities were present in a group of schizophrenic patients appeared in 1976 in the medical journal *Lancet*, reported by Eve Johnstone, Timothy Crow, and other collaborators working at Northwick Park, a prominent research center in Britain at that time. The 1976 *Lancet* study was greeted with doubt. At that time many people believed that schizophrenia was a "psychological illness" that was produced by a bad family environment. As Figure 6–7 illustrates, the doubters were wrong. Approximately 75 percent of the studies published using CT scanning have shown that there are significant group differences in the size of the ventricular system in patients with schizophrenia, as compared with normal controls.

Since enlargement of the ventricles probably reflects some type of broadly defined brain injury (either failure of the brain to grow and develop correctly or loss of nerve cells or dendrites), these findings suggest that patients with schizophrenia have had some type of injury to their brains that is related to their illness and that is probably physical (e.g., viruses, nutrition, birth injuries, abnormalities in genetic programming). It should be stressed, however, that

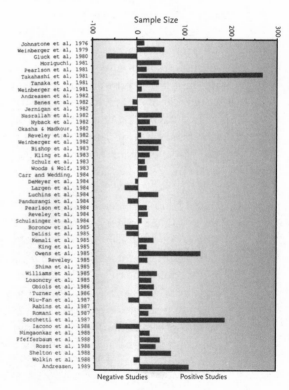

Figure 6–7: A Summary of CT Studies of Schizophrenia

these results are neither specific nor diagnostic. Ventricular enlargement is seen in other illnesses, such as Alzheimer's disease, and it is not present in every person who has schizophrenia. The major importance of this finding is that it has directed medical science to the study of the brain in schizophrenia by indicating conclusively that brain abnormalities are found in some patients who suffer from schizophrenia.

Once that discovery was firmly established, the next step was to attempt to find out how and why brain abnormalities occur. For this effort, magnetic resonance and the functional imaging techniques have provided the most important contributions, since they offer better anatomic resolution and a better opportunity to study physiological and neurochemical dysfunctions.

Ventricular enlargement is a nonspecific indicator of brain injury. Most people at the time thought it was due to an atrophic process—a loss of tissue after the brain had developed normally, as occurs with normal aging and in dementias such as Alzheimer's disease. Could there be a more specific area that is also involved in schizophrenia? One brain region that has been of great interest in the study of schizophrenia has been the prefrontal or frontal cortex. David Ingvar's early SPECT work had pointed to the frontal cortex, but frontal lobe size could not be adequately studied with imaging techniques until MR became available. Since MR has high enough resolution to visualize sulci, and since the frontal cortex is defined by one of the most prominent sulci in the brain, the central sulcus, one could identify that sulcus, separate the frontal lobe, and measure its size.

In 1986 our group reported the first quantitative MR study of schizophrenia. Not surprisingly, the frontal lobes were decreased in size, as was the whole brain and the intracranial cavity. Since skull size is partly driven by brain growth, and since the skull grows very little after around age five, we suggested that the brain abnormalities seen in schizophrenia might be due to a failure in brain development rather than to an atrophic process. There was a red herring here, however, that we duly noted in the discussion. Our control group in this early study contained a substantial number of doctors and other highly educated hospital personnel, so the result could be explained by a confounding factor such as education. And indeed, when we later collected a new sample of patients and compared them to a sample of educationally impaired controls who were equivalent to the patients, the two groups did not differ in frontal or brain size, although the new patients continued to be different (i.e., smaller frontal and brain size) when compared to the original group of controls. The take-home lesson was important. Educational level may affect one's research results, since it may be related to brain size. The later IQ studies confirmed this (IQ and educational level are very highly correlated).

All this work demonstrated that MR is indeed a powerful tool that can measure size differences that have functional significance—a gratifying fact and also an important wake-up call about the importance of using impeccable study design. In later MR studies we and most other groups decided to match patients and controls on the education level of their parents. When this was done, in conjunction with more sophisticated atlas-based techniques for measuring the frontal lobes, decreased frontal size was again found in both chronic patients and patients who were evaluated within a few months after the onset of their symptoms. Since these early MR studies, many interesting findings about other brain regions have been reported in schizophrenia, such as abnormalities in the temporal lobes or the thalamus.

The early studies of schizophrenia were like the halting steps of a toddler learning a new skill, but they pointed in directions that have been followed in many more sophisticated studies. These studies made it clear that MR could be used as a probe to address some very basic questions about the brain and schizophrenia. Here are some examples of the lessons learned from early studies and their related questions.

What mechanisms might produce the abnormalities noted?

Think about neurodevelopment.

Think about whether some specific brain regions might be more affected than others.

Study the course of the changes over time to determine whether gray matter is lost in people with schizophrenia at a more rapid rate than occurs with normal aging.

Study patients at the time of onset, before the chronic disease process has exacted whatever toll it takes. Use twins to think about gene/environment interactions.

The adventure has been an enriching and exciting one, and its story is continued in more detail in chapter 8, which describes schizophrenia.

In addition, the early studies of schizophrenia made it clear that MR could and should be used to ask and answer similar questions about a variety of other mental illnesses. MR has been used to study brain structure in mood disorders, dementia, alcoholism, anorexia nervosa, autism, anxiety disorders, and most other major illnesses. The results of some of that work are described in later chapters.

Using MR to Measure Brain Function and Chemistry
Functional Magnetic Resonance

Functional magnetic resonance (fMR) burst on the scene and soon became an exciting new application of MR in the early 1990s, pioneered initially by Seiji Ogawa at Bell Laboratories. A clever team of investigators at Harvard, led by Jack Belliveau, Bruce Rosen, and Mark Cohen, decided to push MR technology to the limit and use it to measure blood flow in the brain during mental activity—which had only been done previously with SPECT and PET. The novel idea was to build images that were not based on the hydrogen proton, but on the chemistry of hemo-globin, the large molecule in our blood that carries the oxygen that we use to generate energy to walk or run or think. When hemoglobin gives up an oxygen molecule for the worthy purpose of fueling one of our body's activities, it turns into deoxyhemoglobin. Deoxyhemoglobin, like the hydrogen proton, is paramagnetic! So, theoretically, it too could be manipulated with pulse sequences in the MR magnet to produce an image. The problem is that there isn't very much of it. Our bodies are rich in hydrogen, but we don't have a lot of hemoglobin, so the signal pro-duced might be too weak to detect. Again, the wonders of high-speed computers came to the rescue. If you don't have a lot of signal, you have to figure out a way to collect more of it, and more often. If the amount of numerical data in a regular structural MR scan is mind-numbing, fMR may induce a near-comatose state. But investigators have developed a variety of ingenious methods to deal with this data overload.

The early fMR studies of human cognition were a bit like the dancing bear. What was impressive was that the method would work at all. Part of the trick of getting a measurable signal was to have the subjects repeat the same task over and over. Unfortunately, the MR scanning environment is not exactly friendly. Subjects have to have their entire bodies placed inside a narrow hollow tube, which tends to make people feel claustro-phobic. When the subject is inside the tube, it is difficult to deliver a stim-ulus or figure out that he is doing the task he is supposed to be doing. (He could be thinking about tonight's date instead of counting backward from 100, as he has been instructed to do.) The scanner makes loud rhythmic noises when it delivers pulses, and these may interfere with an effort to study anything that involves hearing or listening. Because the method requires repetition of the stimuli or task, and these repeated measurements have to be from the same location in the brain, a tiny movement by the subject will mess up the data.

So initially investigators explored familiar territory. For example, if a person looks at an intense visual stimulus such as a flashing checkerboard, the primary visual cortex should have an increase in blood flow. If that did not occur, the technique would be considered invalid. And voila! It did light up—first in the early Harvard studies and then in other centers around the United States and the rest of the world. Finger movements should produce increased flow in the hand region of the motor cortex. And they did. The changes were very small—a maximum of 5% (produced by visual stimulation, in comparison to a 25–40% increase with this particular stimulus as measured by PET blood flow studies). But they were there, and they were reproducible.

The technique steadily improved. At first people looked at a single slice of brain, choosing it according to the activation expected. But soon people were sifting through the entire brain, just as had been done with PET for many years. As this happened, they began to see the activation of several remote regions at the same time by the same task—a clue to neuroimagers that they are visualizing functional circuits, or regions that work together to perform the task. For example, when the whole brain was studied, it was possible to observe simultaneous activation of the left motor region and the right cerebellum when subjects did a complex motor tapping task with the right hand—a well-recognized anatomical circuit that coordinates motor activity. When subjects switched to the left hand, the opposite pattern was seen—right motor strip and left cerebellum. An example of this pattern is shown in Figure 6–8.

It was a short step from this point to the study of more complex cognitive tasks. Cognitive neuroscientists have shown that fMR methods can be used to study working memory, spatial attention, and many other complicated mental activities. The progress has been so rapid that fMR is now used more widely than PET as a tool for studying mental functions. It has been a short step from the study of normal function to the study of diseases. Many studies have now been done to examine brain function in mental illnesses.

fMR has many advantages over PET. While maintaining a PET center is expensive, and few medical centers have one, every medical center has an MR scanner. MR is thus a widely accessible technology that permits investigators without access to PET to conduct research studies. Other advantages include slightly lower cost, the absence of ionizing radiation, greater repeatability of tasks/scans, and better temporal resolution. However, it also has limitations in comparison with PET. Advantages of PET include the ability to provide absolute flow values, the lower sensitivity of PET imaging to movement artifact, and artifacts in fMR that degrade the signal in some brain regions (e.g., orbital frontal cortex, anterior lateral cerebellar cortex). The PET imaging environment is also quieter and less confining than the fMR scanner. The rhythmic noises produced during fMR scanning make it particularly inappropriate for studies that examine rhythm and timing. Further, patients with mental illnesses tolerate the PET environment much better than the confining MR environment.

Magnetic Resonance Spectroscopy (MRS)

Magnetic resonance spectroscopy (MRS) is a familiar technique to the limited group of people who studied physical chemistry in college. This technique is used to take a sample of tissue, place it in a magnetic field, and use the changes induced to determine the spectra of the chemical contents of the tissue. Any compound that contains a paramagnetic ion can be studied: P^{31}, C^{13}, F^{19}, Na^{23}, Li^7, and of course, hydrogen. Unlike the other imaging techniques, this one does not produce "pictures of the brain," although it does produce quantitative measurements. The pictures that are displayed through MRS are the spectrum of a particular group of chemicals. MRS is a powerful tool for doing in vivo studies of brain chemistry: determining the composition of brain tissue in order to assess whether some pathological process is occurring. It can also be used to examine the density and distribution of some neurotransmitters or drugs used for treatment.

Some of the earliest MRS work studied the phosphorus spectrum. This spectrum was chosen in part because it produces one of the cleanest and most interpretable MR spectra, but also because phosphorus is an important component of cell membranes and of energy metabolism. Nerve cells are surrounded by a protective cell membrane that is like a sandwich: the "bread" on each side is composed of phospholipids, fatty chemicals that provide good insulation and protection. The "meat" in the middle holds the receptors and other components that regulate neuro-transmission. There are two main classes of phospholipids: phosphomo-noesters and phosphodiesters. The monoesters are building blocks: when cells are growing or functioning normally monoesters are prominent. The diesters are breakdown products that reflect neuronal decline or damage. The technique has now been used to study many interesting disorders and health issues, such as childhood and mood disorders, schizophrenia, and the effects of caffeine on the brain.

Using Functional Imaging
to Visualize and Measure Thoughts and Emotions

Chapter 4 provided a tour through the functional systems of the brain: language, memory, attention, executive function, emotion. For more than one hundred years we depended primarily on the lesion method to understand how the brain works. Typically, one or two scientists or physicians observed a single case and made valuable inferences about the systems of the brain, based on the abilities that were lost. Tan taught Paul Broca that "we speak with our left hemispheres." H.M. taught Scoville and Milner that we use our anterior temporal poles to retain or store our memories. Phineas Gage taught Harlow that the prefrontal cortex was important for making mature social judgments and appropriate emotional responses. If we could learn so much from examining a single interesting case, how much more might we learn if we could study large numbers of people and understand how individuals vary! And if we could directly study the intact brain as it is creating speech, learning and remembering, or formulating emotional judgments!

Functional imaging now gives us this opportunity. Just as the genome is being mapped with a precision and payoff that would leave Darwin and Mendel gasping for air, so too Broca and Harlow would be stunned to see their observations extended and mapped using brains from thousands of living human beings.

PET images are very similar to MR in the sense that they collect information about what is happening in chunks of brain tissue (voxels) and turn them into picture elements (pixels). From a lay person's point of view, functional images are much prettier to look at and make a more impressive impact because they are by tradition displayed in color, and they show brain function rather than structure. People like to feel that they are actually seeing what part of the brain "lights up" in a bright yellow or red hue when healthy normal volunteers are doing a mental task, or which parts are blue and inactive during the same task when patients with a particular mental illness perform it. People also find it interesting to see which parts are overactive when patients have abnormal brain activity, such as hallucinations or obsessional thoughts.

Early PET studies of normal cognition and disease processes used the "glucose method" (i.e., F^{18} fluorodeoxyglucose as the tracer) and measured the metabolic activity of the brain. While powerful, this method was limited by the very long time frame that was required to collect the image (nearly one hour) and by the fact that only one set of images could be obtained, and therefore only one mental task could be studied.

When scanning equipment improved enough so that the label could be changed to O^{15} and the tracer to H_2O^{15}, PET scanning changed dramatically. This tracer is used to measure blood flow, which is tightly correlated with metabolism, since blood flow increases in response to increased metabolism. Because the half-life of H_2O^{15} is so short (only 2 minutes), multiple back-to-back studies could be done in the same subject. Investigators could develop elegant designs, choosing a group of tasks that differed very subtly from one another, and could begin to map how small differences in mental activity affect different brain regions. Three groups took the lead in pioneering the use of the "water method," the Washington University group led by Marc Raichle, the Montreal group led by Alan Evans, and the Hammersmith group in England (now at the Queen's Square Neurological Institute) led by Richard Frackowiack. Other groups with access to PET quickly jumped on this fast-moving train as well. The expertise needed for designing PET research shifted from physicists and chemists to cognitive psychologists, psychiatrists, and neurologists.

Marc Raichle did one of the first pioneering studies using the "water method" to map mental processes. Not surprisingly, he chose to examine language—a system that had already been mapped by the lesion method. The results of this early study are frequently displayed on book covers, because the images are so attractive and the results so reassuring. The design of the study was as follows. Scans were obtained while subjects looked at a single word on a screen. During one scan they were simply asked to look at the word. During the next they were asked to read the word aloud. During a third scan they were asked to give a "use" for the word (e.g., knife—cut). The brains of the subjects responded in a way that would be nicely predicted from years of observing patients who developed aphasia after a stroke. The first scan activated the visual cortex, the second Broca's area (the site for producing speech), and the third Wernicke's area (where verbal associations occur).

As functional imaging has matured and as fMR has been added to the repertoire of tools, both the kinds of questions asked and the answers are steadily changing. The goal is no longer to light up a specific region activated by the task, but rather to map the complex distributed functional circuits that the intact brain uses when it thinks and feels. PET has shown that, contrary to the old teaching that most of our brain goes unused, large parts of our brains are in fact busy most of the time, and busy in a very dynamic way.

For example, in one of our PET studies we compared brain activity during two different tasks: remembering new faces, and recognizing whether a face was male or female. Both of these are very important tasks for human beings. Almost immediately after birth, human beings begin the process of learning to identify others around them on the basis of complex variations in the pattern of facial features. Human faces have more common than differentiating features: two eyes and eyebrows, a nose, and a mouth, all in a recurrently consistent spatial pattern. Identifying a face as familiar or unfamiliar or placing it in a category such as male or female requires a capacity to make very subtle discriminations that are probably crucial for individual survival. This capacity is developed early in the human infant, who quickly learns to recognize the mother's face. One must also learn to use the face to identify a person as a family member, a friend, an enemy, a member of a group or tribe, a member of the opposite sex (with requisite rules of behavior), or as a current or past (or future) sexual partner.

In this study we found that two very different functional circuits were used, depending on whether the task was remembering a new face (seen 60 seconds before the scan) or identifying a newly seen face as male or female. The differences are portrayed in Figure 6–9.

Remembering the new face appears to be done primarily by a circuit that runs from the right frontal cortex back to the cerebellum on the other side—the "mental coordination" circuit described in chapter 4. Other parts of the circuit include a smaller area in the left frontal lobe, the right parietal lobe (shown in many previous lesion studies to be related to facial recognition, and probably involved in coding the details of the memory traces for the faces), and the anterior cingulate gyrus (probably reflecting the need to focus attention, a challenge that frequently activates the cingulate gyrus).

This study and others have substantially expanded our understanding of how the brain remembers beyond the relatively simple story taught by the case of H.M. Combined with other studies that we and others have done, it is clear that the frontal lobes are actively involved in both the encoding and retrieval aspects of memory. In this particular study patients were principally engaged in retrieving the memory of the faces they had seen, and the primary site that they were using was in the right frontal cortex, while a smaller region in the left cortex was probably working to do the ongoing encoding. (This pattern of "left encoding, right retrieval," originally proposed by Tulving, has been seen repeatedly in studies of memory.)

What about gender recognition? Interestingly, but not surprisingly, this more primal and less intellectual task evokes activity in the more primal limbic areas of the brain! It also leads us to regions that have never previously been discussed as important brain regions. One area activated when the subjects recognized the faces as male or female was at the base of the frontal lobes in the very middle—an area called the straight gyrus because it is very straight. The olfactory bulbs that we use for smell run right below it. It is primitive cortex, but its function has never been identified. It is used in this task, however, and it must also play a role in memory. Combining this study with what we have seen in other studies, we suspect that the straight gyrus is used when we encode or retrieve intimate and personal memories—the sorts of things that define individual identity. Surely identifying whether another person is of the same or the opposite sex is one of the most personal and intimate things that we do. The other region that is very active during gender recognition is in the left inferior medial temporal lobe—running along its base in the direction of the amygdala. This is also a more primitive brain region—part of the limbic system. Finally, subjects also activated regions in the visual cortex (probably because they were carefully scanning the faces) and in a left medial parietal area known as the precuneus, which appears to be used for visual processing of many different kinds of objects.

The facial processing circuit is just one of the many functional circuits that are being mapped. Several entire books could be written about the many interesting things that functional imaging has taught us about how various connected regions in the brain work together to plan, decide, recognize, recall, feel pain, experience sadness, recognize danger, or the other nearly countless things that we do with this amazing two pound mass of busy tissue. But functional imaging has great potential for examining other topics as well.

For example, we learned from MR that we lose brain tissue as we age, particularly cortical tissue. Does this make any functional difference? Can we see that functional difference if we study the aging brain with PET or fMR?

In another study we examined the relationship between blood flow and aging in healthy normal subjects who were simply lying quietly with their eyes closed—a condition that is often called "rest." Since introspection makes it obvious that our brains never actually rest (probably not even when we sleep!), in other studies we have examined what people actually think about when lying with their eyes closed by debriefing them after the study. Usually they describe a type of episodic memory. People think about past experiences and events of the day or the week, or they make plans about what they will do at some later point in time. The order of these thoughts is somewhat uncensored or random, and this state may be as close as we come to observing the "free association" that occurs during the process of psychoanalysis. With an affection for acronyms, we have designated this state as "Random Episodic Silent Thought," or REST.

First we examined the relationship between age and overall blood flow in the brain. Unlike the general decline in cortical thickness that occurs with aging, we did not observe any overall decline in brain blood flow with age. Apparently our brains maintain more or less the same level of general metabolic activity as we age, at least up to the fifties (the age of the oldest person in this particular sample).

But does aging affect any specific subregions? Here the answer is a definite yes.

As we age, nerve cells in some regions of our brain become less numerous or efficient, which we can visualize with PET as a negative correlation: the older we get, the lower the flow in some brain regions. For example, a decline in blood flow in the visual cortex is correlated with age at a -.62 level, which we subjectively experience as we stretch our arms to hold reading material in order to adapt to our failing visual acuity . . . and finally give up and buy reading glasses. Declines also occur with age in the cingulate region, which is important for alertness and arousal, reaching the -.68 level. Other regions have positive correlations, indicating that blood flow increases in these regions as we age. In this instance, the region may also be experiencing "nerve failure," but the failing cells that remain are working harder to keep up. For example, that key memory region, the hippocampus, is correlated with increasing age at a .5 level, accounting for 25% of the variability (applying the rule of squaring the correlation coefficient to determine how much variance it accounts for). The hippocampus is the first region to be affected in Alzheimer's disease. Does it begin to "go" relatively early, even in a fairly young sample like this one, as we begin to have to search harder to retrieve names or recall where we left our keys, but before the symptoms of actual disease start to emerge? There are clearly differential changes in blood flow in specific regions of our brains as we age, which correspond to recognized changes in functional capacities. These specific changes occur relatively early—even in the late forties and early fifties. The changes appear to become more diffuse with even greater age, according to other studies. This pattern mimics the extension of neural plaques in Alzheimer's disease, where they begin in the anterior temporal lobe and slowly expand to involve the majority of the cortex as the disease progresses.

Can functional imaging be used to examine brain plasticity and the relationship between genetic influences and environment? Again, the answer is yes. In addition to comparing the similarities and differences in brain structure in identical twins, we can also visualize the functional activities of their brains. Since function is even more plastic than structure, we would expect even greater differences. We observed both pairs of twins while they were doing practiced and novel memory tasks, involving both words and faces. An example of the PET scans from one pair of twins is shown in Figure 6–10.

Simple visual inspection indicates that they vary in pattern of blood flow (which indicates which parts of the brain are being used most actively during a task), and they also vary in amount of flow (which indicates how hard the brain has to work overall). In this particular pair, one twin had always been brighter and a better student than the other. This is the twin with lower flow. For her the memory task was easier, and her brain therefore needed to use less energy to perform it. Like most identical twins, this pair *looks* very similar in face and body. Their brains—the most dynamic and plastic organ in the human body—appear to be the most different part of their bodies.

What has PET taught us about the functional activity of the brain in mental illnesses?

Functional imaging studies of mental illness were launched by using SPECT to study schizophrenia, introducing the concept of hypofrontality. The first PET study, reported by Monte Buchsbaum in 1982, also confirmed this result using the glucose method. Impaired frontal function has now been found in numerous studies, using all types of imaging techniques, including MR, SPECT, fMR, and PET. Very importantly, studies indicate that "hypofrontality" is already present in people who are hospitalized for schizophrenia in its beginning stage (first-episode patients). While performing a variety of different mental tasks, patients are unable to activate their frontal cortex as much as healthy normal volunteers. In addition, patients with schizophrenia who have been chronically ill are also unable to activate their frontal cortex while off medication, and their degree of hypofrontality is similar to the never-treated first-episode patients. Treatment with the newer "atypical" medications may improve metabolic activity in the prefrontal cortex, as compared with the older conventional neuroleptics.

From these imaging studies we have been able to reach a number of important conclusions: that the frontal lobes function poorly in untreated schizophrenia, that this malfunction is not caused by treatment (since it occurs in never-treated first-episode patients), that it is not due to chronic illness, that it may be associated with prominent negative symptoms, and that it may be partially reversible with medication. Studies of this type permit us to continue to identify brain regions involved in schizophrenia, and as well they suggest that the medications used to treat the illness may help reverse underlying anatomic, neurochemical, or metabolic deficits. Ultimately, we hope that studies of this type will assist in finding more effective medications that can reverse deficits more completely.

The story of schizophrenia is even more complicated than this, however. We know from the memory studies described earlier, as well as many other functional imaging studies, that the brain works by using distributed circuits involving multiple nodes. It is unlikely that a complicated illness such as schizophrenia could be explained by the involvement of a single brain region, even one as important as the frontal cortex. As will be described in more detail in chapter 8, which describes schizophrenia in detail, recent studies using more complex tasks indicate that schizophrenia is probably a disease characterized by functional misconnections that involve widely separated but important regions such as the frontal cortex, the thalamus, and the cerebellum.

As a prologue to that chapter, while the regions activated by faces are still on our minds, let's have a quick look at the regions where patients with schizophrenia differ from normal controls when remembering the new faces. Where would you guess that they would have problems in a task like this, which challenges them to "learn to meet new people" and remember who they are? What parts of the brain might be working less well?

The answer is illustrated in Figure 6–11, which displays the brain regions where the patients have lower blood flow than the normal controls. The areas of lower flow shown in this figure include that interesting region used for personal and intimate memory, the straight gyrus, the parahippocampal gyrus (part of the hippocampal complex), and the cerebellum. Other regions with diminished flow include the thalamus, the cingulate gyrus, and orbital and medial frontal regions. The distributed circuits that are used to remember, to focus attention, and to coordinate mental activity are all dysfunctional. In particular, the limbic portions of the brain are affected.

Using PET and SPECT to Monitor Treatment

If the brain is dysfunctional in mental illnesses, surely the most important task is to figure out how treatment can reverse the dysfunction. And, indeed, one of the most exciting and fruitful applications of in vivo imaging is to study neurotransmitter systems in the brain and the effects that medications have on them.

The study of neurotransmitters in mental illnesses is particularly difficult because specific tracers must be found to do the task. Both glucose and water are relatively generic workhorses: metabolism and blood flow are relevant to just about everything. If one wishes to study dopamine or serotonin or some other neurotransmitter system, one must find a specific compound that will label that transmitter, or its precursor, or its breakdown products, or a receptor that it occupies. The last choice—labeling receptors—has proven to be the easiest method for most studies, but it is still quite difficult.

The principle that has been used is to identify a drug that works by *occupying* a receptor (roughly equivalent to taking up a chair in a waiting room or restaurant). It need not be a drug that is actually therapeutically useful. Ideally, however, it should be quite specific. That is, it should affect only one neurotransmitter system, and only one class of receptors within that system—e.g., dopamine type 2 (D2) or serotonin type 2 (5HT2). Such highly specific compounds are not that easy to find, and then they must be labeled with an isotope such as F^{18} or C^{11}, and in a manner that does not change their chemical structure so much that their functional activity is changed. When this task is accomplished, the compound is injected prior to or during a PET scan, and its distribution is visualized in the brain. Essentially, the PET study gives us a picture of the location of the receptors that the compound has labeled. The next task is to figure out how to use the method to conduct measurements, since the goal of PET research is to obtain quantitative data.

Methods for measuring receptors were pioneered by three major research groups: the Karolinska group in Sweden, the Johns Hopkins group in Baltimore, and the Brookhaven group in New York. The methods developed by the Karolinska group have now been adapted to SPECT as well and have become the most widely used. The goal of this early work, conducted in the mid-1980s and 1990s, was to measure the number of D2 receptors in schizophrenia, in order to determine whether the chemistry of the illness could be explained by functional hyperactivity within the dopamine system. The reason for suspecting that increased dopamine activity might be the culprit was that most effective anti-schizophrenia medications work by blocking dopamine transmission, and the reason for choosing D2 was that the potency of the therapeutic effect appeared to be neatly correlated with the potency of D2 blockade. So the hypothesis was that patients with schizophrenia would have an increased number of D2 receptors. To test it, however, another difficult task had to be accomplished. The investigators had to find people with schizophrenia who had not yet been treated, since animal studies suggested that treatment caused the number of receptors to increase. (The brain is not dumb. It recognizes that messages aren't getting through, due to receptor blockade, so it responds by growing more receptors!) The Hopkins and Karolinska teams were in a close competition to answer this question, and to complete the task first. The rest of the world watched with interest.

In the end, there was good news and bad news.

The bad news is that the hypothesis was probably too simple, and probably wrong. D2 receptors are not increased markedly in schizophrenia, and D2 (or even dopamine) is not the only system that is injured in this illness.

The good news is that a lot has been learned in the process, and the method that was developed has been applied to even more interesting questions, which have considerable clinical value. Once it was possible to visualize and measure receptors, it was also possible to determine to what degree drugs used to treat schizophrenia block them (called receptor occupancy) and how quickly. The results have been fascinating. This work has been carried forward by the Karolinska group, led by Goran Sedvall and Lars Farde, and a group in Toronto, led by Shitij Kapur and Phillip Seeman.

One of the most interesting observations has been that relatively small doses of medications produce a great deal of receptor blockade. Finding the correct dose of a medication is never easy. Most drug development tests proceed by trial and error. Often a dose ranging study (a study in which three to five different doses are tried, to see which works best for a given illness) is done, but the samples are never as large as those that occur in the real clinical world after a drug has been in use for several years. Doctors often decide from experience that the dose should be lower or higher. Because most psychoactive drugs act slowly, and because doctors, patients, and nurses on inpatient units want rapid improvement, there has been a tendency to use larger and larger doses. For example, when haloperidol was first available in the United States in the late 1960s, the recommended dose was from approximately 2 to 8 milligrams. Over the next decade or two, however, the dose crept up, and some patients received doses as high as 50 milligrams. The PET studies have provided scientific evidence about the effects of higher and lower doses. We now know that the initial recommendations were correct. A dose of two to four milligrams of haloperidol blocks approximately 60–75% of D2 receptors. A good therapeutic dose blocks around 80% of receptors.

Thanks to the PET studies of receptors, we now have a rational brain-based strategy for choosing the correct dose of medication. But why did we need PET to discover this, and why do such low doses work well?

As already discussed in chapter 4, where the functioning of receptors was reviewed, it has been evident for many years that most medications used to treat mental illness do not have any major therapeutic effect right away. They may produce modest improvement, but this is usually due to side effects—sedation in a person who is agitated and overactive as a consequence of her psychotic thoughts. The reduction of the psychotic symptoms, such as delusions, takes several weeks. PET has permitted us to visualize how quickly drugs block the receptors that send the signal on down the neuron to the second messenger system. And the answer is: receptors are blocked quickly. Right after a drug is injected, it passes into the brain, and it migrates to the receptor that it matches, occupying it and closing its door into the receptor neuron. Receptor blockade is almost immediate.

If the therapeutic effects are *not* immediate, which they are not, this tells us that therapeutic effects are a consequence of *more* than just blockade. They require the protein synthesis that is produced through the second messenger system that has been so elegantly studied by Paul Greengard. As long as this is true of psychoactive drugs (and it may be true of the majority of them for a long time), both patients and "health care providers" (i.e., HMOs, managed care systems, insurance companies) will have to be patient. The brain/body cannot recover any faster than it recovers.

Methods for studying neurotransmitter systems with PET are now providing us with a method to create a rational psychopharmacology: a pharmacology based on what we know about how nerve cells communicate, how drugs block receptors, and how the effects of a medication may vary according to the dose. In a sense, PET permits us to do the equivalent of checking blood sugar levels in diabetes. For brain diseases, however, we are interested in "brain levels," and PET permits us to study the relationship between drug dose and clinical response based on a measurement of the levels and activity in the brain. This is a "new world" that will eventually give physicians better information about the most effective doses and will spare patients excessively high doses that may eventually produce unwanted side effects.

Tools for Research or Diagnosis?

Quite understandably, patients and their families have very practical questions about these advances in technology: do they help improve people's lives by making more accurate diagnoses or by guiding treatment?

At this moment, the value for either of those purposes is relatively ited. MR and functional imaging scans cannot be used to make a diag sis, and we have no definitive laboratory markers or genetic tests, even ior Alzheimer's disease. Most of these technological advances are still research tools, useful for probing into the brain or the molecular mechanisms of illnesses. All the data marshalled to date from imaging and electrophysiology are group comparisons. Groups of people with a particular diagnosis are compared to healthy volunteers, and group differences are found. Such studies are very informative in telling us something about the brain mechanisms of an illness—that they may affect the frontal cortex, that they may involve distributed circuits, that they may suggest a neurodevelopmental abnormality. But these studies cannot make any specific predictions about an individual. They can only make predictions about the group. Thus, when neuroscientists or psychiatrists speak about hypofrontality or ventricular enlargement in schizophrenia, they are not implying that every person with schizophrenia will have decreased frontal metabolic activity or big ventricles. Thus these findings are not useful at the moment either as screening or diagnostic tests.

However, these tools can sometimes be useful clinically. Many clinicians believe that a young person who has experienced a psychotic episode for the first time should be examined with a structural imaging technique such as CT or MR, to evaluate for the presence of characteristics such as ventricular enlargement or prominent sulci. These are correlated with a variety of other characteristics that may produce useful information about the long-term course of the disorder. People with schizophrenia who have ventricular enlargement also tend to have more prominent negative symptoms (i.e., symptoms such as apathy and avolition), and a poorer response to treatment. Structural abnormalities may be seen in many other mental disorders, including dementia, alcoholism, anorexia nervosa, and some mood disorders. In some of these illnesses the structural abnormalities worsen over time (e.g., the various dementias), and this can be used to help track the course of the illness. In other illnesses structural imaging techniques can be used to monitor improvement because the abnormalities are reversible. Abstinent alcoholics and anorectics who regain their weight also normalize their brains. Tracking this improvement can be a powerful motivator to help the patient continue the difficult task of remaining abstinent or maintaining a normal weight.

PART III

THE
BURDEN
OF
MENTAL
ILLNESS

CHAPTER 7

UNDERSTANDING WHAT MENTAL ILLNESSES ARE
The Past Is Prologue to Progress

The mind is its own place, and in itself
Can make a Heav'n of Hell, a Hell of Heav'n
—John Milton
Paradise Lost

We human beings moved out of caves and began to cultivate the soil thousands and thousands of years ago. We began to record our history in written language four to five thousand years ago. We do not know when we first began to suffer from mental illnesses, or why this occurred. We do know that mental illnesses have been with us for thousands of years, however, and that they began to be described as early as human beings began to create written accounts of their illnesses. Mental illnesses are described in most surviving early medical texts, such as the Egyptian Ebers Papyrus or the writings of Hippocrates or Galen, going back as early as 1900 B.C. For most of their history, mental illnesses have been recognized as brain diseases that are expressed as changes in thinking and emotion. In early medical texts they are discussed in the same manner as other illnesses, such as dropsy (heart failure) or diabetes. They are described in the Bible, as when King Saul falls into a severe depression. They are portrayed in literature, as when Aeschylus shows us Orestes tormented by the Furies, or Euripides presents the paranoid madness of Medea. Shakespeare and his contemporaries had a whole host of "madness" plays: *The Spanish Tragedy* by Kyd, *The Duchess of Malfi* by Webster, or *Hamlet* and *King Lear* by Shakespeare.

Running counter to the ancient recognition that mental illnesses are diseases of the mind that arise from the brain, however, has been a countertradition of stigmatization and even cruelty. Looking back, it appears that this countertradition arose sometime in the middle of the last millennium, created by a mixture of social forces.

How Did Misunderstanding and Stigmatization of Mental Illnesses Arise?

Between the fall of the Roman Empire and the scientific and philosophical revolutions of the seventeenth to eighteenth centuries, most power, policymaking, and wealth were concentrated primarily in church rather than state. Deviations in belief or behavior were generally viewed with intolerance. The Reformation and Counter-Reformation made the situation even worse, polarizing all of society into a search for heresy and witchcraft. Many unfortunate souls who deviated from the "truth" or the "norm"—defined by the prevailing religious beliefs of their region or country—were perceived as "possessed by the devil" rather than as suffering from an illness beyond their control. Books were written that described techniques for identifying witches, such as the famous *Malleus Maleficarum* (Hammer of Witches), written by two Dominican monks. This text described in detail how to recognize that a person was a witch and possessed by the devil. As we read these descriptions 500 years later, it is clear that the authors were describing mental illnesses such as psychotic depression or schizophrenia. The problem of misunderstanding and misperception was not limited to Roman Catholicism or clergymen, however. No less a leader than the King of England and Scotland (James I) wrote a similar contribution from the Protestant perspective. His book, *Demonologie,* published in 1611, is an almost identical argument and policy statement, simply written from another point of view.

A second social theme that has led to misunderstanding about the nature of mental illness has been the desire to "get them off the streets," or the NIMBY (Not In My Back Yard) syndrome as we know it today. Most of us who are blessed with being healthy or wealthy do not particularly enjoy being confronted by suffering, poverty, or disfigurement. For centuries, the solution for this problem has been to create "institutions" which have gone by various names such as poorhouses, madhouses, orphanages, and prisons. These institutions have housed a mixture of "les miserables." It is not necessarily socially right to mix together people who are murderers, felons, innocent victims of an unjust legal system, homeless children whose parents have died, mentally retarded children, victims of physically disfiguring illnesses such as Elephant Man syndrome, and victims of mentally disfiguring illnesses such as schizophrenia. But it has been done for centuries out of social convenience. When this mixture of people is grouped together and hidden away, it is easy to see how "bad" and "mad" could be confused. The current concern about the tension between the benefits of deinstitutionalizing the mentally ill and the problems of homelessness are simply echoes of problems that people have been confronting for centuries.

These problems and misunderstandings have been counteracted, however, by powerful counterforces.

One was the reemergence of medicine as an empirical scientific discipline around two hundred years ago. As medicine moved from superstition and quackery to science, the understanding of mental illnesses improved, and they began to return to their rightful place as diseases of the mind and brain. Of course, some remnants of that old tradition of superstition and fear sometimes lie thinly disguised below the surface. In contemporary society we may no longer burn our Joan of Arcs at the stake, but many still make disparaging remarks about "crazy people." Increasingly, however, people are recognizing both the importance of mental illnesses and their biomedical origins.

The medical specialty of psychiatry and the modern conceptualization of mental illness grew out of the Enlightenment of the eighteenth century—that exciting new progressive era that led to a belief in the fundamental dignity of all human beings. It produced great ringing statements such as "all men are created equal" or that all are entitled to the rights of "life, liberty, and the pursuit of happiness." In this context, social reformers began to inspect the dimly lit world of "institutions" and to see the need for a great deal of enlightenment there as well. In Europe and the United States a series of reformers began to advocate on behalf of the mentally ill and the mentally retarded and to argue for the creation of medical facilities in which such people could receive humane care. This movement produced the first generation of psychiatrists and the second-oldest of the medical specialties. (The oldest is surgery, which arose from the distinction between physicians and barber-surgeons, the latter being the forerunners of modern surgeons.)

Physicians who became intrigued by the challenges of caring for the mentally ill began to set up hospitals and asylums. Some of these great figures include Benjamin Rush, who founded Pennsylvania Hospital and signed the Declaration of Independence; Philippe Pinel, a leader of the French Revolution who reformed the Salpêtriere in Paris; Vincenzo Chiarugi, a compassionate physician who reformed institutions in Italy, and many more. These men, who were trained as general physicians who cared for diseases affecting the entire body, made a conscious decision to focus primarily on those diseases that affected the mind/brain/spirit. For the first time, a subgroup in medicine was identified based on a shared interest in a single system or organ of the body. The term "psychiatrist" emerged to identify those doctors who were healers of the mind or spirit (psyche = mind/spirit, iatros = healer). These early psychiatrists forged a new and revolutionary way of thinking about diseases of the mind. First and foremost, diseases of the mind were illnesses and should be accorded the best and most humane medical treatments available. Second, the various signs and symptoms could be studied using the techniques of the scientific method, which were being laid down in other fields such as mathematics and physics. Third, people suffering from mental illnesses were unfortunate victims of disease, not criminals or social derelicts.

These early psychiatrists were heroic and visionary figures. I look back on them with admiration. The altruism that they manifested in their daily lives is too often forgotten and too infrequently emulated. Here in the United States, thirteen of the early psychiatrists joined together more than two hundred years ago to create an association of "Superintendents of the Hospitals for the Insane," the forerunner of the current American Psychiatric Association. Their primary goal was to share knowledge and information about how to provide better care for their patients. Benjamin Rush was one of those original thirteen. One of their successors, Amariah Brigham, founded a journal in which they could share information by publishing scientific articles and clinical observations. Brigham, himself an orphan who pulled himself up out of poverty to become a prominent physician, published this journal at his own expense. It first appeared in 1844 and celebrated its 150th anniversary in 1994. It is now the oldest continuously published medical specialty journal in the United States. I am proud to be a direct descendent of the traditions of Amariah Brigham and to serve as the eleventh Editor-in-Chief of the *American Journal of Psychiatry*.

How Did We Discover Specific Types of Mental Illnesses?

The patients whom Rush and Brigham treated in those early mental hospitals—or "asylums" as they were often called—shared one feature in common. Their minds were not working very well, although their hearts, lungs, and muscles usually (though not always) were. But they had a variety of different symptoms. Some were experiencing terrifying intrusions from outside forces that were tormenting them—experiences that would have been called "possession by the devil" one or two hundred years earlier. Some were expressing confused ideas that seemed irrational. Some were anxious and fearful. Some were hearing voices. Some were silently absorbed in a world of their own, totally uncommunicative and even immobile. Some were depressed and guilt-ridden. Many had combinations of these various mental symptoms.

How could those early doctors organize the patients into various homogeneous groups in order to plan treatments? Or should they all be considered the same—all suffering from a single disease that affected the mind and represented a single mental illness? Did all these different manifestations of mental illness have a single physical cause? Or were there different causes? These were fascinating questions, discussed by early psychiatrists in the pages of Brigham's *American Journal of Insanity* (later renamed the *American Journal of Psychiatry*). Similar discussions were also occurring in Britain, Germany, and elsewhere. In general, most of the early psychiatrists believed that mental illnesses were due to specific disturbances in brain function, which led to different patterns of symptoms. For example, Amariah Brigham observed in 1844:

> Insanity is often but an effect of a slight injury or disease of a part of the brain, and in many cases only a few of the faculties of the mind are disordered. From this we infer that the brain is not a single organ, but congeries of organs. . . . Thus each mental faculty has an especial organ, and therefore certain faculties may be disordered by a disease of the brain, while others are not affected.

Using subgroups that reach back to classical times, these first psychiatrists recognized different disease categories such as melancholia, mania, and delirium. Then they began to discuss the different neural causes, using the knowledge of brain science available at the time.

As microscopes and stains for various kinds of cells became available, the early psychiatrists used these tools to improve diagnosis and classification. Soon one group of patients could be clearly identified. They suffered from mental illness due to infection from a bacterium, known as *Treponema pallidum*, also referred to as the spirochete because of its corkscrew-like shape, or *T. pallidum* for short. This infection, known as syphilis, can run through three classic stages. The first one (primary syphilis) occurs when the person develops a venereal lesion, a small painless ulcer or blister, which sometimes passes unnoticed. The second (secondary syphilis) occurs a few weeks later when the bacterium spreads throughout the body, producing fever, chills, and other symptoms. At this time it may also move into the brain, but during the second stage mental symptoms are minimal to absent. Then the disease becomes latent, although the silent carrier may still infect others. People who have a heavy infection of *T. pallidum* in the brain may eventually develop the third stage, tertiary syphilis. This occurs in approximately 15–30% of people who were initially infected. Ten to twenty years later they manifest symptoms such as paranoia, grandiosity, and mental confusion. Friedrich Nietzsche and Paul Gauguin were among the more famous victims of tertiary syphilis.

Fortunately, by the early twentieth century, treatments were developed that arrested the illness in the secondary stage, thereby preventing tertiary syphilis—with its terrible sentence of later madness. A European psychiatrist, Julius Wagner Jauregg, developed a treatment for tertiary syphilis. He noted that infecting people with malaria seemed to eradicate the symptoms in a majority of patients. Although primitive, this treatment was in fact a major breakthrough for a dire disease, and Julius Wagner Jauregg received a Nobel Prize for his discovery. Other treatments were also developed, such as mercury and arsenicals. Isak Dinesen, who described becoming infected with syphilis by her husband in her novel, *Out of Africa* (later a Redford/Streep Academy Award–winning movie), is a famous example of successful treatment during the secondary phase and prevention of tertiary syphilis.

Ultimately, neurosyphilis was eradicated with the development of antibiotics, and most people now no longer even think of it as a mental illness. One hundred years ago, however, it accounted for many of the patients in mental hospitals. As recently as the 1960s and 1970s, a screening test for syphilis (known as the VDRL) was standard procedure for patients admitted to a psychiatric facility with symptoms of psychosis. The delineation of this specific mental illness—ranging from clinical description to laboratory tests to effective treatment—is psychiatry's biggest success story so far. Research psychiatrists still consider the "syphilis achievement" to be the kind of Holy Grail that they seek. Ultimately, we would like to replicate the same process for all mental illnesses: clinical description, identification of cause, finding a treatment for the symptoms, and development of a preventive measure.

As tertiary syphilis was being delineated and treated, psychiatrists were attempting to achieve the same success for the many other patients who populated mental hospitals in the early twentieth century. A German psychiatrist, Emil Kraepelin, made contributions to this effort that laid the foundations upon which psychiatry stood for the remainder of the twentieth century and upon which we still stand today.

Kraepelin was a "quadruple threat" psychiatrist—a great clinician, research scientist, educator, and leader. He was driven by an intense curiosity about how the mind works. He obtained his early training from a prominent psychologist, Wilhelm Wundt, who was interested in how learning occurs, how associations are made, and how memories are formed. Kraepelin rose quickly to become the head of a series of important departments—first in Dorpat (currently Tartu in Estonia), subsequently in Heidelberg, and finally in Munich. He saw numerous patients and recorded detailed observations on their signs, symptoms, and outcome. Relatively early in his career, he began to summarize these observations and to publish them as a textbook of psychiatry. Over the years, he made multiple revisions as his observations accumulated and improved. The last several editions of his textbooks, as well as copies of lectures that he gave to medical students, are classics that are as true and accurate today as they were when he wrote them one hundred years ago. (I was lucky enough to get some rare copies of English translations of these textbooks during my residency training and to learn psychiatry from this brilliant clinician and scientist.) Kraepelin recognized the importance of neuroscience long before the word was invented. He himself, following the tradition of Wundt, would today be called a cognitive neuroscientist, since much of his own experimentation was on the effects of alcohol on learning and memory.

By the time he became head of the department in Munich, he had assembled around himself a group of psychiatrist-neuroscientists whose names have become legendary. Figure 7–1 shows Kraepelin in 1900, seated on a boat with three of his colleagues, two of whom (Alzheimer and Nissl) are now viewed as great figures in neuroscience. Kraepelin's department in Munich included Brodmann, the creator of Brodmann's maps of neurons in the cerebral cortex. It included Nissl, who discovered the Nissl stain for visualizing neurons. It included Alzheimer, who applied Nissl's techniques to study brain tissue in patients with memory problems and personality changes and discovered that some had tangled-appearing

Figure 7–1: Kraepelin and His Colleagues: Alzheimer, Kraepelin, Gaupp, and Nissl on a Boat Trip Together in 1900

neurons and areas of sludge that he called plaques. Kraepelin and Alzheimer noted that these findings tended to occur primarily in older patients with dementia, as opposed to the younger patients whom Kraepelin designated as having early dementia or "dementia praecox." Kraepelin started referring to the illness in the older people as "Alzheimer's disease." The name stuck.

Since Kraepelin was responsible for delineating the alternate syndrome, dementia praecox, I have always thought it unfortunate that we have not named this companion syndrome Kraepelin's disease. Ironically, Alzheimer published only a few papers, while Kraepelin was a prolific author, and yet almost no one has ever heard of Kraepelin. Kraepelin's tombstone bears the inscription: "Though his name will be forgotten, his work will live on." I suspect that if he is looking down upon us from another spiritual universe, he is gratified that the second half of this prophecy has come true.

The Kraepelinian synthesis created the nosological structure that we still continue to use. It recognized one group of disorders, the dementias, that occurred in late life and that were progressive. Alzheimer identified a specific neural mechanism that serves as a marker for the disorder that we call Alzheimer's disease: plaques and tangles. A second group of patients had similar symptoms, but with an earlier onset. Kraepelin considered these to be a second category, which he named dementia praecox and which we now call schizophrenia. Kraepelin also delineated a third group of disorders, which he referred to as manic-depressive illness. Unlike people with Alzheimer's disease or dementia praecox, those who developed manic-depressive illness had transient periods in which they were extremely ill, but eventually recovered fully. Their age of onset varied from youth to mid to late life, and they had prominent disturbances in mood as well as occasional disturbances in cognition and personality. Because these patients could fluctuate between highs and lows, Kraepelin referred to their condition as manic-depressive illness. Kraepelin's three illness categories are still used and continue to be three of the major disease groups that psychiatrists treat. We have made only modest modifications to the basic Kraepelinian nosological system.

Kraepelin and his scientific allies collected postmortem tissue from people suffering from all of these types of mental illnesses. However, they were unable to find characteristic lesions in either schizophrenia or manic-depressive illness. We now recognize that the abnormalities were not visible using the neuropathology tools available at the time, since schizophrenia and manic-depressive illness are disorders of neurochemical transmission and neural connections. Neurotransmitters had not yet been discovered, and methods for tracing connections or visualizing mental functions in vivo had not been created.

While Kraepelin and his team were defining dementia, schizophrenia, and mood disorders in Munich, a Viennese psychiatrist was focusing on another group of disorders known as neuroses. Paradoxically, Sigmund Freud began his career much as Emil Kraepelin did. He was intrigued by the exciting new techniques of neuropathology and himself attempted to invent a stain for nerve cells. He examined the effects of drugs on brain and behavior and experimented with the psychotropic properties of cocaine. He studied neurology in Paris, training with a great French professor, Jean Martin Charcot. Charcot was particularly interested in the use of hypnosis to treat unexplained physical complaints such as paralysis.

When he returned to Vienna, Freud was unable to obtain a university position, probably because of anti-Semitism. He went into private practice, where he specialized in treating the same types of physical complaints that were a Charcot specialty. Viennese ladies and gentlemen with unexplained paralyses, seizures, and lapses in consciousness arrived in his office. He found hypnosis to be helpful, but he also experimented with a new technique that he began to call "free association." Placing his hand on the forehead of the patient, who relaxed on the examining couch, he would invite her to describe spontaneously whatever came into her mind. Interesting things came tumbling out: past memories and hidden desires, often sexual in nature. This was an era when Victorian prudishness was still very much alive, requiring that even table legs be referred to as "limbs." The free associations revealed an unanticipated level of sexual repression in Freud's patients. His ongoing observations led him to formulate a rich repertoire of psychodynamic theories that explained his patients' various signs and symptoms.

Freud's work complemented that of Kraepelin, which emphasized dementias and psychotic disorders. Many of Freud's patients suffered from the illnesses that we now refer to generically as anxiety disorders. They include obsessive-compulsive disorders, panic disorder, posttraumatic stress disorder, and dissociative disorders. Freud explained these various conditions using the concept of the distribution of psychic energy—what came to be called "psychodynamic theory." Freud's terms and concepts are still widely used in the understanding and treatment of anxiety disorders. One of the major efforts in contemporary neuroscience is to understand how psychodynamic psychotherapy and other psychotherapies produce clinical improvement by changing associative memories at the neural level.

Four Stages in the Progress of Understanding Mental Illnesses

The story of syphilis, with which we began, illustrates the overall pattern of medical progress in understanding and conquering almost all kinds of biomedical diseases. This progress consists of four stages, which are summarized in Table 7–1.

The first step, isolating a syndrome, refers to seeing a pattern in symptoms and their changes over time that suggests that their cluster may define a specific disease. The word "syndrome" literally means "running together." In the case of syphilis, the pattern was relatively clear, at least once doctors began to perceive it. Although we now look back on the understanding of syphilis as a great success story, the process of identifying and eradicating it was a long biomedical journey because of the complex presentation of the disease and the fact that it was sexually transmitted. (Primary prevention, or asking people to avoid the possibility of infection by refraining from sexual intercourse, is nearly impossible!)

Syphilis is sometimes referred to as "the first gift of the New World" to Europe. Christopher Columbus and his sailors apparently picked up the infection during their nocturnal revels. The disease was first medically described shortly after the triumphant return of the Columbian expedition, when an epidemic of sores appeared on genital organs (primary syphilis), followed by flu-like symptoms a few weeks or months later (headache, fever, sore throat, skin rash, aches and pains—the signs of secondary syphilis). Months or years later the signs and symptoms of tertiary syphilis began to appear, although it obviously took some time to connect the delayed phase with the earlier infection.

By 1530 an Italian physician, Girolamo Fracastoro, was able to write a description of the disease by recounting the clinical course of a fictional character whom he named Syphilus, thereby giving the disease its name. Medical literature during the subsequent century continued to delineate more clearly the three phases of the illness and the mixtures of signs and symptoms that might occur. Syphilis is a very complicated illness because

TABLE 7–1

The Four Stages of Medical Progress

Isolating a specific syndrome
Identifying its pathophysiology
Finding a treatment to reverse the pathophysiology
Finding a way to prevent the pathophysiology from arising

of its three stages, which may occur at varying intervals in different individuals. It is confusing because it affects so many body systems, ranging from skin to brain to blood vessels to bones and joints. It can be mistaken for a variety of other psychiatric and nonpsychiatric illnesses, and it has therefore been designated as "the great imitator." It is also confusing because only about 30% of people who are initially infected develop tertiary syphilis (also known as neurosyphilis). Yet despite all these puzzling variations in pattern, syphilis was a well-recognized and well-defined medical syndrome within 50 to 100 years after its introduction to Europe.

Identifying a pathophysiology is the second stage in medical progress. Again, the case of syphilis is illuminating, since it reveals that "pathophysiology" can occur at many levels. The term "pathophysiology" is used in medicine to refer to the mechanisms by which a disease arises and (in the case of chronic diseases) continues to progress. Because syphilis was sexually transmitted, doctors figured out almost immediately that the pathophysiology suggested a "contagious disease." Of course, until the microscope permitted visualization of microorganisms, the very specific nature of contagious diseases could not be identified. In the case of syphilis, the actual organism, *Treponema pallidum*, was not discovered until 1905. By inference, however, physicians recognized that the disease must involve spread of the infection from the initial painless sore in the genital area to other parts of the body, leading to the various symptoms of secondary and tertiary syphilis. Part of the pathophysiology was the germ's particular affection for nerve tissue, but the spirochete may also produce damage to the heart and to the skin. The late-stage skin lesions are disfiguring painful sores known as "gummas."

Finding a treatment to reverse the pathophysiology is the third stage in medical progress. Sometimes this occurs before the pathophysiology is actually fully understood. For example, increasingly effective treatments were found for syphilis, such as mercury and malarial treatment, but they could not be recognized as eradicating the germ until the germ itself was found. The treatments were conspicuously more successful if used during the secondary stages, presumably because they reduced the number of active T. pallidum bacteria and therefore prevented them from being able to do the later damage of tertiary syphilis. Once antibiotics became available, they were an obvious treatment for syphilis, since it had been known for nearly forty years that the disease was produced by a specific germ. The only trick was finding the best antibiotic to attack *T. pallidum*. This turned out to be penicillin, which is now the standard treatment for all stages of syphilis. Fortunately, this treatment is so successful that both the germ and the disease that it produces have largely been eradicated in the Western world.

Finding a way to prevent the pathophysiology from arising is the fourth step in medical progress. Epidemiologists speak of prevention at multiple levels. Primary prevention refers to preventing a disease altogether. In the case of syphilis, this has already occurred in North American and European countries, since people no longer carry the germ and infect one another. If primary syphilis does occur (for example, when a traveler has a "fling" in a country where syphilis is still present), prevention is still possible through penicillin treatment when the lesion of primary syphilis is first observed (as long as it is noticed. Women often have internal lesions that are not visible). This early treatment prevents further pathophysiology—i.e., the secondary and tertiary stages. Other examples of primary prevention include vaccination for infectious diseases such as small pox and the use of preventive measures such as condoms to prevent the spread of HIV or other sexually transmitted diseases.

Tertiary syphilis was once a mental illness that accounted for approximately 25% of the patients in psychiatric hospitals. It was a terrible scourge for 350 years and has now essentially disappeared. In fact, it is a tribute to this medical success story that the average reader of this book probably did not realize that neurosyphilis was once a very important mental illness.

But where do we stand with all those other mental illnesses that are still around? For syphilis the path from syndrome to identifying pathophysiology took three hundred years, while it took another fifty to find a successful treatment and implement prevention. How long will it take for schizophrenia, dementia, depression, and other mental illnesses?

Mental Illnesses as Syndromes: Does This Make Them Myths?

The definition of most mental illnesses is still at the syndromal level. That is, they are defined by a clustering of signs and symptoms, in combination with a long-term course, just as syphilis was for nearly three hundred years. We have steadily improved our clinical definitions for most of the major mental illnesses, successfully delineating them into categories that differ in outcome and response to treatment. Further, we have accumulated a great deal of information about the extent to which they run in families and (in some cases) about changes in brain structure and function. Because we cannot yet point to a specific lesion or a specific cause such as *T. pallidum*, some critics (most notably Thomas Szasz of the University of Syracuse) have argued that mental illnesses must be myths.

To place our current understanding of the nature of mental illnesses in perspective, two things must be considered. First, the four steps described in Table 7–1 do not necessarily occur sequentially, even though they seem to form a logical progression. Sometimes we find a successful treatment before we understand how it works. Digitalis, for example, was used for several centuries to treat heart failure, well before we understood the principles of how the heart muscle contracts in response to the bundles of nerve fibers in its command center. Sometimes we recognize a syndrome and have a reasonably sound understanding of its pathophysiology, but remain frustrated in our search for successful treatments or preventions. Cancer and Huntington's disease are good examples of this problem. Sometimes a single discovery, such as the development of antibiotics, permits a dramatic breakthrough in a previously difficult disease, as happened with syphilis when penicillin was discovered. Although we like to think of science as rational and logical, we cannot always predict how, why, and when breakthroughs will occur.

We can summarize our progress in understanding mental illnesses by seeing how we score on a "report card" that provides grades for each of the four steps. The report card (reflecting how things stand in 2001) is shown in Table 7–2. (It will be interesting to see how the grades improve by 2010 or 2020.)

The details explaining these grades are summarized in the four following chapters. However, the report card illustrates that progress in understanding mental illness is uneven, complex, unpredictable, and serendipitous. Sometimes we are able to work forward from our knowledge of the syndrome and its pathophysiology to developing improved treatments. In the case of the dementias, we have learned a great deal about the genetic mechanisms of Huntington's disease, and a considerable amount about

TABLE 7–2
Progress in Understanding Mental Illness: A Report Card

Disease Category	Syndromal Definitions	Pathophysiology	Treatment	Prevention
Neurosyphilis in 1900	A	C	D	D
Neurosyphilis in 2000	A	A	A	A-
Dementias	A-	B+	D+	D+
Schizophrenia	B+	B-	C+	D
Mood disorders	B+	C	A	D-
Anxiety disorders	B+	C	B+	C-

the brain and genetic mechanisms of Alzheimer's disease, which comprise two of the major dementias. Our progress on treatment to date has been relatively poor, however, hence the grade of "D+." The grade is D+ rather than D because our knowledge of the genetic mechanisms of Huntington's disease permits genetic testing and prevention if people who carry the Huntington's gene do not have children. Many hope that this grade will rise rapidly very soon, due to the huge efforts being expended to develop a treatment for Alzheimer's disease. For some illnesses, such as schizophrenia, we have good syndromal definitions and have made considerable progress in understanding the pathophysiology, largely through the application of neuroimaging techniques, which suggest a neurodevelopmental pathophysiology that leads to misconnections between brain regions. Treatments are improving, but still have a long way to go. For the anxiety and mood disorders, we know less about pathophysiology, but have a relatively good score card for treatment, due to the development of effective medications.

Subsequent chapters tell the story of how science has brought us to our present levels of success, and the hope that the report card will improve markedly during the next several decades.

The Diagnostic and Statistical Manual (DSM): How Syndromal Definitions Are Created

The syndromal definitions used in modern psychiatry are summarized in a carefully developed diagnostic manual, which is currently in its fourth edition. It is known as DSM IV, which stands for the fourth *Diagnostic and Statistical Manual* of the American Psychiatric Association. DSM IV is an enormous tome, consisting of 886 pages of definitions, criteria, glossaries, and other materials. It forms the basis for diagnosing mental disorders of all types, including childhood disorders, substance-related disorders, dementias, mental disorders secondary to a general medical condition, eating disorders, sleep disorders, adjustment disorders, and so on. Each of the disorders is given a brief description that covers topics such as diagnostic features, prevalence, course, and familial pattern. After the general description, specific diagnostic criteria are provided.

This manual has become the standard diagnostic method for teaching medical students and psychiatric trainees. Psychiatrists seeking board certification, the "quality assurance" indicator that a psychiatrist has adequate training and knowledge, are tested for their knowledge of DSM definitions. Treatment guidelines for both medications and psychotherapy are based on DSM. The manual is also widely used by other professionals, such as psychologists, neurologists, and lawyers.

Although it is an influential book, it is important to realize that DSM has limitations as well as strengths. Because it was developed with considerable care and contains very explicit rules, it is sometimes accorded undue reverence—as if it somehow contained divine revelations about the definitions and boundaries of mental illnesses. Like other standard references, such as dictionaries and encyclopedias, it was in fact written by human beings, based on the best knowledge available at the time. It should not be accorded the same respect as the Bible, the Koran, or the Talmud. Unlike those religious documents, DSM was written by committees of physicians and scientists. It undergoes regular revisions as the knowledge base of psychiatry changes. The most recent version of DSM IV was published in 2000, and it will probably be revised again in another ten years. The Bible, Koran, and Talmud do not seem to need such frequent revisions!

The idea of creating a standard diagnostic manual for use by psychiatrists arose after the Second World War. For the first time, psychiatrists from all over the United States (and even from all over the world) were brought together in clinical settings and discovered that they did not always speak a common diagnostic language. The concept of how to define schizophrenia or depression was not precisely the same in Peoria as it was in Manhattan or in London. Some psychiatrists had been trained to conceptualize an illness quite broadly and to diagnose it in large numbers of patients, while others had a narrower concept. After World War II, the Veterans Administration found itself confronted with decisions about psychiatric disability in veterans and recognized the need for national standardization. This led to a *Veteran's Administration Diagnostic Manual*, which was the forerunner of the first *Diagnostic and Statistical Manual* (DSM I), which appeared in 1952, followed by a later revision (DSM II) in 1968. The early DSMs were modest documents, which contained relatively concise and telegraphic definitions. Here, for example, are the definitions of schizophrenia from DSM I (1952), DSM II (1968), DSM III (1980), and DSM IV (2000).

Warning: don't spend a lot of time reading them carefully (unless you plan to become a psychiatrist or a lawyer specializing in forensic psychiatry). They are full of technical jargon that won't make a lot of sense. Just go for the gestalt. Get a sense of how the style of the definitions has changed over time.

DSM I - Schizophrenic Reactions

This term is synonymous with the formerly used term "dementia praecox." It represents a group of psychotic reactions characterized by fundamental disturbances in reality relationships and concept formations, with affective, behavioral, and intellectual disturbances in varying degrees and mixtures. The disorders are marked by a strong tendency to retreat from reality, by emotional disharmony, unpredictable disturbances in stream of thought, regressive behavior, and in some, by a tendency to "deterioration."

DSM II - Schizophrenia

This large category includes a group of disorders manifested by characteristic disturbances of thinking, mood, and behavior. Disturbances in thinking are marked by alterations in concept formation which may lead to misinterpretation of reality and sometimes to delusions and hallucinations, which frequently appear psychologically self protective. Corollary mood changes include ambivalent, constricted, and inappropriate emotional responsiveness and loss of empathy with others. Behavior may be withdrawn, regressive, and bizarre. The schizophrenias, in which the mental status is attributable primarily to a *thought* disorder, are to be distinguished from the *major* affective illnesses that are dominated by a *mood* disorder.

The major change between DSM II and DSM III was the use of diagnostic criteria to define each of the mental illnesses in the manual.

DSM III – Diagnostic Criteria for a Schizophrenic Disorder

A. At least one of the following during a phase of the illness:

1. Bizarre delusions (content is patently absurd and has no possible basis in fact), such as delusions of being controlled, thought broadcasting, thought insertion, or thought withdrawal.

2. Somatic, grandiose, religious, nihilistic, or other delusions without persecutory or jealous content.

3. Delusions with persecutory or jealous content if accompanied by hallucinations of any type.

4. Auditory hallucinations in which either a voice keeps up a running commentary on the individual's behavior or thoughts, or two or more voices converse with one another.

5. Auditory hallucinations on several occasions with content of more than one or two words, having no apparent relation to depression or elation.

6. Incoherence, marked loosening of associations, markedly

illogical thinking, or marked poverty of content of speech if associated with at least one of the following:
    a. Blunted, flat, or inappropriate affect
    b. Delusions or hallucinations
    c. Catatonic or other grossly disorganized behavior
B. Deterioration from a previous level of functioning in such areas
as work, social relations, and self-care.
C. Duration: Continuous signs of the illness for at least six months
at sometime during the person's life, with some signs of the illness at
present. The six-month period must include an active phase during
which there were symptoms from A.
D. A full depressive or manic syndrome, developed after any psychotic symptoms, or was brief in duration relative to the duration of
the psychotic symptoms in A.
E. Onset before age 45.
F. Not due to any organic disorder or mental retardation.

If those criteria sound extremely complex or convoluted to the
average reader, that is because they are! The definition was made
considerably simpler in DSM IV.

        DSM IV – Diagnostic Criteria for Schizophrenia
A. *Characteristic symptoms*: Two (or more) of the following, each
present for a significant portion of time during a one-month period
(or less if successfully treated):
    1. Delusions
    2. Hallucinations
    3. Disorganized speech (frequent derailment or incoherence)
    4. Grossly disorganized or catatonic behavior
    5. Negative symptoms, i.e., affective flattening, alogia, or avoli-
tion. (Note: only one criterion A symptom is required if delusions are bizarre or hallucinations consist of a voice keeping up a
running commentary on the person's behavior or thoughts, or
two or more voices conversing with each other.
B. *Social/Occupational dysfunction*: for a significant portion of time
since the onset of the disturbance, one or more major areas of functioning such as work, interpersonal relations, or self care are
markedly below the level achieved prior to the onset (or when the
onset is in childhood or adolescence, failure to achieve expected
level of interpersonal, academic, or occupational achievement).
C. *Duration*: Continuous signs of the disturbance persist for at least

six months. This six-month period must include at least one month of symptoms (or less if successfully treated) that meet criterion A (i.e., active-phase symptoms) and may include periods of prodromal or residual symptoms. During these prodromal or residual periods, the signs of the disturbance may be manifested by only negative symptoms or two or more symptoms listed in criterion A present in attenuated form (e.g., odd beliefs, unusual perceptual experiences).

D. *Schizoaffective and mood disorder exclusion*: Schizoaffective and mood disorder with psychotic features have been ruled out because either (1) no major depressive, manic, or mixed episodes have occurred concurrently with the active phase symptoms or (2) if mood episodes have occurred during active-phase symptoms, their total duration has been brief relative to the duration of the active and residual periods.

E. *Substance/general medical condition exclusion*: The disturbance is not due to the direct physiological effects of a substance (e.g., a drug of abuse, a medication) or a general medical condition.

F. *Relationship to a pervasive developmental disorder:* If there was a history of autistic disorder or another pervasive developmental disorder, the initial diagnosis of schizophrenia is made only if prominent delusions or hallucinations are also present for at least a month (or less if successfully treated).

As the differences between the definitions of schizophrenia in DSM I, II, III, and IV indicate, the definitions of mental illnesses have undergone changes in both style and content since the first standardized manual was developed in 1952. Why did these changes occur? Do they represent progress, or just change for the sake of change?

The decision to move from the brief descriptive definitions of DSM I and II to the diagnostic criteria of DSM III and later manuals was driven by a very real need for improvement in the precision of definitions. During the 1950s and 1960s, multiple research studies had indicated that psychiatric diagnoses tended to be inconsistent and unreliable. The criticisms came from multiple fronts. Powerful films, such as *One Flew Over the Cuckoo's Nest*, suggested that people who were simply angry or rebellious might be misdiagnosed with an illness such as schizophrenia and subjected to inhumane treatments as a consequence. Evidence for this came from a much-publicized article, called "On Being Sane in Insane Places." Several research investigators presented themselves to psychiatric hospitals, complained of relatively minor hallucinatory experiences (e.g., hearing a voice saying "thud"), and were hospitalized for evaluation for several weeks before they were found to have nothing seriously wrong and discharged. This study was widely discussed when it appeared in the 1970s, since it suggested that some psychiatric facilities were not identifying people who were sanely feigning mental illness.

Several famous multinational studies compared diagnostic practices around the world and noted serious inconsistencies. For example, the International Pilot Study of Schizophrenia compared diagnostic practices in twelve countries and found that schizophrenia was more frequently diagnosed in the United States and the Soviet Union than in other nations throughout the world. Similarly, a comparative study was done between the United States and Great Britain. In this study, British and American psychiatrists evaluated the same patients. The American psychiatrists were more likely to diagnose schizophrenia and the British psychiatrists mood disorder.

The goals that motivated the development of DSM I and DSM II were clearly not being achieved. The simple descriptive definitions were not leading to diagnostic consistency. The boundaries for some disorders, such as schizophrenia, were relatively broad in the United States, while others were relatively narrow (e.g., mood disorders). Psychiatric diagnosis, as conducted during the 1960s and 1970s, needed a critical reappraisal that would lead to improved precision and reliability. Specifying the precise steps—diagnostic criteria—was the obvious solution.

The main function of diagnostic criteria is to improve the agreement between two clinicians who are evaluating the same information or the same patient. Psychometricians refer to this as reliability. Reliability can be quantitatively measured using indices of agreement such as various types of "correlation coefficients." In general, reliability levels of .6 or .7 are considered to be good, and levels of .8 or .9 are excellent to outstanding. A variety of research studies have shown that the introduction of diagnostic criteria raised reliability coefficients considerably. Therefore, they have substantially improved clinical practice, since psychiatrists throughout the United States and the remainder of the world are using the same objective definitions of the various mental illnesses. In addition, diagnostic criteria have improved psychiatric research, since they ensure that genetic and imaging studies are using similar definitions. Since diagnoses are standardized, patients from different institutions can be pooled together in research studies in order to produce larger samples that will have more statistical power. The ability to collect large samples is particularly important in genetic studies.

Patients have also benefited from the increased objectivity provided by diagnostic criteria. Twenty or thirty years ago, psychiatric diagnosis was a bit of a mystery, and many people were not quite sure exactly what "mental illness" meant. Since the development of DSM III, the entire process of defining mental illnesses and making diagnoses has become both objective and public. In fact, anyone who wants to can buy a copy of DSM IV or even look up definitions of disorders on the Internet. Patients themselves can look up diagnostic criteria and decide whether they think they have a particular diagnosis. Whether they identify themselves as patients or consumers, people with mental illnesses are placed on a much more equal footing with their doctors. Having access to knowledge about the diagnostic process and the definitions of mental illness has empowered them.

The "Downside" of Diagnostic Criteria

As the saying goes, it is hard to have gains without also having pains. While the more objective definitions of mental illnesses have produced undeniable benefits, they have also introduced some new problems.

One problem is that some people take the DSM definitions too seriously. DSM definitions are based on the best evidence available. In addition to syndromal clustering of signs and symptoms and a characteristic course, specific disorders have been delineated from one another because they have different patterns of family transmission or different responses to treatment. The scientific basis of DSM is credible. But it is not infallible. Because DSM has become institutionalized in training programs and quality assurance testing programs, it is revered too much and doubted too little. Both physicians and the lay public must recognize that, while progress with syndromal definitions has been good, we still have a great deal to learn about the pathophysiology of most mental illnesses. As we learn more, definitions and classifications may change. Furthermore, research studies must experiment with the power of non-DSM definitions to probe the deepest levels of how mental illnesses are caused. For example, people who develop Alzheimer's disease at an early age have a much worse outcome than those who develop it later. Could it be that early and late Alzheimer's disease are two distinctly different disorders with different pathophysiologies? Many research scientists are concerned that DSM criteria may limit creativity and flexibility in thinking, which may inhibit progress in understanding the underlying mechanisms of mental illnesses.

A second criticism of the DSM criteria is that they may have sacrificed validity in order to achieve reliability. The evolution of the criteria for schizophrenia illustrates this problem very nicely. Historically, and probably correctly, schizophrenia was defined by Kraepelin and Bleuler as a multisystem disorder affecting multiple mental functions. In particular, it produced changes in emotional responsiveness and the ability to think clearly. These concepts were well reflected in DSM I and DSM II. When DSM III was written, however, concerns about overdiagnosis of schizophrenia and poor reliability led to an emphasis on symptoms that were easily defined because they were more objective than subjective. Specifically, the definition emphasized hallucinations (hearing voices) and delusions (a variety of false beliefs, such as being controlled by outside forces or persecuted). The definition of schizophrenia became more reliable with the new DSM III criteria, but the essence of its concept may have been lost in the process. Psychiatrists who trained after 1980, most of whom learned only the DSM III or later definitions, are able to ask a seemingly endless series of questions about hearing voices, but in the process they may miss or even misunderstand the broader intrapsychic problems that people with schizophrenia experience. At the level of research, understanding the pathophysiology of schizophrenia must explain more than simply delusions and hallucinations. It must also explain why people with schizophrenia experience changes in their ability to think clearly or in their emotional responsiveness. These symptoms, which were revived in DSM IV through the concept of negative symptoms (in an effort to improve validity), are core features of the illness. Although more difficult to define reliably, they may have greater clinical significance (i.e., validity) than delusions or hallucinations.

A third objection to overdependence on DSM definitions and diagnostic criteria is that they may dehumanize clinical care. Whether they are suffering from cancer, a heart attack, or a depression, all patients would like their doctors to think of them as human beings first and as "cases of disease" second. Multiple pressures impinge on the modern doctor, forcing him or her to diagnose and treat quickly. DSM facilitates speed, since diagnoses can be made by using relatively simple checklists containing a series of questions. Therefore, it conveniently matches the needs of HMOs and managed care organizations. But it may not meet the emotional and psychological needs of individual patients. Each person who walks into a doctor's office is a unique being who lives in his or her own individual economic and social world and who happens to have a medical complaint of some type. He or she is a *person* with signs and symptoms, not a package of signs and symptoms that just happen to be occurring in a person. Too often, the DSM criteria encourage physicians to jump immediately into inquiring about specific signs and symptoms without taking adequate time to get to know the patient as a unique individual who lives in a particular economic, social, and spiritual world. Not only can this process be dehumanizing for the patient, but it may have negative medical consequences, if the physician prescribes a medication that the patient cannot afford or a treatment that the patient cannot accept for religious or cultural reasons.

A Prologue to Progress

The remaining four chapters in this part summarize the status of our knowledge concerning four major groups of mental illnesses: schizophrenia, mood disorders, dementias, and anxiety disorders. Each of the four chapters summarizes the syndromal definitions of these various disorders, our current state of knowledge about the pathophysiology, and the treatments that are currently considered to be optimal. As summarized in the report card, the syndromal definitions of these conditions are solid. Nearly all diagnoses are based on what is referred to as "the clinical history," which consists of signs and symptoms and their pattern of waxing and waning over time.

We have laboratory tests for only two diagnoses, Alzheimer's disease and Huntington's disease. The laboratory test for Alzheimer's disease, which checks for the presence of the characteristic plaques and tangles, can only be made by examination of the brain after death, so diagnoses of Alzheimer's disease in living people are provisional. The progression to laboratory tests is typically done on the basis of understanding the pathophysiology. The Huntington's test, for example, identifies the genetic abnormality that defines the illness: a characteristic mutation on Chromosome 4. Genetic and imaging tools are at present being applied actively to all the other disorders, but it is unlikely that any will be so simple that they will be easily diagnosed by a single laboratory test at any time in the future.

Most mental illnesses are "complex" in pathophysiology. That is, their causes and mechanisms must be understood from multiple perspectives, running the gamut from genes to environment. The story of pathophysiology told in the next four chapters is likely to progress rapidly over the next several decades. As this occurs, scores on the report card are likely to improve. The good news, however, is that we already know a great deal—far more than we did twenty years ago. Further good news is that we have good and steadily improving treatments for most of the major mental illnesses.

Each group of illnesses is introduced through the story of a particular patient. Anyone who wants to have a genuine understanding of mental illnesses must appreciate the impact that they have on human lives. To miss the human aspect is to miss the whole point of this book. The dignity and value of individual human beings need not and should not be lost as we integrate our scientific understanding of their brains, minds, and spirits.

CHAPTER 8

SCHIZOPHRENIA
A Mind Divided

> I felt a Cleaving in my Mind—
> As if my Brain had split—
> I tried to match it—Seam by Seam—
> But could not make them fit.
> The thought behind, I strove to join
> Unto the thought before—
> But Sequence raveled out of Sound
> Like Balls—upon a Floor.
> —Emily Dickinson
> "Poem 937"

Scott's mom and dad, Sue and Phil, were puzzled and frightened. Scott, a nice-looking kid despite his rather scraggly blond hair and baggy, ill-fitting clothing, was no longer taking care of his appearance. He showered infrequently and seemed to have stopped brushing his teeth. Even his younger sister, Laura, was beginning to say he was outside the normal range for cool teenage scruffiness. He was also spending lots of time alone in his room and seemed to have lost interest in hanging out with his friends. Was he on drugs? Was he depressed? What could be wrong?

If I tell them what is happening, they will probably think I am losing my mind and make me see a psychiatrist, Scott thought to himself. God, it's scary. Maybe I should tell them.

Everyone is making fun of me all the time. Mom and dad wouldn't believe it if I told them how bad it really is.

You can't imagine what it is like. My whole life has changed. It is like I woke up and found myself living in hell. Almost everyone around me is like a demon who is tormenting me. I get to school in the morning, and everyone is staring at me. A bunch of kids will be standing in the hall. I can tell that they are talking about me. I hear them saying my name, while they're standing around laughing and making jokes. I can just tell. Really, they say "See, Scott just walked in the door. He looks like he has been jerking off all night long. He's Scott the jerk-off jerk." Then they laugh like it's the funniest thing in the world. Those lines "Scott the jerk-off jerk" keep running through my head. I hear it all the time. It's like the kids at school are talking to me even when they are not around. Sometimes they say other nasty things too, like, "Scott, you are an asshole" or "Scott, just get lost."

Kevin and Clyde, my two best friends, haven't done anything to stick up for me. In fact, they have turned against me. It would be OK if they just decided not to be my friends and left me alone. But they want to torment me too . . . I don't know why. I never did anything to them. We used to have such a good time, getting together to fix up our car, listening to music, and stuff like that. Now they are sending electrical signals from the car battery. It shouldn't be that strong, from a 12–volt battery, but they can actually give me electrical shocks on my skin. I don't know how they transmit the current through the air, even through walls. But they do. Sometimes they even hit my nipples. I don't try to see them any more. If I told anyone that this was happening, they probably wouldn't believe it. It really burns and hurts. Sometimes I hear them talking to me too, saying the same kind of shitty things that the other kids at school are saying.

I have to keep this secret. Otherwise people may even lock me up. I may just have to run away somewhere to escape.

There isn't any point in studying any more. I can't keep anything straight in my head. It is really weird. I start to have a thought, and then things just don't connect right. It is like my brain is babelized. I try to do an assignment, and at first a million thoughts occur to me. Then they get all jumbled up and my mind goes blank. A terrible empty blank. Like standing on the edge of a cavern and being about to fall into it.

I'm so confused. I am so scared. I can't describe this to anyone. I have to deal with this alone. I used to be one of the guys. But now I'm an outcast. Why is this happening? What did I do? I've completely lost control over who I am, what I do, what I think. God, I've got to do something to stop what is happening to me, but I don't even know what it is. Could it really be that I am losing my mind, like one of those people in One Flew Over the Cuckoo's Nest? *If I am, then I should just kill myself. I don't want to keep living if it is going to be this bad the rest of my life.*

Sue and Phil just couldn't figure it out. Scott had been such a joy most of his life, and now he had become withdrawn and even a little hostile at times. No kids are perfect, but both Scott and Laura were fun and a great source of pride—good students, lots of interests, and many friends. What a change now. Scott was sullen, apathetic, and even dirty and smelly.

Scott, now seventeen, had been an adorable little boy. Sue was tiny, only four foot eleven and under 100 pounds, and so Scott's birth was a little hard on her. The labor lasted almost 24 hours, and they nearly did a C-section. But finally they managed to deliver an 8 pound 1 ounce baby boy by using forceps. Scott's head looked like a banana during the first week or so, but its shape gradually became normal, and he was otherwise a very alert and bright-eyed little guy. Sue and Phil would look at him in his crib and wonder at the tiny miracle they had produced together. They had no worries that he might have problems—just the opposite, in fact. He sucked vigorously, cried loudly, held his head up strongly, and kicked his arms and legs actively. He did almost everything earlier than the little boy next door, who was born just a week later. He crawled at six months and toddled at nine. He seemed a little behind in speech, saying only a few words until around 18 months, but then he started talking to them in complete sentences. It was as if he had wanted to take it all in and do it right, rather than sounding like a baby.

Phil was thrilled to have a son, and he played all sorts of games with him. As Scott grew older, he became accomplished at both physical and intellectual games. He learned to swim when he was only two, to play checkers when he was three, and chess when he was five. He was the star pitcher and leading batter on every baseball team that he joined, right on through high school. He also picked up tennis fairly well and learned to play a solid game against his father and his friends. He was remarkably ambidextrous, which was an asset in both baseball and tennis. People could never be sure whether they would have to deal with a right-hander or a lefty.

Although Phil and Sue were pleased with Scott's athletic achievements, they most enjoyed his playful and loving nature. He was a little blue-eyed towhead who looked like Dennis the Menace, but behaved like Anthony Angel . . . most of the time anyway. When he was only one-and-a-half, they developed a tuck-in ritual that they kept up for many years. Sue would say, "love you," as she turned to leave the room, and Scott would call back, "no, love you first." Soon, "love you first" became a family expression that Phil, Sue, Scott, and (eventually) Laura all used, as a way of expressing how much they cared for one another.

Scott was their golden boy. They watched him grow from an adorable toddler to a rambunctious preschooler to an alternately grave or giggly grade-schooler. They watched his body move from chubby to muscular as he matured from childhood to adolescence. He was a handsome and appealing kid at each stage. They watched his mind expand and grow from wondering where butterflies go in winter to wondering how airplanes fly. He never got less than a B in any course in school and was always in the upper quarter of his class. They expected big things from him. He could become a lawyer, an engineer, a veterinarian, a pilot, a leader in business—anything he wanted.

It all started to fall apart the summer before his senior year at North High. He had just broken up with his girlfriend. She was his first serious girlfriend, and for the previous year they had seemed to be madly in love with one another. Phil and Sue never knew exactly what went wrong with the relationship, and Scott didn't seem to want to discuss it, even though he was usually pretty open with them. They wrote it off to "disappointed love" and "teenage growing pains" when he started to spend more time in his room alone—at first listening to music, but later just staring silently at the wall, sometimes for three or four hours. They were more surprised when he also stopped hanging out with his buddies Kevin and Clyde. Those three kids had been inseparable since junior high—riding bikes, playing catch, hanging out at the mall, ordering pizza late at night, and eternally dismantling and reassembling a 1982 Pontiac Firebird that resided in Clyde's garage. When they finally asked Scott about Kevin and Clyde, he gave them a funny look and said, "I told them that I didn't want to hang out with them any more. It's because they don't like me any more. They aren't really my friends after all." That seemed very strange to Phil and Sue.

Scott seemed to be having a personality change—going from a friendly outgoing kid who was a self-starter to a lethargic lump. He had a summer job as a checker at the supermarket. He had had to work his way up to the checker job by loading groceries in cars for the previous year, and he was pleased with the extra pay and responsibility. But he actually quit the job in mid-July and didn't even tell Phil and Sue about it for a week or two. The only explanation that he gave was that "the customers made fun of the way that I looked and talked." Now Scott did look a little funny to his graying parents, but lots of other kids had the same funny look, and Phil and Sue believed that the way a kid acts is more important than how he wears his hair or jeans. Scott, their athlete, was wearing his blond hair in long golden ringlets and donning baggy, torn trousers every day. They worried more when he stopped washing that golden hair and when it became greasy and even bad smelling. He seemed to have lost all interest in his appearance. He slept late in the morning, and spent his time aimlessly. In fact, he seemed to have withdrawn into a different world. When school started in the fall, he had difficulty getting up and was reluctant to attend classes.

At first Phil and Sue thought it was just adolescent moodiness. Then they wondered about drugs. Normally respectful of his privacy, they decided to check his room when he was away. No signs of drugs. Not much of anything that gave them any clues—a few car magazines, a few rock magazines, and a few copies of *Playboy*. They didn't know what to do or what to think. They decided to watch carefully and to give Scott his space and time, at least for a little while.

Then one night Scott disappeared. He left for the evening around 8 p.m., mumbling something about going out for a walk to "think things over." He had a kind of preoccupied and wild expression on his face. They didn't ask him where he was going, because they just assumed that he would walk for an hour or so and return home. He had done that many times before. When 10 p.m. arrived, they started to worry. By midnight they were frightened. Scott had always been good about telling them what he was doing, and it was unlike him to disappear like this. At 1 a.m. they decided to call the police for help. However, the police could do nothing—kids stayed out late all the time without telling their parents . . . law enforcement officers can't be surrogate parents. Then around 1:30 they got another call. Scott had been found. A driver had seen him on a bridge and then seen him jump over the side. His body was found on the railroad tracks below. He had been taken by ambulance to Mercy Hospital, still alive but probably seriously injured.

Saving Scott's life and setting his two shattered legs were the first priority. As he awakened from the anesthetic and began to talk to Phil, Sue, and Laura, the rest of story came out. They listened in horror as he explained that Kevin and Clyde were devil worshipers who had been transformed into demons, as had all the rest of the kids at school. They were mocking him all the time, telling him that he was worthless, calling him names, tormenting him by sending electrical shocks to his body. Sometimes he could even smell sulphur when he was around Kevin and Clyde. That night they had told him exactly how to "get lost." He should go to the Winston Street Bridge and jump off. He was just following their orders. He was powerless to resist the control they were exercising over him.

What had happened to Scott? This was the sort of thing that happened in movies, not to real people. It couldn't be true. Scott was a good kid, a normal kid, a promising kid. This was just delirium, due to the pain of the broken legs, the confusion caused by anesthesia. It couldn't be *mental* illness.

But Scott's account of inner torment turned out to be only too true.

The psychiatrist who cared for him during the six weeks that he lay in traction on the orthopedic ward gently explained Scott's illness to them. After the doctor confirmed from them that there was no history of drug abuse and no strong evidence for a mood disorder, he concluded that Scott must be suffering from schizophrenia. Scott had described an abundance of the painful inner experiences that plague people who develop schizophrenia—all the symptoms that doctors consider "psychotic," such as hearing voices when no one is around or feeling that others are tormenting them, controlling them, or sending them signals and messages through indirect methods. He also had the telltale signs of a change in personality, inability to think clearly, and social withdrawal.

Scott's "broken brain," the psychiatrist explained regretfully, would be tougher to heal than his broken legs. The broken brain of a person with schizophrenia cannot be put in traction and casts. Fortunately, medications can help a great deal, especially with psychotic symptoms. New ones had been developed recently that would be especially helpful. The doctor assured them that Scott's delusions and hallucinations would probably diminish markedly within a few weeks after medications were begun and that they might disappear altogether. Phil and Sue were very relieved. But . . . the doctor also explained that schizophrenia is a complex brain disease that affects many aspects of its victim's life and personality. It might be hard to get the "old Scott" back completely—the sunny cheerful kid who approached life like a tennis ball on its upward bounce. Schizophrenia could sometimes drain away emotions and drive, and medications were less successful in transfusing these mental traits back in again. But love, attention, and support from everyone in the family would certainly help.

✻   ✻   ✻

The next few years turned out to be rocky and rough for the whole family. As predicted, Scott's psychotic symptoms vanished for nearly six months. Scott was able to return to school, reluctantly, after Christmas. He had enough credits so that he could even graduate on schedule with his class. Kevin and Clyde were great, once they realized that Scott's strange behavior had been due to a mental illness. They stopped by the hospital often, and by the house after he returned home, and they tried hard to keep him on track in every possible way—getting him to class, taking him out on weekends, working on the car together. Scott tried hard too. But something had changed. He just wasn't very motivated. His mind wandered. He didn't concentrate very well. He graduated, but everyone agreed that he should put off college for another year while he continued to get his life back together. Then the voices and the demons came back again right after graduation, even though he was still on medication. The dose was increased, but he began to be restless and fearful. The doctor decided that, since he had made a serious suicide attempt while psychotic, he should go into a psychiatric hospital for a few weeks until the psychosis diminished.

For the first time, Phil, Sue, and Laura saw a "schizophrenia treatment unit." There, in a group, were approximately thirty people ranging from teens to early thirties—all of them with symptoms similar to Scott's. It was chilling. Would this be Scott's future? Would he, now 18, become like that 28–year-old man, sitting in front of a TV, chain-smoking, and rocking back and forth? Their first reaction was to want to run away, to take Scott somewhere else, to refuse to hospitalize him at all. Scott had the same feeling. Although Scott had come to the hospital because he was pacing and restless and mentally tormented, he felt he couldn't really be like all these other people who were pacing and worried. The doctor explained that these were all people like him, people with families, people who were suffering. He suggested to Sue and Phil that they might, in fact, like to join an organization that was composed of patients with mental illnesses and their families, the National Alliance for the Mentally Ill (NAMI), which had a local chapter that met monthly. They might even be surprised at who they would meet there. They would find a lot of families not very different from their own, fighting the same battle against fear and grief. It turned out to be a great suggestion. NAMI was an invaluable resource for information, advice, consolation, and even hope during the ensuing years.

Scott was able to leave the hospital within ten days, with his psychotic symptoms much improved. But over the next two to three years he had occasional relapses and had to return to the hospital. Phil and Sue discovered that their insurance had placed a "cap" on treatment for mental illness—only 60 days were covered for an entire life! Hardly enough. A social worker assigned to Scott by the hospital explained that Medicaid would cover him if he applied for permanent disability. But that would be like admitting defeat. Scott was going to get better, not be disabled for life. However, as they watched Scott's college fund disappear and Laura's diminish, they finally capitulated. It took nearly a year to process all the paperwork. Still worse, if Scott were to recover enough to work more than half time, he would have to lose his Medicaid coverage. What a disincentive! There ought to be a way to provide good medical care for people with schizophrenia without penalizing them for trying to return to a productive life.

After five years or so, fortunately, the relapses stopped occurring. Scott stabilized and even improved. Eventually he was able to get a half-time job and to take an occasional class at the local community college. He was a clerk at a local garden center, and he actually enjoyed the contact with customers. The family learned to live with his illness, which remained below the surface, still robbing him of drive and energy and impairing his ability to read other people exactly right. He was still a little suspicious and untrusting. The golden boy never returned to his old sunny self. But he was still their Scott, still sweet, lovable, and loving in his own way. Phil and Sue, who had been haunted by worries about how Scott would survive if anything would happen to them, felt their fears eased a bit.

✳　　✳　　✳

Schizophrenia is probably the cruelest and most devastating of the various mental illnesses. An estimated 1% of the population has schizophrenia, which claims its victims at a youthful age and prevents their full participation in society. Schizophrenia also creates an enormous economic burden, costing society billions of dollars annually.

Primarily a disease of young people, it typically strikes during the late teens and early twenties. It is a disease that has many different faces. Some sufferers, like Scott, appear to be perfectly "normal" kids before becoming ill, catching family and friends by surprise as they begin to experience symptoms. Some have subtle indicators that can be identified retrospectively, especially in comparison with brothers and sisters who do not become ill. As children, they may have been less well coordinated, shyer, more anxious, or slower learners. Scott had none of these indicators, although he did have two others that research studies suggest are "predisposing factors." First, his mother had a prolonged labor, and his delivery was difficult. Second, perhaps as a consequence of a subtle head injury incurred at the time of birth, he was ambidextrous—neither strongly right nor left-handed.

Such "early signs" can be misleading when they are a product of hindsight. It is easy to look back and see a higher rate of "clues" after a son or daughter has already become ill. During the past decade, however, clever scientists have looked for such early indicators using what is called a "prospective design." That is, they obtain information that was somehow recorded about the child *before* the illness of schizophrenia had become manifest.

One approach, applied by Elaine Walker and Rich Lewine at Emory University, was to obtain home movies from families who belonged to NAMI and who had at least one child suffering from schizophrenia. All the children in the movie were rated by trained professionals using standardized methods to assess how well they were coordinated or how well they interacted with brothers, sisters, and other children. The raters did not know which of the children would later become mentally ill. Analyzing these "blind ratings," Walker and Lewine found that the children who eventually developed schizophrenia tended to be more awkward both physically and socially.

Another approach, used by Robin Murray and his team in England, was to examine the records from a national health study conducted in England shortly after the Second World War. All the children born during a single week in 1947 were identified and had all aspects of their health measured at regular intervals. This forward-looking and innovative national study has told us a great deal about many illnesses, including schizophrenia. As of the last few years, 37 of these children have developed schizophrenia. Murray and his group were able to look back at health and school records and determine how those who eventually became ill differed as children from those who did not. They have found some interesting things that are what psychiatrists call premorbid indicators—small signs that may be early markers of a predisposition to become ill and that are present before the full illness emerges. Among them is an increased rate of mixed-handedness, or ambidextrousness. Yet another prospective study, conducted by Michael Davidson in Israel, used health records collected by the military. Israel has a policy of universal military service for both men and women, so Davidson was able to examine later development of schizophrenia in a representative group of young people tested during their late teens prior to onset of illness. Problems in relating socially with others were found to exist in those who later became ill. These prospective studies are among the strong evidence that has been accumulating to suggest that schizophrenia is due to an abnormality in brain/mind development that may begin well before the person develops obvious symptoms.

What Is Schizophrenia?

Unfortunately, "schizo" has become a slang term, often used in a derogatory way. Schizophrenia was given its name by a Swiss psychiatrist, Eugen Bleuler, early in the twentieth century. Bleuler wanted to choose a name that would describe the most important features of this illness, which had previously been called dementia praecox by Emil Kraepelin, the German psychiatrist who originally identified it as a discrete mental illness. Kraepelin named the illness dementia praecox because that name means "an illness that affects the ability to think clearly and is persistent and chronic" (dementia) and "an illness that occurs primarily in young people" (praecox).

His colleague in Switzerland, Eugen Bleuler, initially took exception to some aspects of Kraepelin's definition. Like Kraepelin, Bleuler worked in a large psychiatric hospital, the Burgholzli, and saw many patients over long periods of time. Bleuler believed that some patients showed substantial improvement after their initial onset of illness. Therefore, the use of the term "dementia" was misleading, since it suggested that the patients would steadily worsen over time, as typically happens in neurodegenerative disorders. Further, he noted that some patients developed their illness at a later age—in their twenties, thirties, or (rarely) forties. Therefore he suggested that the name "dementia praecox" be replaced with one that was more descriptively accurate. He proposed the name "schizophrenia." This name literally means "splitting" or "fragmenting" of the mind and is derived from classical Greek (schizo = split, fragmented; phren = mind). Bleuler chose this name because he believed that the essential feature of schizophrenia was an inability to think clearly and to link together "associative threads" during the process of thought and speech. His new name gradually replaced Kraepelin's "dementia praecox."

Schizophrenia is an illness that can be difficult to explain or define because patients have so many different kinds of symptoms. Perhaps the most striking thing about schizophrenia is its sweepingly broad injury to a large array of cognitive and emotional systems in the human brain. The signs and symptoms of schizophrenia are diverse; they include disorders of perception (i.e., hallucinations), inferential thinking (delusions), goal-directed behavior (avolition), and emotional expression (affective blunting), to mention only a few. No single one of its many signs and symptoms can be considered to be pathognomonic or defining. Each is present in some patients, but none is present in all. In this respect, schizophrenia differs from most other mental illnesses, which typically affect a single mind/brain system, such as Alzheimer's disease (memory) or manic-depressive illness (mood).

Because the signs and symptoms of schizophrenia are so complex and diverse, an effort has been made to simplify thinking about the illness by subdividing them into natural categories. The most widely accepted subdivision is into "positive" and "negative" symptoms. This terminology is a bit confusing because there is nothing that is positive or good about the positive symptoms—they are unpleasant experiences such as hallucinations. This terminology derives from a nineteenth-century British neurologist, John Hughlings-Jackson. As often happens in science and medicine, Jackson's theories were shaped by new ideas from an unrelated field—in this case, Darwin's theory of evolution. He thought of the human brain as like an onion. It contained more primitive levels at its inner core, with "higher" governing levels encasing and enclosing this inner core. The "higher" levels were added on during the process of evolution. He thought the positive symptoms of psychosis were a "release phenomenon." Thoughts from a lower level of evolution would break through out of the inner core because some higher governing brain region had lost control, and so the person would have hallucinations or delusions. Negative symptoms, on the other hand, were due to a simple loss of function, presumably due to neuronal loss. They were expressed by apathy or disinterest.

My research program has been studying the symptoms of schizophrenia and the best ways to define it for more than twenty years. One of our more useful contributions was the reintroduction of Jackson's ideas in a modern form. We dropped the Darwinian baggage, which was unproved, and redefined positive and negative symptoms in a simple descriptive way. We also emphasized the importance of negative symptoms and their underlying relationship with cognitive and emotional impairments.

This modern reconceptualization defines positive symptoms as an exaggeration of normal functions (the presence of something that should be absent), and negative symptoms as a loss of normal functions (the absence of something that should be present). We classified the array of signs and symptoms as positive or negative according to the kinds of mental functions involved, revealing that the two groups together include nearly all the mental functions that human beings possess. A summary of the symptoms and their corresponding mental functions appears in Table 8–1.

Positive symptoms are typically those that call attention to the illness. People usually are recognized as being mentally ill because their positive symptoms are clear indicators that they suffer from a serious problem that impairs their sense of reality. Scott's positive symptoms are what led to his suicide attempt, for example, and they shocked and frightened his parents when Scott initially described them.

Negative symptoms are often the first signs of the illness to appear. For example, Scott became withdrawn and lost interest in things that he pre-

TABLE 8–1
The Symptoms of Schizophrenia

SYMPTOMS	MENTAL FUNCTIONS
Positive	
Hallucinations	Perception
Delusions	Inferential thinking
Disorganized speech	Organization of language and ideas
Disorganized behavior	Monitoring and planning of behavior
Inappropriate emotions	Emotional appraisal and response
Negative	
Alogia	Fluency of language and thoughts
Affective blunting	Expression of emotions and feelings
Anhedonia	Ability to experience pleasure
Avolition	Ability to start things and follow through
Attentional impairment	Ability to focus attention

viously enjoyed. Positive symptoms tend to respond to treatment more rapidly and readily than negative symptoms. Even when positive symptoms have been reduced, however, it is usually clear that the person is still "not well." The negative symptoms often persist and lead to impairment in many important aspects of life—the ability to work, to return to school, to have close friends, to enjoy hobbies and sports, to have a girl-friend or boyfriend, or to feel close to family members. The negative symptoms can sometimes be mistaken for laziness or bad manners, but they in fact reflect a loss of the ability to start things and follow through on them and a loss of the capacity to experience joy and pleasure in the normal activities of daily living. People with schizophrenia often notice this loss of pleasure and drive, and they find it as troubling as the positive symptoms. In some ways, negative symptoms are even worse because they seem to rob patients of their personality and identity.

What Causes Schizophrenia?

When a young person develops a mental illness, the knee-jerk response is often: "what did the parents do wrong?" Schizophrenia is not a disease that parents cause. Nor is it a disease that parents can prevent or arrest, much to the despair of people like Phil and Sue. Despite parental love and care, the disease strikes, injures, and leaves its suffering victims and their families in pained submission. Schizophrenia is a brain/mind disease. In most cases several causes have conspired to injure the developing brain and mind, but bad parenting is not one of them.

Genetic Influences

Studies showing that genetic factors may contribute to the development of schizophrenia provided the earliest evidence that schizophrenia has a biological basis. As described in chapter 5, our methods for studying genes and their contribution to disease has grown steadily more sophisticated. The earliest genetic work on schizophrenia was based on the simple observation that mental illnesses sometimes run in families—an observation that suggests a role for genes but does not prove this, since familial aggregation could be due to learned behavior and role modeling. In the case of schizophrenia, the actual pattern of familial transmission does suggest a role for genes. If one parent has schizophrenia, there is about a 10% chance that one of the children will develop schizophrenia. If both parents have schizophrenia, then this risk increases substantially to about 40 or 50%. Likewise, chances for developing schizophrenia if one brother or sister has the illness are about 10%, and these increase to about 20% if one parent and one brother or sister is ill with schizophrenia. So there is a modest risk if one family member has the illness, and the risk increases substantially if two or more family members are ill.

During the past fifty years, more sophisticated methods have been used to translate these observations into a more precise study of genetic influences. Comparing the rates of illness in identical and nonidentical twins provides a more direct test of the influences of genes. The higher the concordance rates in identical twins as compared to nonidentical twins, the greater the probability that an illness is genetic, since identical twins share almost exactly the same genes, while nonidentical twins share approximately 50%. The term "concordance rate" refers to both twins having the same illness. More than ten twin studies of schizophrenia have now been done, involving hundreds of twin pairs. They have consistently demonstrated higher concordance rates in identical (monozygotic) twins than in nonidentical (dizygotic) twins—around 40% as compared to 10%. The 10% rate seen in nonidentical twins is similar to that for brothers and sisters. The summary of this comparison is sometimes referred to as the monozygotic:dizygotic ratio, or MZ:DZ ratio, which is a rough index of the degree to which an illness is under genetic influence. For schizophrenia the MZ:DZ ratio is approximately 4:1—clear evidence that genes must play a role.

Doubters who wish to argue that schizophrenia is due to bad parenting or a poor family environment might still point out, however, that twins grow up in the same home, and that identical twins tend to be encouraged to behave in similar ways. Therefore, two gifted scientists, Seymour Kety and Leonard Heston, independently pioneered a new and more powerful approach to isolating the role of genes in schizophrenia. They separated the role of family environment and the role of genes by studying adopted children who grew up without knowing anything about their birth (biological) parents. Both groups of adoptees were reared in families that were considered to be "normal" or "healthy." Kety and Heston compared the rates of schizophrenia in adopted children who had a birth mother with schizophrenia to adopted children who had a birth mother with no evidence of mental illness. Both studies found essentially the same thing. The rate of schizophrenia in adopted children reared apart from their birth mothers was about the same as for children who had a schizophrenic parent and grew up in the same home. It was around 10%. On the other hand, the rate in the children of healthy birth mothers was similar to the population rate, or about one percent.

Although these studies of the rates of schizophrenia in families, twins, and adopted children are often cited as proof that schizophrenia is "a genetic disease," close inspection of this evidence indicates that the story is not really that simple. If identical twins have almost identical genes, and if schizophrenia is a disease caused solely by genetic influences, then the concordance rate in identical twins should come close to 100%. The actual rate of around 40% is nowhere near that high. The various genetic and family studies suggest that genes must indeed play a role, but genes alone do not cause schizophrenia. Other factors, described later, probably need to be added to a genetic predisposition for schizophrenia to develop. This is good news, since it is more difficult to make changes in genes than it is in other predisposing factors in order to achieve the long-term goal of preventing schizophrenia in predisposed children.

As described in chapter 5, scientists are working strenuously to apply the tools of molecular genetics and molecular biology to identifying the genes that may be involved in the development of schizophrenia and other major mental illnesses. The early success in finding a single gene that caused Huntington's disease raised the hope that single genes might be found for illnesses like schizophrenia, but this has not turned out to be the case. Most experts now think that schizophrenia is clearly multifactorial, involving multiple genes, and possibly even different genes in different individuals, as well as many nongenetic or environmental influences. The fact that multiple genes are probably involved is the main reason why the various reports that "the schizophrenia gene has been found on chromosome 5" (or 11 or 22 or elsewhere) have not been consistently repeated. Any single gene can probably only explain a small fraction of the causation of schizophrenia. Although this makes the search for the genetic influences more difficult on a scientific level, it is again good news on the human level. If multiple genes are required and have to co-occur, the chances of developing schizophrenia are reduced, and genetically predisposed children cannot be considered to be "doomed from the womb."

Since an emphasis on genetic influences can lead to undue pessimism, especially in families where at least one member has schizophrenia, another potentially optimistic note should be struck. That is about the relationship between creativity or originality and schizophrenia. The relationship between mood disorders and creativity has been frequently publicized during the past twenty years. When I initiated the objective scientific study of creativity and mental illness some thirty years ago, I expected to find a relationship to schizophrenia. My early studies, begun in the early 1970s, were inspired by anecdotal observations of familial relationships between genius and schizophrenia. James Joyce, one of my favorite writers, had a daughter, Lucia, who suffered from life-long schizophrenia, was treated by Jung, and died in a mental hospital in England. Bertrand Russell, a cultural icon of the twentieth century, had an uncle, a son, and a granddaughter who suffered from schizophrenia. Albert Einstein also had a son with schizophrenia. Further, Leonard Heston (at that time one of my colleagues at Iowa) had observed that a substantial number of the adopted children of schizophrenic mothers pursued creative interests or hobbies, suggesting that there might be a genetic association between schizophrenia and tendencies to be creative or think in original ways.

These observations are coupled with another interesting fact. Although people with schizophrenia often do not marry and do not themselves have children, the disease appears to have persisted down through the centuries and at an equal rate throughout the world. What could explain this? The answer might be that "schizophrenia genes" may also confer some evolutionary benefit that leads them to persist. Having them may transmit some abilities that are useful to human beings, either on an individual or a group basis. Physicians are familiar with other models of this for nonmental illness. Sickle-cell anemia, for example, persists in Africa because it protects against the development of malaria.

Recent films and other events have brought the relationship between genius and schizophrenia back to the forefront. The film Shine, although unfortunately implying that the young artist's illness might have been caused by an overpunitive and aloof father, portrayed the artistic triumph of a musical genius who had a schizophrenia-like illness. In 1994 many of us who have worked with people suffering from schizophrenia rejoiced to learn that the gifted economist John Forbes Nash had won the Nobel prize for his work on game theory. Not only had Nash made major contributions to this and many other branches of mathematics, but he also had been seriously ill with schizophrenia during his thirties and forties, eventually improving markedly and functioning again at a very high level, supported by his loving wife, Alicia. An interesting anecdote in the prologue to a recent biography of Nash highlights the possible relationship between genius and schizophrenia. Hospitalized in a psychiatric facility in 1959, Nash was asked by one of his friends how he, such a rational and logical man, could believe that he was getting messages from aliens in outer space. Nash looked at his friend and replied, "The ideas I had about supernatural beings came to me the same way that my mathematical ideas did. So I took them seriously."

People with schizophrenia indeed perceive the world in unusual and original ways. The ability to do this may lead to erroneous insights that we consider psychotic. On the other hand, this ability may also lead to highly original ideas or observations that turn out to be true. Einstein, the father of a son with schizophrenia and himself a highly eccentric man, is an obvious example. Isaac Newton, who identified the laws of mechanics that laid the foundation for the industrial revolution, was also a solitary and eccentric man who experienced a psychotic episode in his forties. Perhaps giftedness in mathematical, scientific, and abstract creativity is particularly related to schizophrenia. In any case, people who seem to have carried "the schizophrenic tendency" provided the two major scientific contributions of modern physics. We owe the laws of gravity and mechanics and the theory of relativity to the two original and beautiful minds of Newton and Einstein.

Neurodevelopmental Factors:
Are People with Schizophrenia Doomed from the Womb?

How do the various genetic and nongenetic influences actually add up and eventually cause a person to develop schizophrenia? Most clinical neuroscientists now suspect that schizophrenia is a "neurodevelopmental disorder." Something—and probably several different things—has gone wrong in the orderly process of brain development that begins at the time of conception and continues on into young adult life.

Many other neurodevelopmental disorders are well recognized in pediatrics and general medicine. Some of these are due to brain abnormalities that begin during pregnancy, causing their victims to be "doomed from the womb." Down's syndrome, or trisomy 21, is a genetically caused disease that is due to an abnormality in chromosome 21. The genetic mutation leads to a classic syndrome of mild to moderate mental retardation, typical physical traits such as "mongoloid" or oriental-appearing facial features, a sweet and loving personality, and a tendency to show intellectual decline in the thirties and forties. Trisomy 21 occurs with increasing frequency in the children of older mothers, presumably because the genes in the mother's ova become less efficient during cell division with increasing age. (Women have a lifetime supply of eggs in their ovaries; this supply does not increase or get replenished over time.) Older mothers now often obtain genetic testing early in pregnancy to determine whether their developing baby has this genetic abnormality.

At the other end of the continuum is a neurodevelopmental disorder that is completely nongenetic, fetal alcohol syndrome (FAS). FAS occurs when children developing in the uterus are exposed to massive amounts of alcohol because the mother drinks too much during pregnancy. Children with FAS have a low birth weight, small heads, small brains, learning disabilities or mild mental retardation, and behavioral hyperactivity. They have characteristic facial features such as wide-set eyes, flattened noses, or a lack of the characteristic dent in the upper lip. Our work at Iowa studying children with FAS, led by Victor Swayze, indicates that they also have marked and easily visualized abnormalities in their brains that reflect failure of the normal formation of connections in brain regions that occur across the midline (e.g., agenesis of the corpus callosum; see chapters 4 and 6 for more details). Scientists have not yet determined exactly how or when alcohol causes this damage to brain development. Alcohol consumed by a pregnant mother does pass directly to her developing baby and enters into brain tissue, since alcohol is soluble in fats and brain tissue is very rich in fat. Most evidence seems to suggest, however, that an occasional glass of wine is not likely to do much damage, and that the mischief is caused primarily by intermittent binges when very large amounts of alcohol are consumed.

Both of these neurodevelopmental disorders begin before birth. Trisomy 21 lays out the wrong genetic program for brain development, while FAS introduces a toxic substance that interferes with the orderly progression of the genetic plan for connecting brain regions. In either case the damage is done irreversibly before the child is even born, and the diagnosis is usually obvious in the delivery room or during the first few days of life.

Schizophrenia, however, is a different type of neurodevelopmental disorder, and it is unlikely the people with schizophrenia are doomed from the womb due to either genetic or nongenetic factors. Most people who have it were either completely normal or relatively normal at the time of birth. They do not show their first signs of illness until much later. Some people who eventually develop schizophrenia have mild premorbid indicators, but many are as normal as Scott was before he became ill. As described in chapter 4, we know that brain development is an ongoing process that does not end until sometime in the mid-twenties. Unlike Trisomy 21 or FAS, the factors that negatively affect brain development in schizophrenia, probably occur at multiple times. Any one factor could cause the illness if severe enough, but probably in most cases they need to add up. The most critical abnormality must be one that occurs during the late stages of brain development, when the brain does its final "growing up" during the late teens and early twenties. This is a critical time for young people, since they must learn to fly out of the parental nest and live on their own, choose an occupation, and find friends and partners with whom they can share their lives and perhaps ultimately marry.

The scientific evidence suggesting that schizophrenia is a neurodevelopmental disorder affecting multiple stages of brain development is substantial and steadily increasing. This evidence indicates that many different kinds of influences may be involved, and that they may be both genetic and environmental. Genetic factors were just reviewed. Scientists have also documented the importance of several environmental factors. For example, people with schizophrenia are more likely to have been born in the wintertime, a season during which mother and child are more often exposed to a variety of viral illnesses. Higher rates of schizophrenia have been observed among people who have been born during influenza epidemics, also suggesting that viruses may be a factor. Viruses, of which poliovirus or the human immunodeficiency virus (HIV) are notorious examples, are famous for their ability to damage tissue in the nervous system and also for their ability to invade the cells and produce changes in the genetic material. Studies of regions in Europe exposed to severe famine during the Second World War have also shown that malnutrition during pregnancy can contribute to the development of schizophrenia. Like Scott, people who develop schizophrenia are more likely than average to have a history of birth injury or perinatal complications. These may cause a hidden brain injury that sets the stage for the later development of schizophrenia.

Perhaps the strongest evidence supporting neurodevelopmental abnormalities comes from neuroimaging studies. Our research group, as well as

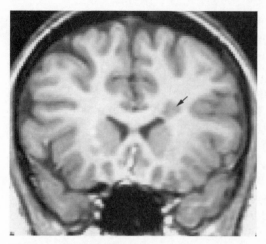

Figure 8–1: MR Scan Showing Ectopic Gray Matter in a Person with Schizophrenia

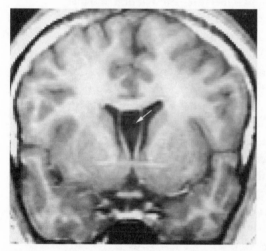

Figure 8–2: MR Scan Showing Enlargement of the Cavum Septi Pellucidi in a Person with Schizophrenia

several others, has been especially interested in using MR to examine the brains of patients with schizophrenia for telltale signs that something went wrong during the process of brain growth. One such abnormality is "ectopic gray matter," or tiny islands of neuronal cells that did not make it to their proper destination when they began the arduous task of neuronal migration to the cortex during the second trimester of pregnancy (see chapter 4). Although this clue is not seen very often, ectopic gray matter is nevertheless more common in people with schizophrenia than in healthy normal individuals. Figure 8–1 shows the brain from a person with schizophrenia who has ectopic gray matter (found in about 5% of males with schizophrenia). Another more common neurodevelopmental abnormality is found in about 20% of males with schizophrenia. Most of us are born with a small gap between the two hemispheres of our brain, known as the cavum septi pellucidi. As our brains mature during early childhood, this gap closes. However, it has failed to close in approximately 20% of men who have schizophrenia, suggesting that something went wrong in brain growth during early childhood or later. The brain of a person with schizophrenia, showing this abnormality, appears in Figure 8–2.

Imaging studies of large numbers of people with schizophrenia studied shortly after onset of symptoms also suggest that some abnormality in brain development has occurred prior to onset. Clinical scientists refer to these as "first-episode studies." MR is used to visualize the living brain in young people who typically are in their late teens or early twenties and who have become ill only recently. Such studies can be very informative, since they let us examine possible causes early in the game, before treatments with medication or the effects of chronic illness may have caused changes. Five or six studies of first-episode patients have now been done by centers throughout the world, ranging from Australia and the Orient to England and Europe to the United States and Canada.

Quite consistently, these studies have shown that first-episode patients have the same types of brain abnormalities that are seen in people who have been more chronically ill. These include enlargement of the ventricles, enlargement of the sulci on the surface of the brain, a general decrease in overall size, and specific decreases in size in crucial brain regions such as the prefrontal cortex or the hippocampus. In addition, the cortex is thinner, although the total number of cells is not decreased, suggesting that the change is due to a loss of the projections from the cell bodies of nerves that permit them to make connections (i.e., dendrites and spines). Such substantial structural brain abnormalities (e.g., overall smaller brain size, increased ventricular size) have probably been present for at least a few years and preceded the development of the illness. The most likely interpretation is that they are additional indicators of a previous problem in brain growth and development that led eventually to the symptoms of schizophrenia.

This inference has been given even more solid support by recent studies of childhood onset schizophrenia conducted by Judy Rapoport's team at the National Institute of Mental Health (NIMH). They have now studied brain growth in large numbers of healthy children and adolescents and compared them to children and young adults who have recently developed schizophrenia. Both groups are being studied repeatedly over time using anatomic brain measurements obtained with MR. This very interesting work has focused on the crucial teenage years, when young people are maturing mentally and socially. Their brains are making the key connections that permit them to grow up and become mature and functional adults. We know from basic neuroscience that this process involves overgrowth of connections, which subsequently need to be pruned back. The NIMH group has shown that during the teenage years (between 13 and 18), brain growth and change occurs in both healthy control subjects and those with schizophrenia, but that the developmental curves are quite different. The progression of brain development in the childhood-onset schizophrenia patients involves a decline in total brain volume, and an increase in the volume of gray matter. These studies add further evidence that the brain changes in schizophrenia may occur at multiple points in time, ranging from early fetal development to late adolescence or young adulthood.

Why does this matter?

Pinpointing the crucial time window for the development of an illness is an important step in figuring out how to prevent it. Determining whether people with schizophrenia are "doomed from the womb" is not limited to asking whether the disease is genetic or not. It also means asking about nongenetic factors such as birth injuries or viral infections. If the key brain injury occurs during fetal development, we have few opportunities to arrest the progression of the illness. If, on the other hand (as is probably the case), maturational brain changes during the teenage years are important, then eventually we may be able to identify what they are and arrest them very early. Most treatment programs currently being developed for Alzheimer's disease and other neurodegenerative disorders are based on this kind of strategy—to identify the neurobiological processes that produce the illness and then to influence or arrest their occurrence or progression. Many of us have similar hopes that we can do the same for schizophrenia.

Schizophrenia as a Disorder of Misconnections in the Brain

So far we have seen how schizophrenia is caused by a mixture of genetic influences and nongenetic influences such as head or birth injuries, viral infections, exposure to toxins and drugs of abuse, hormonal changes, and other factors. These influence the development of the brain during the prolonged period of brain maturation that occurs in human beings and that is probably not completed until the early twenties. But exactly what effect do these neurodevelopmental aberrations have? If we were to go hunting for the locus of schizophrenia in the brain/mind, where would we look? To answer this question, we must think about "functional genomics," or how the influence of specific genes is translated to abnormalities in the functions of the brain and mind.

We would look for the locus of schizophrenia almost everywhere. But it will be hard to find it in any single place.

Some mental illnesses, such as Alzheimer's disease, have characteristic changes in specific cells and cell layers. Other mental illnesses, such as Huntington's disease, affect a single brain region. For schizophrenia, however, a diligent search by many talented neuroscientists has not identified any such specific regional abnormalities or nerve cell lesions. Some skeptics of neurobiological explanations for schizophrenia have suggested that the illness cannot therefore be considered a brain disease. A much more likely explanation, however, is that schizophrenia is a disease that affects the brain in ways that are different from Alzheimer's or Huntington's disease. It does not damage specific cells or even specific regions. Instead it damages the way regions are connected to one another, so that a breakdown in signal transfer occurs, and the messages sent back and forth between various brain regions are garbled or confused. In the language of neuroscience, schizophrenia is a disease that affects distributed neural circuits rather than single cells or single regions. Such disorders are sometimes referred to as misconnection syndromes.

When I talk to patients with schizophrenia, I often begin by asking them what kinds of problems are troubling them the most. They tend to come up with answers like this:

My thinking is confused.

My ideas don't seem to connect quite right.

I have trouble filtering out unimportant information.

I feel bombarded by stimuli.

In short, most people with schizophrenia have a subjective sense that their ability to think and feel has somehow become disorganized, disconnected, or misconnected. The tools of neuroimaging, already briefly described in chapter 6, have allowed us to study how the brains of people with schizophrenia function differently from healthy individuals when they are performing similar mental tasks. These studies have shown us that the subjective experience of "misconnection" or "disorganization" reflects an underlying problem in the ability of distributed brain regions to send messages back and forth efficiently and accurately.

Although computers are not a perfect metaphor for the brain, using them as a model can be somewhat helpful. The typical computer system has to continually link up information from multiple sources, such as a variety of software programs, stored data, and peripheral devices such as printers and scanners. We have all seen how computers can get tied up in knots and crash if we ask them to do too many things at once, or if incompatibilities occur between various software programs or between software and hardware. Performance can become sluggish and inefficient if the information transfer rate becomes too slow or if files become too large to handle. The underlying brain abnormalities in schizophrenia are somewhat analogous: they occur because of problems with integrating multiple components. Therefore, we are not likely to succeed in localizing schizophrenia in any single brain region, although we do observe that some regions are abnormal if measured anatomically or functionally using MR or PET.

Some specific abnormalities in subregions of the brain have been found relatively consistently. Measures of structural size using MR have demonstrated decreases in the frontal lobes, temporal lobes, or hippocampus. More recently our group has suggested that the thalamus and the regions of the cerebellum also have specific size decreases. These subregions are scattered all over the brain, and they are quite numerous. The only way to make good sense of these findings is to conclude that schizophrenia is not "in" any single one of them, but instead occurs because of problems in all of them, caused by the connections or relationships between them. Like a malfunctioning computer, the various nodes in brain networks send information back and forth in a way that causes file corruption, garbled information, or crashes. For a few people with schizophrenia, the neurodevelopmental abnormality producing the misconnections may be due to abnormal wiring on a large scale—e.g., connections between neurons at the level of axons. This is evident in those rare cases of ectopic gray matter. Most of the time, however, the abnormality is at a finer level, involving the sending and receiving of messages by synapses or spines on dendrites.

Although we cannot see the lesions of schizophrenia at postmortem as we can for Alzheimer's disease, we can see the abnormalities in connections between functional circuits using in vivo imaging techniques such as PET. These studies show that people with schizophrenia have trouble transporting signals and information back and forth around their brains in the same way and at the same speed as normal people performing the same task.

In our own PET center, we have now studied healthy volunteers and people with schizophrenia while they perform many different tasks. These have included remembering past experiences, learning and later remembering word lists or stories, responding emotionally to pleasant or unpleasant visual images, sniffing nasty or pleasing odors, focusing attention on specific areas of a video screen, listening to sounds and identifying which ear is receiving them, tapping fingers in a particular rhythm, or judging how long a particular time interval is. That is obviously a long list of very different kinds of mental tasks. They were chosen to examine the diverse kinds of mental problems that people with schizophrenia have, such as focusing attention, encoding and using memories efficiently, or responding emotionally and experiencing pleasure.

Rather remarkably, people with schizophrenia have abnormal patterns of blood flow in all of these tasks, but some regions are abnormal in almost every kind of task. These consistently abnormal regions include the thalamus and the cerebellum. We now suspect that the cerebellum may be malfunctioning as a "metronome" or timekeeper, causing signaling to lose its synchrony and coordination. The thalamus, which functions as a filter or gatekeeper that helps determine how much information should be let in or out of the brain, may be failing to screen information out, so that the system becomes overwhelmed with so much data that the person's thinking becomes confused or sluggish. The net effect is incoordination or miscommunication between key processing centers needed for a given task that are distributed throughout the brain. Figure 8–3 (A-C) illustrates these abnormalities in distributed brain circuits in schizophrenia. These studies from our research center, as well as many from other centers, suggest that schizophrenia is not a disease of any specific brain region, but rather a disease of functional connectivity between distributed regions. For this reason, we need to see the lesion or injury using different techniques from those used for illnesses such as Alzheimer's disease or Huntington's disease. At the level of genetic influences, we must now work to understand how genes regulate the ongoing process of establishing and maintaining connections throughout widely distributed regions of the brain, including all components (e.g., synapses, spines, cells, and key nodes such as the thalamus.)

The Neurochemistry of Schizophrenia

The circuits that link brain regions together use chemical messengers that permit the cells on either end to communicate with one another. Disturbances in the chemical activity of the brain create another "invisible lesion" that cannot be seen with the naked eye or under a microscope. Evidence from several sources indicates that chemical imbalances in schizophrenia are "real," contribute to the development of symptoms, and can be corrected through medications that affect brain chemistry.

Dopamine was the first neurotransmitter to be discovered as a contributor to the symptoms of schizophrenia. Several lines of evidence have suggested that dopamine is important in this illness. First, schizophrenia-like symptoms can be produced through stimulant drugs such as amphetamine, which produce their exhilarating effects by causing large amounts of dopamine to be released. In fact, chronic amphetamine abuse may produce a permanent change in the brain and be yet another nongenetic factor contributing to the development of schizophrenia in predisposed individuals.

A second piece of evidence is provided by our knowledge of how antipsychotic drugs work. Almost all the drugs that successfully reduce symptoms of schizophrenia decrease the activity of the dopamine system in the brain by blocking dopamine receptors, the little patches on nerve cell membranes that are designed in order to receive chemical messages. Nearly all antipsychotics, ranging from older ones such as chlorpromazine to newer ones such as risperidone or olanzapine, lock into these receptors and shut out dopamine molecules that may be trying to transmit messages. As dopamine tone is reduced, the psychotic symptoms are also quieted down. These observations led to the formulation of the "dopamine hypothesis of schizophrenia" by Arvid Carlsson, a distinguished Swedish neuropharmacologist and Nobel laureate, in the 1960s. Direct support for this hypothesis was provided by several different studies of postmortem brain tissue independently conducted by two eminent psychiatrist/pharmacologists, Solomon Snyder and Phillip Seeman. These studies showed a direct relationship between the ability of drugs to block dopamine (type 2) receptors and their antipsychotic potency. Studies also showed an increase in dopamine receptors in limbic brain regions such as the nucleus accumbens in patients with schizophrenia.

The dopamine hypothesis was universally accepted as explaining the chemical imbalance in schizophrenia for nearly thirty years. Simply stated, it suggested that the symptoms of schizophrenia were due primarily to hyperactivity in the dopamine system. In the past decade, however, the plot has thickened, as our knowledge about neurotransmitter systems has increased further and as new antipsychotic and antischizophrenic drugs have been developed. The dopamine hypothesis has not been supplanted, but it has had to deal with the arrival of at least two new younger siblings, serotonin and glutamate. More recent hypotheses suggest a key role for both serotonin and glutamate in the development of the symptoms of schizophrenia, based on multiple recent studies of the neurochemistry and neuropharmacology of the illness. Clinician scientists now think that schizophrenia occurs as a consequence of a much more complex chemical imbalance that includes multiple neurotransmitter systems that interact with and modulate one another. Dopamine is almost certainly a crucial component, but other neurotransmitters also play a role.

What Treatments Are Available?

The development of new medications for schizophrenia is one example of the remarkable progress that has occurred in the treatment of mental illness over the past fifty years. Although we all wish that mental illnesses such as schizophrenia could be cured completely, and we are as yet unable to achieve this, we should pause and take stock of how far we have actually come.

Fifty years ago a person diagnosed as having schizophrenia was faced with three or four options, all of them relatively unpleasant. One was the irreversible surgical treatment, prefrontal leucotomy, developed by the Portuguese neurologist Egas Moniz. Insulin coma and electroconvulsive therapy were other alternatives. These three treatments were palliative measures that worked in some patients. The majority of people with schizophrenia remained chronically ill and required lifetime care because they experienced so much confusion, lack of drive, or psychotic agitation that they were not able to care for themselves independently and meet the demands of ordinary daily living. Literally every other hospital bed was occupied by someone suffering from schizophrenia—50% of the people in hospitals throughout the world.

New Medications and False Hopes

In this context, a new medication, chlorpromazine (with the trade name Thorazine), burst on the scene in the early 1950s. The first experiments were conducted in Paris in 1952 by two French psychiatrists, Jean Delay and Pierre Deniker. They tried the new drug in psychiatric patients with many different diagnoses, including mania, schizophrenia, and depression. They observed that, in addition to calming or tranquilizing, it also markedly reduced the terrifying hallucinations and troubling delusions that were causing patients to be agitated, fearful, and restless. Their observations were quickly confirmed in many other countries, including the United States, and a new era in the care of schizophrenia was launched. The number of inpatient beds needed for psychiatric patients dropped steadily, institutions were closed, and more and more patients began to receive their care in community settings where they could lead more normal lives. Soon other new antipsychotics, many of them with fewer side effects, were added to the armamentarium, such as haldoperidol, which was developed by Paul Janssen in Belgium.

Psychiatrists began to approach the care of first-episode patients with new hope. Many believed that if the first episode of psychosis could be treated aggressively and further relapses prevented with the new medications, schizophrenia could potentially be cured altogether. Gradually, however, these hopes were dashed, and schizophrenia continued to be viewed as a devastating and grim diagnosis. The antipsychotic medications, it turned out, were simply that: they were effective for reducing psychotic symptoms, such as delusions and hallucinations, but they were much less helpful with negative symptoms. Although no longer psychotic, patients continued to suffer from a reduction in the ability to think fluidly, to experience joy or pleasure, or to start tasks and finish them.

People would be admitted to the hospital in the midst of an agitated psychotic episode, improve markedly with antipsychotic treatment, and be discharged with plans to return to work or school. Despite their best efforts, and despite encouragement from family and friends, they often were unable to do the things that used to come easily to them. They would simply lose interest in classes, wander away from a job in the middle of the day, or fall into a pattern of staying in bed or spending all their time watching TV. It became increasingly clear that the negative symptoms were due to a reduction in the ability to function cognitively and emotionally. The antipsychotics did not seem to cure the "schizo-phrenia," the fragmentation of mind and emotion, that was described by Bleuler and used to name this often incapacitating illness. After an era of hope launched in the 1960s with the introduction of antipsychotics, the treatment of schizophrenia failed to improve further. Yes, patients were better, but they were far from well.

The Next Generation of Medications
Things have improved again recently, however, raising hopes that increasing progress will be made in the care of schizophrenia. A "new generation" of antipsychotic medications has been developed, with the hope that they will be truly "antischizophrenic." These are the so-called atypical antipsychotics.

Clozapine was the first of the new atypical medications. Although it had been used to a modest degree in Europe, it was largely abandoned after it was noted to produce a dangerous reduction in the number of white blood cells, a condition known as agranulocytosis. Two American psychiatrists, John Kane and Herbert Meltzer, decided to explore its utility in a group of people with schizophrenia who had remained severely and chronically ill. They were "treatment refractory." That is, they had not responded to any of the antipsychotics that had been tried. In this desperate situation, it seemed worthwhile to try a somewhat risky medication, using weekly checks of white blood count to determine whether agranulocytosis was developing. Gratifyingly, Clozapine worked! Many patients who had been chronically ill improved markedly.

Clozapine is an interesting medication for several reasons. First, unlike every antipsychotic developed up to that point, it did not work intensively on the dopamine system to block the dopamine type 2 receptor, which had been the primary target for drug development. Further, it seemed to work not just for positive psychotic symptoms but also for the negative symptoms of schizophrenia. People with the illness gradually began to regain their interest in life, their ability to enjoy things, and the capacity to think in a clearer, more logical way. Finally, clozapine had minimal extrapyramidal side effects.

In addition to being less effective for improving cognition or reducing negative symptoms, the older antipsychotic medications produced a variety of unpleasant side effects, many of which are related to their blockade of the dopamine system. Extrapyramidal side effects, sometimes called EPS for short, include a rigidity of muscles, a frozen blank expression on the face, a tremor, and a tendency to walk with a shuffling gait. Another unpleasant side effect produced by older neuroleptics is akathisia. This is an intense subjective sense of anxiousness, which causes the person to want to pace and move around restlessly. Some people on traditional antipsychotics also feel depressed. Because these side effects are so unpleasant, many people with schizophrenia do not want to take traditional antipsychotic medications. For them, the treatment is almost worse than the disease itself. Some would prefer to have psychotic symptoms if they have to pay the price of having EPS in order to get rid of them.

The new atypical medications have been a godsend in this regard. Paul Janssen, who had previously developed haloperidol, has devoted much of his life to finding better treatments for schizophrenia, and he synthesized and tested yet another new medication, risperidone, which worked on other neurotransmitters besides dopamine and was the first of the new atypical medications that could be used without the difficult blood-monitoring required for clozapine. Other new atypicals subsequently followed Janssen's lead. These include olanzapine (Zyprexa), quetiapine (Seroquel), and ziprasidone (Zeldox). These drugs have now become "first-line treatments." That is, they are normally the first choice for treating people in their first episode of illness, and they are largely supplanting the older antipsychotic medications for most people who have schizophrenia. The traits that define "atypical" include relatively less D2 blockade, reduced extrapyramidal side effects, ability to improve both psychotic and negative symptoms, and an ability to improve mental alertness. The atypicals are also less likely to produce tardive dyskinesia, a potentially irreversible movement disorder that occurs in 20–30% of patients treated with traditional neuroleptics. This movement disorder is a relatively disfiguring tendency to grimace, twitch, and pace.

The newer atypicals do have their own problems with side effects, however. One of the most troubling is that some of them cause excessive weight gain, partly as a consequence of increased appetite, but partly as a direct effect of the medication. A related side effect is an increased tendency to develop diabetes mellitus. Another problem is an effect on one component of the endocrine system (prolactin, which stimulates breast development).

Cognitive and Psychosocial Rehabilitative Treatment

When the diagnosis of schizophrenia was considered to carry a grim prognosis, few people worried about introducing psychotherapy, psychosocial rehabilitation, or cognitive retraining. Because people with schizophrenia who have been treated with the new atypicals are more alert and interested, both patients and families now hope that the addition of various psychotherapeutic interventions to the newer medications can substantially improve outcome. Programs that emphasize this aspect of treatment may soon become a mainstay, as novel approaches are developed. These treatment programs will probably have two components.

First, people with schizophrenia may need help with learning how to organize the activities of everyday life, which psychiatrists call psychosocial rehabilitation. This illness strikes its blow at a critical period in development, the teenage and young adult years. This is the time when people are just becoming emancipated from the home. Therefore, young adults with schizophrenia may need to learn to pass through the natural maturational processes of learning how to set up an independent home, look for a job, or plan a routine normal day. People with schizophrenia are sometimes shy or fearful of being around others, and so they may need help in figuring out how to join group activities and make friends. Partial hospitalization, group programs, and outpatient therapies may all be helpful.

Cognitive relearning is a second approach. Treatment programs implementing this strategy are in their infancy. Their essence is to focus on the fundamental cognitive abnormalities that characterize schizophrenia. They are designed to help patients learn to focus attention more precisely, to solve problems more efficiently or more rapidly, to monitor ideas and speech more effectively, and to improve both motor and mental coordination. These programs are founded on the concept of neuroplasticity, described in chapter 4: "neurons that fire together wire together." The hope is that if people with schizophrenia can go through extensive retraining similar to that used to rehabilitate stroke patients or train children with specific hearing or learning disabilities, they will gradually rewire their brains so that new connections are formed, and they can then learn to think and function more clearly and effectively.

CHAPTER 9

MOOD DISORDERS
Riding the Emotional
Roller Coaster

> For aught we know to the contrary, 103 degrees or
> 104 degrees Fahrenheit might be a much more
> favorable temperature for truth to germinate and
> sprout in, than the more ordinary blood heat of 97
> or 98 degrees.
> —William James
> *Varieties of Religious Experience*

Marcia was puzzled. Hal was really acting strange. He was doing things that were completely out of character for him.

Marcia and Hal were getting ready to celebrate their fifteenth wedding anniversary. It had been a great fifteen years. True, Marcia had given birth to two kids and no longer weighed 105 with a 22–inch waist, but she still was a pretty good-looking 112. Hal was the one who should be able to use the pregnancy excuse! On Marcia's cooking he had ballooned from 140 to 175 and acquired a loveable paunch.

Their fortunes had ballooned as well. Hal was a developer in Santa Fe, and he had a real knack for it. Although the market had leveled a bit during the late 1980s and early 1990s, things had been booming for the last five years. Hal himself had really done well, even during the "soft" times. Hal knew what he was doing and cared about doing it exactly right. People respected his meticulousness. Hal could anticipate better than most, and he worried about all the right details. He could look at 120 acres of land, and where other people saw a wasteland, he could envision roads that took interesting turns to well-sited houses looking at the Jemez or Sangre de Cristo mountains. He wasn't one of those shoddy developers. He also worried about the community and the environment. People trusted him. They knew he was honest, that he wouldn't compromise quality or standards to make an extra buck or two. In the wild whirlwind of Santa Fe real estate, a guy like Hal was a community treasure. Lots of locals were selling out to the Almighty Dollar, and carpetbaggers were also coming in from both coasts . . . quite literally to capitalize on what could be done in the still unspoiled Southwest. These days Hollywood had discovered Santa Fe. The O'Keefe and the Los Alamos traditions were bowing before the onslaught of people looking for a quick dance with a wolf. Hal and Marcia were old Santa Fe—although they were good business people, they were also dedicated to preserving Santa Fe's unique history and ecology.

But now Hal was behaving strangely.

Hal had always been the more outgoing one in the pair, but now he was talking constantly. Last night over dinner, he talked a mile a minute, describing to her a new development that he was going to create in the hills on the northwest side, where all the houses would cost over a million dollars. Their market would be celebrities and second- or third-homers. He was going to go way out on a limb to get the funds for the land. He would call the area Principe de Paz, to honor the importance of world peace in the third millennium. That in itself was a great idea, he explained. He had big plans for recruiting buyers. First, he would buy a plane or two himself. He had researched that, and he could pick up a fairly nice used plane for $150,000. Eventually he'd like to move up to a very nice new twin-engine, and ultimately a Learjet. Second, he was going to take flying lessons so that he could fly people in himself, soaring over the land sites so that they could see the locale from an eagle's view. Third, he would include stables and a horse farm—people were getting tired of cars and wanted to get back to nature.

Parts of the idea seemed fine to Marcia. What worried her was that he was working so impulsively and so grandiosely. Hal was usually very systematic. Yes, he was a developer, which involved some risk, but up to this point it had been reasonable, and not so great that they would be over-extended. Normally, Hal got some land that would hold ten to twenty houses in the $300–400,000 range and built them slowly, one or two houses at a time. Most were spec homes, but Hal would occasionally do a more expensive custom home that cost a half million. They really couldn't afford to get into such a large-scale development, not to mention extra overhead like airplanes. Their kids were now 11 and 13, and both Hal and Marcia had always agreed that setting aside college funds was a top priority. This plan would use *all* their college savings. And . . . it wasn't like Hal to want Michelle Pfeiffer or Tom Hanks to move to Santa Fe.

Even the way he described it all was strange . . . he talked much faster, and it was all kind of disjointed. He hadn't been sleeping. He had been staying up until around 1 or 2 a.m., working at his desk, and then getting up again at 4 or 5 a.m. Maybe sleeplessness was messing him up. Last night he crawled in bed around 1:30 and insisted on making love even though she was sound asleep. He woke her up to do it again around 4:30. Now that was a bit much of a good thing too!!

Hal had always been energetic and hardworking, but this Hal seemed like a domesticated animal who had turned wild. What was happening?

Hal looked in the mirror, having jumped out of bed at 5 a.m. He felt GREAT. *That was GREAT last night with Marcia.*

He surveyed his appearance. He liked what he saw. Although not quite as handsome as Tom Cruise, he was still a real studmuffin. He peered into his glacier-blue eyes and gave himself a big smile. Irresistible. His teeth were white and even. Teeth tell a lot about people. His eyes moved down to the cleft in his chin. That was one of the things that women loved. He himself liked the rugged effect of his chin too. He turned to look at himself in the full-length mirror on the back of the door. It was a well-sculpted body. Women liked that part of him too. He chuckled.

Maybe I ought to change my name to Dick. Marcia couldn't resist me last night. I think I'll introduce myself to people today as DICK Davis, also sometimes known as Hal...Here I am, the greatest land developer in the southwest. AND the greatest studmuffin. Land . . . land . . . land. . . . grand . . . grand land . . scan . . . scan the land from the air . . . grand land . . . scan the situation . . . scam . . . . do a land scam. Don't call it a scam. Land scam. Building scam. Don't call it Principe de Paz. Call it Principe de Dick. Dick's division.

Maybe not quite enough water. Got to get it through anyway. I have the charm to do it. Will make millions out of Principe!! Millions!!!! Got to make a few more connections. (Erections??!!!) Maybe I can sell a lot to Michelle Pfeiffer. Look up her email address on the internet. Send her a picture of me. Get her to come over for a visit. Fly over and pick her up.

Hal glanced back in the mirror again. Yes, the eyes were blue, the teeth were white, and the chin had a cleft. But there was a bit of gray in the temples and some white hair on his chest. That sculpted body DID have a slight paunch. Well, never mind. He still felt GREAT.

What about the development plan? Was it too big? No, it was worth the risk. It would make them a fortune. He and Marcia would be secure for life! A few compromises? Who would notice? Besides, that's the way of the world.

One last look. He looked GREAT. Ready to take over land development in the entire Southwest eventually!

Over the next few weeks, Marcia's concerns turned out to be well founded. Hal went from pleasantly expansive to downright "off the wall." They were a close couple, and they always discussed new ventures in advance and agreed on a strategy. Suddenly, however, Hal was acting like a free agent. He took out several huge loans from their local bank—far more than ever before, and far more than she was willing to risk. He somehow convinced the president, Jack Selini, that Principe was a good bet and that the dude ranch and the private planes would be the wave of the future. Hal and Marcia now owed millions in construction loans. Hal had decided to wait for a while to take flying lessons himself, but he was expanding the office staff to include a pilot, an office manager, and two additional secretaries. Their lawyer (also convinced of the value of "the plan" based on Hal's enthusiasm and his solid reputation) had helped write two-year contracts with excellent fringe benefits, generous bonuses, and special options for purchasing lots at Principe. That was necessary to recruit the additional talent.

When Marcia questioned Hal about these decisions, he began yelling at her. How could she, his best friend, doubt his judgment and criticize him? Especially when there were other doubters. Two rival developers had gotten wind of the plan. They were going to try to block it. In fact, they seemed to be in cahoots with each other, unusual because they typically quarreled with each other. But there seemed to be a conspiracy growing against him, Hal confided. They also knew that he was in communication with Michelle Pfeiffer, Tom Hanks, and Oprah Winfrey on the Internet. They had been tapping into his email and reading his messages. He could tell because he was getting special banners on the screen that said things like "Principe sucks!" They would just flash briefly, but he *knew* where they were coming from. Hal described all of this in an animated way, nearly shouting, and sometimes not quite finishing his sentences. It was strange. He was just *too happy*. All these responsibilities should worry him. Normally, a small loan was enough to give him nightmares. But he was excited, enthusiastic, and *way* overconfident. This sudden personality change was scary.

Having lost the allegiance of the family banker and lawyer, Marcia decided to try the family doctor. She had decided that Hal was sick . . . maybe even sick in the head.

Dr. Schwartz listened to the story over the phone. He was sympathetic, but he explained that he would have to see Hal in order to figure out what the problem might be. He asked a few questions about Hal's family and made a thoughtful "Hmmm" when Marcia mentioned that Hal's sister suffered from depression.

"This could be manic-depressive illness," he said.

That was bad news, but at least it explained why Hal was acting so strange. Dr. Schwartz suggested two options: an appointment with him, which Hal might find more acceptable, or an appointment with one of the local psychiatrists. Even as they spoke, Marcia realized that either would be a tough sell.

That evening Marcia tentatively brought up her concerns. Hal listened for only a minute or two and flew into a rage.

"You think I'm CRAZY?! How COULD you? I'm the smartest developer around. Look what I talked Jack Selini into lending me. And I've been in communication with all those Hollywood people. I'm a star myself."

Marcia mentioned her doubts about whether he was *really* communicating with them. Maybe it was their secretaries or assistants. And she really doubted that "the competition" was flashing the banners. He was not himself. He was talking fast. Sometimes he didn't make sense. He wasn't sleeping. He was preoccupied with sex.

That was the last straw for Hal. He tore out the door, jumped in the car, and drove out of the driveway, tires screeching. Marcia wondered what would happen next.

❖   ❖   ❖

In a few hours, she found out. She received a call from the police. Hal had been picked up for speeding. When the officer stopped him, Hal jumped out of the car and started arguing. When the officer asked him to get in the patrol car, Hal slugged him, dashed to his car, and tried to drive away. The car received a bullet in the tires. Hal received handcuffs and an arrest for assaulting an officer. He was now at the police station. He was loud, argumentative, and abusive. He had given his name as Dick Davis, although his wallet indicated that his name was Harold Davis. Did they have the right home? He also said that he was a prominent celebrity— that he had built homes for Ted Danson, Michelle Pfeiffer, and Tom Hanks. He was going to be starring in a film with Michelle, with whom he had an "especially special" relationship. He had hinted that other important things were going to happen—something about the Prince of Peace coming to reside in Santa Fe, the city of holy faith. Would Mrs. Davis be willing to come to the police station to help them figure out what was going on?

Marcia got there as fast as she could without speeding.

Hal looked distraught, disheveled, defensive, and a bit demoralized. His hands were still cuffed. His shirt was torn. He was surrounded by three burly guys in uniforms, who obviously had no patience with anyone who would slug an officer. Marcia felt sorry for Hal. She could also tell that he could easily be spending the night in jail. And that she was going to have a tough time convincing the police officers that Hal wasn't so bad . . . that he was maybe sick instead of bad. She never thought she would be choosing between such dreadful possibilities. Her Hal wasn't either one.

She felt close to tears.

She took a deep breath, looked the officers in the eye, and started explaining Hal's recent personality change in a tiny cracking voice. She told them that Hal really was a builder/developer in Santa Fe, and a very good one. He was still fairly small, but well-known and respected within the building trade. Normally, he was a conscientious man, well liked, and trusted by everyone. He was also usually quite gentle. He rarely lost his temper, and he had never hit anyone before in his life.

But something had happened to him just lately. Everything about him seemed to have changed. He had become overconfident, developed expansive plans, borrowed way too much money, and even been crabby and mean to her. It had scared her so much that she had contacted the family doctor, who suggested that Hal might have manic-depressive illness. In fact, Hal's "joy ride" was precipitated by Marcia's candid confrontation about a possible mental problem.

The policemen were far more sympathetic than Marcia expected. This was apparently not the first time that they had dealt with someone who was manic. In fact, they seemed to know more about the problem than Marcia did. They suggested that Hal should be taken to a psychiatric unit in Albuquerque, where he could be held for 48 hours and evaluated. If mania was the problem, the charges would be dropped.

Hal was surprisingly silent as he listened to Marcia. But the mention of a psychiatric unit and a "48–hour hold" pushed the wrong button.

"There's nothing wrong with me," he yelled. "You can't do this. Call my lawyer. Call my banker. They'll tell you how important I am. You will be changing the course of history. Michelle will be furious!!"

Marcia had almost never fought with Hal. She rarely disagreed with him. She loved him and cared about him. Deciding to send him to a psychiatric unit against his will was the hardest and bravest thing that she had ever done. But she did it anyway. She knew that he might hold it against her for the rest of their life together. It might ruin her marriage. It might discredit Hal in the community. It might scare their kids. But she knew it was the only reasonable alternative, between two that both looked bad.

And the officers agreed with her. Hal was taken to the psychiatry ward in Albuquerque in handcuffs, protesting all the way.

Marcia . . . and eventually Hal . . . were pleasantly surprised. The first couple of days were pretty rocky, since Hal was distressed to learn that other people thought his expansive plans and irritable behavior were "crazy." He was immediately diagnosed as suffering from mania, and he reluctantly agreed to accept treatment. He was given medications that sedated him slightly, producing a dramatic reduction in his anger and agitation, and also lithium, a mood stabilizer. Within a few days angry grandiose Hal turned into guilty remorseful Hal, as he realized how overextended he had become, and how much he had frightened Marcia. Luckily, the grand schemes for Principe de Paz were still rather preliminary, and so he had not gotten in so deeply that he was ruined. Later, he and Marcia were able to look back gratefully on his "joy ride" and even chuckle about gentle Hal slugging a cop. If he had not done that, who knows how long it might have taken for him to be forced to obtain treatment?

Over the next decade Hal and Marcia learned to live—and live successfully—with the emotional roller coaster that Hal sometimes had to ride. As often happens to people who suffer from mania, Hal also had occasional episodes of depression. They both quickly learned the value of catching episodes early and seeking appropriate medical treatment. Although they had to explain Hal's problem to Jack Selini and their other professional associates, Hal's reputation as a builder was not impaired. In fact, it continued to improve. Honesty about his transient mood problems actually seemed to help, since his friends in the Santa Fe community became even more loyal supporters.

On their twenty-fifth wedding anniversary, Hal and Marcia were actually able to make a real champagne toast to a real Principe de Paz—one of the best designed areas of Santa Fe, developed and built principally by Hal Davis.

* * *

Way back in the twentieth century, sometime in the 1970s, a clever wag wrote a Broadway theatrical called, "Stop the World. I Want to Get Off." People who suffer from mood disorders know that feeling only too well. Subjectively, they feel as if someone picked them up and put them on an emotional roller coaster that just speeds ahead, and they have no brakes to slow it down, no accelerator to speed it up, and no conductor to hear them cry, "I want to stop and get off." Plummeting down the hill creates feelings of terror and dread. Going up the hill may feel like an exciting surge upward, but over time it is combined with the recognition that it will come to an end and will be followed by another frightening downhill ride.

It is sometimes unappreciated that billions of people are on this emotional roller coaster. Mood disorders are the most common of the mental illnesses. Approximately 1% of people throughout the world suffer from the most severe form, bipolar disorder or manic-depressive illness. This was Hal's problem (and therefore Marcia's). Depending on the narrowness or breadth of the definition, another 10–25% will have at least one episode of depression at some time in their lives. Jim, who lived through the "waking nightmare" described in chapter 2, suffered from depression. According to World Health Organization statistics, depression is second only to cardiovascular disease in economic cost to society in developed countries, outranking cancer. If we examine the cost worldwide, depression ranks fourth (behind respiratory infections, diarrhoeal diseases such as cholera, and diseases occurring near the time of childbirth).

Almost everyone you know has a friend or family member with mood disorders, or has suffered from this illness herself or himself. Once people begin to turn to one another and discuss their experiences with these illnesses, they find themselves looking into a crowded cauldron of human suffering. Further, mood disorders can be fatal. Approximately 15% of people with mood disorders commit suicide, leaving behind friends and family who wonder what they could and should have done.

What Are the Mood Disorders?

Mood disorders, which are also sometimes referred to as affective disorders, are illnesses that primarily affect the emotional coloring with which a person perceives the world. The two main forms of mood disorder, depression and mania, are on opposite poles from one another on the emotional scale, as shown on the thermometer of mood. When a person's emotional temperature is normal, the mood is said to be neutral. During mania, the world is viewed through rose-colored glasses because the mood is abnormally elevated or "high." During depression, the coloring of life is blue or sad. Some people move only from the neutral point on the thermometer to the negative end of the scale and suffer only from the blueness of depression. Jim, whose story was told in chapter 2, illustrates this type of mood disorder. Others, like Hal, ride a steeper emotional roller coaster and experience both highs and lows. Because they move back and forth between the two poles of mood, their illness is referred to as bipolar. Those who experience only depression are sometimes referred to as unipolar. "Manic-depressive illness" is the older name for this group of disorders. Some people prefer to continue to use it because of its historical importance. There are two milder forms of mood disorders, where the drift on the emotional scale moves to less severe extremes. A person with a full major depression might score in the 7 to 10 range, while a person who only gets down to a 3 or 4 is said to have dysthymia. The milder form of bipolar disorder is referred to as cyclothymia. The classification of mood disorders is summarized in Table 9–1.

Doctors have recognized mood disorders as medical illnesses and treated them accordingly for centuries. As documented in the writings of Hippocrates, the "father of medicine," Greek physicians included both mania and melancholia in the compendium of illnesses that they treated. Somewhat later, the physician Galen developed the "theory of humors." This theory persisted well into the eighteenth century and continued to influence medical treatment. The theory of humors suggests that the body is composed of four fluids (i.e., humors): blood, phlegm, yellow bile, and black bile. When we are in a healthy state, the four humors are in balance. If the quantity of any of the humors rises excessively and causes an imbalance, diseases occur as a consequence. Depression or melancholy, for example, is due to an excess of black bile. Melancholia literally means black bile (melan = black, cholic = bile).

The Thermometer of Mood

-10	-5	0	+5	+10
Dysphoric		Neutral		Euphoric
Depressed				High
Blue				Elevated
Sad				Expansive

TABLE 9–1

Classification of Mood Disorders

Depressive Disorders	Bipolar Disorders
Major depression	Bipolar, manic phase
Dysthymia	Bipolar, depressed phase
	Bipolar, mixed or rapidly cycling
	Cyclothymia

According to the theory of humors, mild variability in the balance of fluids leads people to have different types of personalities or temperaments. If they were slightly high in blood levels, they might be sanguine or optimistic. If they have a mild excess in black bile, they might tend to be somber or melancholic. If phlegm is relatively increased, they would be phlegmatic or laid back. If yellow bile is prominent, they would be choleric or irritable. The theory of humors provided the medical rationale for a variety of treatments, since reducing excessive fluid could ameliorate diseases. Unfortunately, blood is the most accessible of the body fluids, and so bloodletting became a very common treatment, which persisted well into the nineteenth century, particularly as a treatment for infections and fevers, which were characterized by redness and flushing. Even as recently as the early twentieth century, physicians used leeches to reduce patients' "excessive" blood. Fortunately for people suffering from melancholy, black bile was harder to find and remove!

Not only have mood disorders been understood as major medical illnesses for centuries, but they have also been recognized to be associated with giftedness or creativity. As early as the fourth century B.C., Aristotle commented: "Those who have become eminent in philosophy, politics, poetry, and the arts have all had tendencies toward melancholia."

Riding the emotional roller coaster may be a little scary, but it also seems to endow people with the ability to feel deeply and to express these feelings in ways that have been useful to society. Many religious leaders have struggled with periods of despondency or depression, such as Maimonides, Martin Luther, St. Augustine, St. John of the Cross, or Gerard Manley Hopkins. Problems with mood have been extremely common in artists and writers, such as John Keats, Leo Tolstoy, Ernest Hemingway, William Styron, Robert Lowell, and Sylvia Plath. Mood disorders have affected great leaders and politicians, such as Oliver Cromwell, Abraham Lincoln, and Teddy Roosevelt. These people, and many others who may be less famous but no less important, have turned their capacity to feel deeply into a creative reservoir from which they have given many gifts to human society. As the motto of the National Alliance for the Mentally Ill states: People with Mental Illness Enrich Our Lives.

Mood disorders usually occur in episodes that come and go. Some of the more debilitating mental illnesses, such as schizophrenia and dementia, tend to be chronic. People with mood disorders, however, spend much of their lives hovering around the neutral point on the emotional temperature scale. When at that point, or even when in the mildly elevated range, they function very normally. Jim, described in chapter 2, had only a single episode of depression, and his mother had only two. People like this are in fact all around us and are "perfectly normal," even though they have had a mental illness from which they have fully recovered. A few people suffer from very chronic depression. Multiple episodes are more common in people who are bipolar, but they too are "perfectly normal" between episodes, or perhaps a bit more energetic and upbeat than average. Because of their emotional sensitivity and responsiveness, people with mood disorders often make good friends, companions, and co-workers. People who spend a great deal of their lives around the +1 or +2 level of the emotional temperature scale are usually fun, energetic, lively, and interesting.

The Depressive Episode

When a person falls into a depressive episode, however, it is a miserable experience. Having symptoms of depression is no fun at all. Table 9–2 lists these symptoms.

The core abnormality is in emotional tone, but people with depression also have problems with motor activity and cognition. As described later in this chapter, we think that the emotional changes are primary and that the motor and cognitive symptoms occur in association with the emotional changes. But, as discussed in other chapters in this book (e.g., chapters 3 and 4), these subdivisions are somewhat arbitrary at both the conceptual and the neural levels.

Anyone looking through this list of symptoms is likely to respond:

TABLE 9–2
Symptoms of Depression

Emotional
Depressed mood
Diminished interest or pleasure in most activities
Somatic
Weight loss or gain
Insomnia or hypersomnia
Psychomotor agitation or retardation
Fatigue or loss of energy
Cognitive
Feelings of worthlessness or excessive or inappropriate guilt
Diminished ability to think or concentrate, or indecisiveness
Recurrent thoughts of death or suicide

"Hmmm. I've had lots of those symptoms. I wonder if I have sometimes been depressed."

Just as none of us in fact has a "perfectly normal" 98.6 temperature in our bodies all hours of the day and all days of the week, so too our mood temperature may fluctuate within normal limits. Some people may have their temperature setting centered on −1 or +1 and be normal and healthy, but those who are set around −1 are likely to be more self-critical, to look at the world a bit more pessimistically, or to have sleep disturbances. Further, all of us dip into the -2 or -3 range for a few days when something goes wrong. A dip can be produced by a critical remark from a colleague, the breakup of a relationship, an illness in a family member or friend, a cold or a case of the flu, worries about a rambunctious child or an aging parent, or a bad grade on an exam. Many different kinds of things affect our emotional tone.

The difference between these normal fluctuations in mood and a diagnosable illness depends on three things: the number of symptoms, their severity, and their duration. The criteria that psychiatrists use to diagnose a major depression, as specified in their *Diagnostic and Statistical Manual* (DSM IV), require that at least five symptoms be present most of the day, nearly every day, for at least two weeks, and that depressed mood or markedly diminished interest or pleasure be among these five symptoms. Further, the symptoms must be an obvious change from the person's usual state. These criteria provide a useful way to place a somewhat arbitrary dividing line between "mild but normal blues" and "pathological blues."

The first two symptoms on the list reflect the core emotional abnormality in depression. A person who is feeling depressed often describes feeling sad, blue, down in the dumps, despondent, or even despairing. Instead of recognizing an emotion of sadness, some people notice a change in their ability to feel pleasure or be interested in things that they normally enjoy. Listening to that favorite song may not produce the usual surge of joy when the trumpets join in, or the invitation to play tennis or bridge may be declined because it just feels unappealing. Favorite hobbies may be abandoned. Spontaneous inspirations to contact a friend to share lunch or a cup of coffee may dry up. Friends and family may note that Grandma no longer smiles as much at the antics of the grandkids, that a spouse has lost interest in everything including sex, or that an adolescent is retreating to his room on Friday night instead of going out with the gang.

The second group of symptoms on the list is sometimes referred to as vegetative or somatic. These symptoms are important markers, particularly for more severe types depression. Typically, a person with severe depression has a decreased appetite, which leads in turn to weight loss. But appetite may be increased in atypical or milder depression. Likewise, insomnia is an indicator of severe depression, but people with milder or atypical depression may sleep too much. Insomnia is sometimes described as *initial*, *middle*, or *terminal*. Initial insomnia is trouble falling asleep, and it is usually considered significant if the person tosses and turns for more than about a half hour before dozing off. Middle insomnia is the tendency to wake up in the middle of the night and to remain awake for an hour or two, usually followed by a fitful sleep. People with terminal insomnia wake up one or two hours before their usual time and are unable to fall back asleep again. As anyone who has experienced insomnia is aware, it is an unpleasant symptom, particularly because it is usually combined with a tendency to ruminate or worry during the periods of sleeplessness.

Psychiatrists use the symptoms of insomnia and decreased appetite (also referred to as anorexia) in order to identify the most severe form of depression, which is called melancholia. This used to be called endogenous depression, because it comes on suddenly without any precipitants and seems to grow from within (endo = inside, genous = to grow) for no apparent reason. In addition to terminal insomnia and anorexia, people with melancholia complain of other symptoms. One is a pervasive loss of interest or pleasure, including a subjective sense that their change in mood is different from normal (e.g., different from feelings that occur in response to a personal loss). Another indicator of melancholia is a change in mood state during the day, with a tendency to feel worse in the morning and better as the day goes on (known as diurnal variation). Other symptoms of melancholia include excessive guilt and severe psychomotor retardation or agitation. People with melancholia usually are able to identify their depressed mood state as a very distinct change from their normal self, and they often complain with surprise that the change has not been precipitated by anything in particular: "There is no reason for me to feel this way!" The good news about this relatively severe form of depression is that it tends to respond quite well to medications, and people do return to the normal self that they used to have, sometimes with an offset as distinct as the onset.

People with depression may also notice changes in their level of motor activity. Feeling slowed down, referred to as psychomotor retardation, is more common. The depressed person may walk more slowly, have less spring in her step, or think and speak more sluggishly and briefly. People with severe psychomotor retardation may sit in a chair or lie in bed all day. When asked how they feel, they may complain that their energy is drained or that they are just too tired to do anything. Rather than being slowed down, some people may have "psychomotor agitation." The person may be restless, fidgety, and unable to sit still. She may wring her hands, tug at her clothes, drum her fingers, or speak in a rapid staccato way about how terrible everything has become. Subjectively, people with psychomotor agitation may describe their mood as tense or anxious, in addition to having some coloring of sadness or despair.

Although a disorder of emotion, depression also has a variety of cognitive symptoms. Sometimes people who are depressed complain of having trouble concentrating or thinking clearly. They feel they can't keep their mind on their studies, focus well on their work, or even keep track of what is going on while watching a football game or reading "escape fiction." A depressed person often feels worthless and may completely lose confidence in himself, so that he is reluctant to go to work, take his exams, or volunteer to participate in a favorite activity such as sports or music. He may ruminate guiltily about things he has done in the past that he should have done better or should not have done at all. Sometimes depressed people become focused on particular "misdeeds" that they think they have done and that they can never rectify. Often, when the person finally describes the misdeed, it turns out to be a small and forgivable act, at least in proportion to the level of guilt. Some people with depression become preoccupied with death and consider the possibility of suicide.

A person who becomes depressed can have cognitive symptoms so severe that he is psychotic (i.e., has delusions or hallucinations). These psychotic symptoms are usually quite different from those of a person with schizophrenia, in that they are consistent with the person's depressed mood. The hallucinatory voices may come from God and tell the person that he is going to be banished to eternal torment for his misbehavior, or they may be the voice of a teacher or friend repeatedly telling him that he will always fail at everything he does. Delusional thoughts also typically focus on worthlessness and guilt. Sometimes the delusions are turned inward, so that the person thinks that he is being punished in some way, such as feeling his internal organs are rotting away from a fatal disease. Sometimes they are turned outward, so that he is going to be "found out," through police surveillance and monitoring, or tormented by the electronic signals being sent from the angry neighbors next door.

Suicidal thoughts are a particularly troubling symptom of depression, because this is an illness that has a very high suicide risk. The risk is tied to the severity of the depression. Among people with severe depression, approximately 15% will die by suicide. However, the risk should not be minimized for those who have a milder depression. Suicidal thoughts are considered to be a medical emergency and usually require either hospitalization or very close monitoring and supervision.

How should a friend or family member react when a person close to them appears to be suffering from a depression and to be at risk for suicide? Urban legend has spread several unfortunate rumors that should be firmly ignored. One is that "people who talk about it never do it." The other is that "you can precipitate someone into committing suicide by asking them about it." Fact demonstrates exactly the opposite. The risk for suicide goes up among those who have expressed suicidal thoughts, and it goes up even higher for people who have made previous suicide attempts. The best strategy for dealing with a concern about suicide is to discuss it openly with the friend or loved one, using a simple straightforward question, such as, "Have you ever thought about taking your life?" Quite often, the other person feels relieved to be asked and responds well to the consolation. A person with suicidal thoughts who is not under medical treatment should be encouraged to seek it promptly. A variety of follow-up questions help determine the seriousness of the risk, such as whether or not the person has formulated a specific plan or has access to drugs or weapons that could be used. Needless to say, weapons should never be left lying around the house if a person living there is known to have suffered from depression.

Suicide risk in adolescents and young adults is especially worrisome, since rates are rising along with the general rates of social violence. Teenagers may make suicide attempts without realizing how easy it can be to complete the task. They also may be more reluctant to discuss their thoughts and feelings, particularly with their parents. In such instances, worried parents may share or explore their concerns with the teenager's closest friend. All suicide attempts should be taken seriously, even if they appear to be an obvious "cry for help." That cry should never be minimized or treated as unimportant.

In addition to the classical depressive syndrome described above, several other milder forms have been identified. While most people with classical depression respond well to treatment and have a good long-term outcome, milder forms may be a bit more chronic.

Atypical depression is at the opposite pole from melancholia. People who have this type of depression are particularly sensitive to slights and rejections. Unlike the person with melancholia, whose mood does not respond by brightening up when something pleasant occurs, people with atypical depression respond intensely to both positive and negative experiences, but are especially sensitive to negative ones. This "rejection sensitivity" makes them particularly vulnerable to having difficulties in their love life, since they are easily hurt. They tend to have many partners and frequent breakups. Other symptoms are also opposite from melancholia. Instead of weight loss, insomnia, and anorexia, people with atypical depression may have weight gain, hypersomnia, and increased appetite. They are prone to getting "the munchies." They sometimes complain they feel as if their arms and legs are made of lead, or that they just can't get out of bed and move around. People with atypical depression respond less well to most treatments, but they often do better on the selective serotonin reuptake inhibitors (SSRIs), the newer antidepressant medications such as Prozac. They are also likely to do better with a mixture of psychotherapy and medications than medications alone.

Dysthymia is a mild but persistent mood disorder. By definition, using DSM IV criteria, a person with dysthymia must have had a disturbance in mood that has been present for at least two years. People with dysthymia tend to be chronically miserable. They feel as if their mood temperature is permanently set somewhere around −4 or −5. Sometimes they complain, "I have been depressed since the moment I was born," or "I've been depressed my entire life." People with dysthymia tend to have symptoms that are similar to those of typical depression, such as decreased energy, feelings of worthlessness, or trouble concentrating, and they are more prone to have symptoms from the "cognitive" group combined with the "emotion" group. The criteria for dysthymia require that only two symptoms are present, but they must have persisted more or less continuously for at least two years. Some people with dysthymia have episodes when their mood worsens and they develop a relatively more severe major depressive syndrome. When this major depression clears, they then return to their chronic state of dysthymia. When people move back and forth between these two forms of depression, their condition is referred to as double depression.

The Manic Episode

Mania is at the opposite pole from depression. Like Hal, a person with mania has a mood temperature that has been reset at the 6–10 level. People with mania can be fun to be around, at least for short periods of time, because their mood is cheerful, enthusiastic, and expansive. In fact, if people with mania did not have the other symptoms associated with it, they would be living in an ideal state, since they are intensely happy all the time. Sometimes, however, people with mania are simply irritable, or they lapse into irritability when contradicted or when their energetic plans are thwarted. When this happens, the happy and enthusiastic person can be much less fun to be around.

The story of Hal, which began this chapter, illustrates many of the symptoms of mania, which are summarized in Table 9–3.

In contrast to the self-doubt of depression, people "suffering" from mania have boundless self-confidence. Their combination of confidence with expansive mood leads them to formulate great plans to expand their business ventures, found new organizations, lead religious movements, or write books or music. They are often convinced that they have special abilities or powers, such as the capacity to predict the future, read people's minds, or perform extraordinary physical feats. When their grandiosity reaches extreme levels, they may think they are a religious figure such as the Messiah, a famous athlete such as Michael Jordan, or a rock star such as Prince.

The manic episode may begin as a gradual change, but typically the onset is relatively abrupt. Although a person in the midst of a manic episode may have little insight, family and friends recognize that the person is showing a distinct change. In extreme cases, the person's usual personality is unassuming and soft spoken, and so the contrast between the person's "normal self" and the "manic self" is striking. More frequently, however, people who suffer from mania have mood temperatures that are set slightly higher than zero, and often more in the +1 or +2 range.

TABLE 9–3
Symptoms of Mania

Persistently elevated, expansive, or irritable mood

Inflated self-esteem or grandiosity

Decreased need for sleep

More talkative than usual or pressure to keep talking

Flight of ideas or subjective experience that thoughts are racing

Distractibility

Increase in goal-directed activity or psychomotor agitation

Excessive involvement in pleasurable activities that have a high potential for painful consequences

During a manic episode, the person acquires an enormous surge of energy. His need for sleep decreases, and he will often work late on some new project, fall into bed for a few hours, and awaken at four or five in the morning, feeling refreshed and ready to jump out of bed and return to work again. Unlike the insomnia of major depression, which is troubling and annoying, a person with mania usually enjoys the decreased need for sleep and does not feel at all fatigued. The conveniently extended day of the manic person is spent planning and initiating projects. Frequently, however, the person is much better at planning and initiating the projects than completing them. After a week or two has been spent in a manic episode, a person suffering from mania may have begun writing a book, taken his computer apart in order to install extra memory (without putting it back together again), and begun shopping for a new and better home. Like a juggler, a manic person enjoys having multiple balls in the air. Unlike a skilled juggler, however, he cannot sustain the feat, and after a week or two is surrounded by a litter of uncompleted plans. Sometimes, the increase in energy becomes so extreme that the person is physically agitated. When this happens, he may pace frantically, or even become physically aggressive if thwarted or irritated.

A person in a manic episode is usually very talkative. He may approach both friends and strangers, describing a new idea or a recent experience with great enthusiasm. Speech tends to be rapid and may be quite loud. It can be difficult to get the person to stop talking. When a question is asked that could be answered with a brief response or even a "yes" or "no," the reply can go on for five or ten minutes. Attempts to break in are ignored, and the manic person may even shout down someone else who tries to interrupt his long monologue. Psychiatrists call this pressured speech. Behind the pressured speech is a "flight of ideas." If asked, the person may say that his thoughts are racing rapidly or that millions of ideas are occurring to him all at once. As he tries to express this wealth of ideas, his speech may skip from topic to topic so that it is sometimes incoherent or barely makes sense. Sometimes he entertains himself with puns, rhyming, and word play. (Psychiatrists call this clang associations.) Sometimes the disorganized pattern of speech is quite similar to that observed in schizophrenia, with the difference that a manic person tends to speak rapidly and enthusiastically. As the mania begins to clear and speech is slowed down, however, the "thought disorder" of mania and schizophrenia can be nearly indistinguishable.

If the increased energy and richness of thought could be harnessed and focused, a time of mania could be a time of great creativity and productivity. Unfortunately, however, a person who is manic is riding a horse that is out of control. He is easily distracted, so that both thoughts and projects are broken off in the middle, as another exciting new object or idea comes into view, causing the person to veer off in yet another new direction. The net result is lots of activity with few or no useful accomplishments.

The most damaging symptom of mania is the one somewhat euphemistically referred to as "excessive involvement in pleasurable activities that have a high potential for painful consequences." People who develop a manic episode begin to think and behave in ways that would be quite out of character for them when their emotional thermometer is near its zero point. A person who is normally a devoted and faithful spouse may become annoyingly hypersexual or may engage in sexual indiscretions with another person. Huge bills may be run up during spending sprees. Everyone sitting in a bar may be treated to drinks. Friends or employees may be given expensive gifts. Being in the middle of a mania is a bit like being in the middle of a "happiness binge"—lots and lots of fun until one wakes up the next morning with terrible financial and emotional headaches.

Manic episodes usually begin relatively abruptly, building up over a few days and then running their course. If untreated, a manic episode may last anywhere from a few days to a few months. Although a person with mania usually has very little insight that something is wrong while in the midst of the manic episode, he or she usually feels terrible when trying to pick up the pieces of a shattered life after the episode is over. Mania can ruin marriages, families, fortunes, and careers. Because "coming down" from a mania can occur with a very bumpy landing or even a crash, many people with bipolar disorder cycle from mania into a depressive episode almost immediately or within a month or two. The tendency to become depressed can partially be explained on a psychological basis, as a person looks back insightfully at the silly, dangerous, or inappropriate things that have been done. Very likely, however, cyclic neural activity in the brain is also driving the process.

While few people experience a depressive episode and then later develop mania, nearly everyone who experiences a manic episode eventually suffers from a depression. The illness is truly "bipolar." Fortunately, most people with bipolar mood disorder enjoy long periods of normal mood. Some may have the pleasant experience of living at around +1 on the emotional thermometer—the psychological equivalent of living in a particularly nice climate. Some have the misfortune, however, of having difficulty achieving a stable neutral mood and instead cycle rapidly and repeatedly. Psychiatrists call these people rapid cyclers. The rapid cycler rides the worst emotional roller coaster of all—moving from one extreme to another over the course of a few days or sometimes even a few hours.

When people are at a level of around +5 on the emotional thermometer, they are said to have hypomania—a state of mind just short of extreme manic euphoria and poor judgment. A person who is hypomanic is talkative, energetic, and ebullient. He or she may have mildly inappropriate behavior, such as being overly friendly to strangers or overly affectionate to friends. Instead of the wild shopping sprees of mania, the person simply spends a bit too much money somewhat impulsively. When people combine episodes of major depression with periods of hypomania, they are said to have bipolar II disorder.

Cyclothymia is the mildest form of bipolar disorder. A person with cyclothymia fluctuates to a mild degree on either side of the emotional thermometer, sometimes rising to a +3, and sometimes falling to a −3, with the shifts occurring during periods of time that last from a few days to a few weeks. Cyclothymia may be more a variant of normal personality than a true disorder, since people with cyclothymia rarely seek treatment in health care settings.

What Causes Mood Disorders?

If you ask the average person, "Why do people get depressed?" you are likely to hear the response, "because bad things have happened to them, or because they have more problems than they can cope with." If you ask the same question about dementia or schizophrenia, you are likely to hear, "because something has gone wrong in their brains that causes them to have trouble thinking clearly or remembering." Because depression is a disorder that primarily affects emotions, and because our emotions are acutely sensitive to everything going on in the world around us, we are naturally more likely to assume that mood disorders are triggered primarily by recent personal experiences. A few people might opine that depression is caused by "abnormalities in brain chemistry," and those who are familiar with mania would almost certainly give this explanation for it.

In fact, like most mental illnesses, mood disorders are probably produced by a mixture of factors, spanning the range from personal experience to brain chemistry. The relative contribution of these factors varies from one person to another.

The Role of Stress and Personal Loss

All of us think we understand stress. It is being in situations where we feel pressured, overwhelmed, and unable to keep up. We get "stressed out" during exam periods, work deadlines, planning marriages, getting divorced, trying to discipline unruly kids, trying to deal with critical parents, trying to recover from a cold or the flu, learning that a loved one has a serious illness, buying a house, selling a house, worrying about money, worrying about holidays, caring for a new baby, caring for an aging parent . . . the list could go on and on. Surviving within the complex social and economic structure of the twenty-first century is stressful indeed. It is a wonder that we manage to cope at all.

Stress is experienced mentally, but it also affects us physically. Our body contains a finely tuned system, the endocrine system, that is designed to help us adapt to and cope with stress. This stress feedback system is shown below.

When stressful experiences in our daily life get registered in our cerebral cortex and limbic system as emotionally troubling, those parts of the brain send an alert to other parts of the body so that we are prepared to cope. The ultimate targets are our adrenal glands, paired endocrine glands that sit on top of our two kidneys (ad-renal = above the kidneys). The adrenals produce the life-sustaining hormone cortisol, which plays a key role in regulating our sleep and appetite, our kidney function, our immune system, and our overall well-being. An adequate supply of cortisol is fundamental to our survival. Rarely, a person will develop a disease in which the adrenals no longer produce cortisol, known as Addison's disease. President John Kennedy was one of its more famous victims. If they are not given regular supplies of replacement cortisol, as Kennedy was, they will die from the cortisol insufficiency.

Because the endocrine system is so important, it sends the message through a series of checkpoints that include the hypothalamus and pituitary gland, which use the messengers corticotrophin releasing factor (CRF) and adrenocorticotropic hormone (ACTH). As cortisol production is increased by the adrenals, the hypothalamus checks on the levels, so that it can sense when they have become high enough or are becoming too high, and in turn can send a message back through the pituitary telling the adrenals to now "slow down." Charles Nemeroff of Emory University has made significant contributions to understanding the role of these endocrine functions in mood disorders.

If the function of the stress response system is measured in people in the midst of a mood disorder, many show indications that its autoregula-

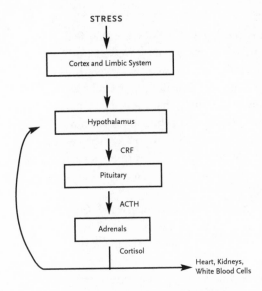

Figure 9–1: The Hypothalamic-Pituitary-Adrenal Axis

tion is disrupted. For a time, checking the autoregulatory mechanism for cortisol production was proposed as a potential "laboratory test" for depression. That test was known as the Dexamethasone Suppression Test or DST. The essence of the test is to give dexamethasone, a synthetic cortisone-like agent, which should "turn off" the signal from the hypothalamus to the adrenals that tells them that cortisol levels are sufficiently high. In the normal healthy state, cortisol production has a characteristic pattern during the 24–hour day, rising at around 5:00 or 6:00 in the morning in order to prepare us to get up and confront the coming day. Cortisol levels tend to peak around midday and then gradually decline, so that we can slow down in the evening and fall asleep naturally at 10:00 or 11:00 p.m.

As the DST studies indicated, many people with mood disorders have a cortisol regulation system that is not working quite right. Administration of dexamethasone did not shut down cortisol production as it should. Instead, many people with mood disorders continue to produce high levels of cortisol all day long. From Nemeroff's work, we now know that CRF is likely to be the basis of overactivity of the hypothalamic-pituitary-adrenal cascade shown in Figure 9–1, and that this is the reason for overproduction of cortisol. Most research and new drug discovery now focuses on CRF. The pattern of hypercortisolemia probably explains why so many people with depression have problems with their sleep cycle, especially insomnia. The DST has now been largely abandoned as a laboratory test, since it produces too many false positives. Other problems besides mood disorder show the same pattern of non-DST suppression, such as excessive dieting, anorexia nervosa, or recent illnesses such as colds or flu. Many different things can throw this delicate feedback loop out of kilter, which explains why so many different things can be potential contributors to the development of a depressive episode. A series of stressful life events induces a biological reaction (i.e., an outpouring of cortisol) which, once initiated, may overreact and trigger or exacerbate a mood disorder.

Several different kinds of animal models have been used to examine the relationship between stress and depressive mood disorders. One early group of studies grew out of the observation by child psychiatrist Rene Spitz that young infants who did not receive adequate maternal love and cuddling developed a severe apathy that he called anaclitic depression. Working at the University of Wisconsin, Harry Harlow and William McKinney examined the effects of separation from the mother on monkeys. They observed that young monkeys developed a syndrome similar to human depression if they were reared either in total isolation or in wire cages where they could see other monkeys but had no physical contact. If the monkeys grew up in this environment during their first six to twelve months of life, they were unable to respond normally when subsequently put in a more normal living environment. Instead they reacted with depression and demoralization, cowering in a corner and avoiding opportunities to play with other animals.

Martin Seligman at the University of Pennsylvania has studied the relationship between stress and depression in rats through a similar "learned helplessness model." Exposing rats to painful stimuli such as shocks, from which they are unable to escape, produces a similar syndrome of demoralization and withdrawal. As with monkeys, the response continues when the stress is no longer present. As in humans, the rats who have developed learned helplessness display a disregulation in the brain-adrenal feedback loop; they have high levels of cortisol that do not respond to suppression with dexamethasone. Further, their demoralization and apathy responds to the same treatments that improve depressive episodes in human beings, such as tricyclic antidepressants.

Another interesting perspective on the relationship between stress and mood disorder arises from the observation that the rate of depression appears to have risen rapidly. A team of investigators from Iowa, Columbia, Harvard, Washington University in St. Louis, and Rush Medical School in Chicago joined together in a major project called the Collaborative Study of Depression. We pooled our efforts and extensively evaluated a group of more than 1,000 patients who suffered from mood disorders, as well as more than 3,000 of their family members. Interested in determining whether an early age of onset was related to worse outcome, we divided the patients into "cohorts" based on their birth date (i.e., current age at the time of evaluation). We discovered something known as a "cohort effect." That is, people in the

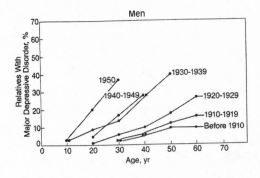

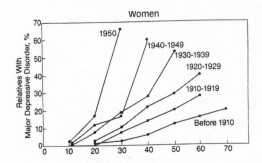

Figure 9–2: Increasing Rates of Depression in Baby Boomers: Differences in Frequency of Depression in Cohorts Broken Down by Birth Decade and Gender

younger age ranges reported a steadily earlier age of onset than those in the older cohorts.

Figure 9–2 shows the graphs from this study, using data about rates of depression in the family members, since the sample is very large and therefore likely to be representative of people in the country as a whole. The graphs show "cumulative risk," or the likelihood that people will become seriously depressed at some time in their lives. People have been divided into six groups, based on age at birth: before 1910, 1910–1919, 1920–1929, 1930–1939, 1940–1949, and after 1950. People in the last two groups are mostly baby boomers. The sample is also divided into men and women, so that gender differences in the prevalence of depression can be examined. The graphs plot the likelihood of developing a major depression based on age cohort and gender. There is a striking difference between the age cohorts. The curves plateau for people born before 1930, but they keep climbing for people born after 1930. For people born after 1940—the baby boomers—the curves point nearly straight up. Women are more likely to become depressed than men.

The prediction curves from this study are somewhat frightening, suggesting that more and more baby boomers are likely to experience a major depressive episode as they grow steadily older. In fact, unless the prediction curves level out at some point, the majority of them are certain to have a depressive episode eventually.

This demographic trend for increased depression in the baby boomers may reflect another interaction between stress and mood disorder. The "boomers" have experienced more intense competition than any generation in recent history, simply because of their large numbers. Beginning with competition for college admission, they went on to compete for jobs in an economy that has converted from using human beings to using computers and other machines. They were promised an "Age of Aquarius," and instead live in a world where economic materialism and violence are rampant and on the rise, where idealism has been replaced by greed and cynicism, and where prominent national leaders repeatedly have their clay feet exposed by sex, lies, and audiotapes. Although it is not possible to prove scientifically that economic and social pressures are producing the high rate of depression observed in this cohort, a purely genetic or biological model of depression cannot explain the cohort effect. It is another piece of evidence suggesting that environmental stresses can contribute to the development of mood disorders.

When the relationship between stress and mood disorders is discussed, however, cause and effect are sometimes difficult to separate. Importantly, the experience of depression or mania is itself a psychological stressor. Once a person has had a depressive or manic episode, self-confidence is eroded. Knowing that both mania and depression can recur, many people experience fearful anticipation, while they also look back and recall how painful and difficult the mood disorder was. Further, these illnesses can cause changes in a person's life that have long-lasting effects, such as divorce, loss of a job, or bad grades. These disappointments can color future perceptions as well as future potential for success. Obviously, the best antidote to this cause-and-effect situation is successful treatment and vigilant attempts to prevent further episodes.

Genetic Factors

As is the case for schizophrenia, the dementias, and many other mental illnesses, mood disorders tend to run in families. Mood disorders have been studied with the full range of techniques that permit clinical scientists to disentangle the contributions of nature and nurture.

The first, and simplest, line of attack on the question is to determine whether first-degree relatives (parents, brothers and sisters, and children) have increased rates of mood disorder in comparison with the general population. Many studies of this topic have been done, and nearly all show an increase in mood disorder, especially bipolar disorder. The rates of illness in relatives vary, depending on how broad or narrow definitions are. Approximately 20% of the parents of people with mood disorder also suffer from a mood disorder, while the rates in their brothers and sisters and children are as high as 30%.

Scientists studying the genetics of mood disorders are particularly interested in the question of whether or not the bipolar and unipolar forms "breed true," as has been discussed in chapter 5. That is, geneticists wonder whether people with bipolar illness have only bipolar illness among their relatives, while people with major depression have only major depression. If the two forms of mood disorder do breed true, then this suggests that they are discreet illnesses genetically. If they overlap, on the other hand, then they may be related illnesses. This issue is especially important to the search for genetic mutations.

The answer to the question is not simple or conclusive, however. For example, people with bipolar illness do have a higher rate of bipolar illness in their relatives than unipolar patients. For example, in the Collaborative Study of Depression, 10% of the relatives of bipolar patients had bipolar disorder, while 24% had unipolar disorder. Thus, even the relatives of bipolar patients have a great deal of unipolar illness. On the other hand, 5% of the relatives of unipolar patients had bipolar disorder, while 29% had unipolar disorder. The most likely interpretation of these findings is that, while there is some evidence for breeding true, there is also some overlap in the genetic vulnerability for these two forms of mood disorder.

Twin and adoption studies have been done as a complement to the family studies, in order to determine the extent to which mood disorders are genetic in addition to familial. The concordance ratio of identical (monozygotic) to nonidentical (dizygotic) twins is quite high, just as it is in schizophrenia. The average rates of illness in both types of twins are relatively higher, reflecting the fact that mood disorders are more common than schizophrenia. If one identical twin has a mood disorder, then there is a 65% chance that the other twin will have a mood disorder; the concordance rate is 14% in nonidentical twins. This yields an overall MZ:DZ ratio of 4:1, which indicates that mood disorders have a strong genetic component. Adoption studies also support this conclusion.

Mood disorders were among the first of the major mental illnesses to be studied with the techniques of genetics using linkage methods. In a study of the Old Order Amish that was published in *Nature* and received wide press coverage when it appeared in 1987, linkage to the short arm of Chromosome 11 was supported. Later, when more families were added to the sample, the linkage scores were no longer significant. In subsequent studies, linkage to several other genes has been reported, including Chromosome 18 in several different loci, Chromosome 21q, Chromosome 4p16, Chromosome 6, and the X chromosome. Several candidate gene studies have also been positive. So far these studies have primarily examined candidate genes for aspects of neural transmission, particularly the receptors for dopamine or serotonin. Molecular genetics studies have focused more heavily on bipolar disorder than on unipolar major depression, largely because bipolar disorder is a relatively severe disorder that is easily diagnosed.

The fact that mood disorders are linked to creativity was mentioned earlier in this chapter. Interestingly, however, my studies of creative writers at the University of Iowa Writers' Workshop also suggest that the tendency to be creative and to have mood disorders co-occurs within families. That is, not only is the predisposition to mood disorder familially transmitted, but so too is the inclination to creativity. In a pair of studies that were the first to use modern scientific techniques to look at the relationship between "genius and insanity," I studied 30 writers and a control group matched for age, sex, and education who had professions outside the arts. I began this study in the context of my interest in the relationship between schizophrenia and creativity, based on the family histories of people such as Joyce, Einstein, and Russell, as described in chapter 8. I expected that the writers themselves would be relatively free of mental illness, but that they would have an increased rate of schizophrenia in their first-degree relatives.

As I interviewed one writer after another, I was astonished to find that the majority of them suffered from mood disorder, not schizophrenia. This was completely counter to my working hypothesis!

Furthermore, their relatives also suffered from mood disorders! Their relatives also had a very high rate of creativity. In fact, there was scarcely a family that did not have a first-degree relative with either mood disorder or creativity, and most had both.

The statistics from the thirty writers and their families, compared with the thirty controls and their families, are quite amazing. Thirteen percent of the writers were bipolar I, 30% were bipolar II, and 37% suffered from major depression. Thirty percent also suffered from alcoholism. All these rates were significantly higher than in the control group. Turning to their first-degree relatives, I found that they too had high rates of mood disorder in comparison to normal controls. Three percent had bipolar disorder, 15% major depression, and 7% alcoholism. Further, their rates of creativity were also increased in comparison with the controls. Twenty percent of their relatives were creatively successful. Interestingly, the type of creativity exhibited in the relatives of these writers was not limited to literary creativity. Instead, family members were dancers, painters, inventers, musicians, and photographers, as well as writers. It appears that the trait running in these families is a primary one reflecting personality and cognitive style, which predisposes people to being creative, but also makes them susceptible to mood disorder. Cognitive and personality testing of this sample indicated that they tended to be more intellectually open, adventuresome, curious, and questioning. This way of approaching the world may permit them to perceive in novel ways and to be generally more inventive, but it also makes them more vulnerable to rebuffs and mood swings.

Several other investigators, including Kay Jamison, Ruth Richards, and Hagop Akiskal, have subsequently confirmed the general direction of these findings. These multiple replications have created a sea change in thinking about the relationship between "genius and insanity." Many types of creativity are clearly related to mood disorder. The possible connection with schizophrenia has largely been forgotten or abandoned. However, this is probably an overreaction because (as described in chapter 8) schizophrenia has struck prominent scientists and mathematicians and their families as well. Creativity is no doubt a complex trait. While the more "human" forms of creativity may be associated with mood disorder, the association with mental illness may differ for creativity in highly abstract fields such as mathematics or physics.

The Neurochemistry of Mood Disorders

Juli Axelrod, a gifted neuroscientist who has spent most of his career at the National Institute of Mental Health (NIMH), set the stage for understanding some aspects of the neurochemistry of mood disorders by studying the mechanism by which tricyclic antidepressants exert their therapeutic effects. He observed that the recently developed drug imipramine blocked the reuptake of norepinephrine in transmitter neurons, as shown in Figure 9–3.

Axelrod was awarded the Nobel Prize for this discovery in 1970. Juli is a sweet and modest man, and the exciting announcement caught everyone by surprise, including him. Legend has it that, as the press converged on NIMH for a conference, the director of the institute turned up in Axelrod's lab to ask if there was anything that he needed. Seated in his embarrassingly small and crowded lab space, Axelrod replied, "Do you think I might now be allowed to have a parking space for my car?"

Axelrod's pivotal discovery provided the foundation for what later came to be known as the catecholamine hypothesis of mood disorders. The net effect of preventing reuptake of norepinephrine was to increase the amount of this neurotransmitter available at the synapse and to therefore up-regulate the overall tone of the norepinephrine system. This led several psychiatrists to formulate the hypothesis that norepinephrine in

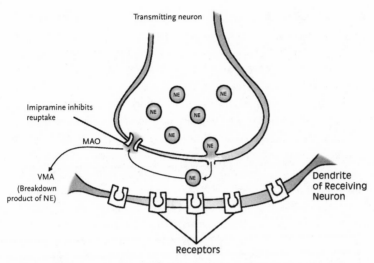

Figure 9–3: Mechanism of Action of Tricyclic Antidepressants: Blockage of Reuptake

particular might be dysfunctional in mood disorders, including Joseph Schildkraut, William Bunney, John Davis, and Arthur Prange. The clearest formulation came from Joseph Schildkraut:

> Some, if not all, depressions are associated with an absolute or relative deficiency of catecholamines, particularly norepinephrine, at functionally important adrenergic receptor sites in the brain. Elation conversely may be associated with an excess of such amines.
>
> (*American Journal of Psychiatry*, 1965; 122:509–522)

As shown in the figure, further support for this hypothesis came from the fact that another group of medications that were also effective in treating depression, the monoamine oxidase inhibiters (MAOIs), also increase the amount of norepinephrine available to receptors. They effect this by inhibiting the breakdown of norepinephrine through MAO. Two major groups of drugs seem to be increasing norepinephrine tone and correcting a functional deficit. As described in chapter 4, norepinephrine is widely distributed throughout the brain, suggesting that it serves a general regulatory function. In addition, it has well-recognized targets in the hypothalamus, a site that the neuroendocrine studies of cortisol have suggested might be disregulated. A specific site such as this might be the target upon which the antidepressant drugs are acting.

As other new tricyclic antidepressants were introduced, however, clinician-scientists noticed that some of them had effects on another neuro-transmitter, serotonin. In fact, while imipramine (Tofranil) was fairly selective as a norepinephrine uptake inhibitor, the next potent antide-pressant, amitriptyline (Elavil) affected both the serotonin system and the acetylcholine system. It increased serotonin transmission and decreased acetylcholine transmission. Clinically, it was evident that there were indi-vidual differences in response to antidepressants. Some people seemed to do better on a drug like Tofranil, while others responded better to Elavil.

As a consequence, the catecholamine hypothesis was eventually sup-plemented by another, the serotonin hypothesis. Like norepinephrine, serotonin is widely distributed throughout the brain and also appears to have a general modulating effect. These observations eventually led to the development of the group of drugs known as selective serotonin reuptake inhibitors (SSRIs), which primarily affect the serotonin system. Prozac is the most famous of the SSRIs. These drugs work by the same type of mechanism as the tricyclics, but have their effect solely on serotonin. That is, they block reuptake and therefore increase the amount of serotonin functionally available at the terminal. As more and more SSRIs have been developed, they have become the treatment of choice for many people suffering from depression. However, not everyone responds to them, and therefore tricyclics also continue to be used.

What does this pattern of response suggest about the neurochemistry of depression? One theory is that there are several different forms of depression, one of which is characterized by abnormalities in the norepi-nephrine system and another of which is characterized by deficits in serotonin. Much more likely, however, is the possibility that these two neurotransmitters, in conjunction with others as well, such as acetyl-choline and dopamine, work together to achieve an overall balance. The simplistic formulations that "schizophrenia is a dopamine disease" or that "mood disorder is a norepinephrine or serotonin disease" are almost cer-tainly wrong, given what we now know about the complex interaction between chemical systems in the brain. As other chapters in this book indicate, serotonin has been implicated in other illnesses such as schizo-phrenia and norepinephrine in other disorders such as anxiety. The one neurotransmitter = one disease approach to conceptualizing mental ill-nesses, while intellectually convenient because of its simplicity and ease of understanding, is just too easy to be true.

Neural Circuitry and Mood Disorders

Our internal experiences of sadness, depression, joy, and elation are cre-ated as neurotransmitters send messages back and forth between brain regions that somehow connect many tiny bits of information from many domains of experience. At a clinical level, a psychiatrist might say that a person develops depression because an early life experience, such as loss of a parent, makes her vulnerable to subsequent rejection or loss. A full-blown depression occurs when a sufficient number of events co-occur. The person who had the early loss may also carry an abnormal gene or two, which predispose to depression. The depression itself is finally triggered when she goes off to college, is isolated from high school friends and family, and gets several bad grades on midterm exams. The feeling of emotional suffocation or despair, which the well-known writer Sylvia Plath described as being suffocated under a bell jar in her autobiographical account of her own depression, sets in. The chain of events just described is not unlike that portrayed in *The Bell Jar*, since Plath lost her father at an early age and developed her first depression while in New York, away from friends and family, struggling to succeed as a *Mademoiselle* guest editor.

The subjective experience of depression is the expression of the interplay between our brain regions that register and interpret our current emotional experiences in the light of our past emotional experiences. Exactly how the brain does this is currently being mapped by neuroscientists interested in the neural substrates of emotion, largely through functional imaging techniques. An obvious candidate region to study is the limbic system, described in chapters 4 and 11 and long recognized as a key region for emotional perception and processing.

Functional imaging studies have suggested, however, that the limbic system has broader cortical connections than was previously surmised. When healthy volunteers are studied in "emotional challenge experiments" such as internally playing a personal script of a sad experience, they use a group of widely distributed cortical regions in addition to the "usual suspects" in the amygdala or hippocampus. Neuroscientists are now studying the "extended limbic system" as a neural basis for emotion. They are also recognizing that "emotion" is not a single thing, but is composed of many types, such as pleasure, fear, or sadness. The neural circuitry of depression resides in this extended system. Regions at the base of the frontal lobes, the deep enfolding between the frontal and temporal lobes known as the insula, and the anterior cingulate gyrus are key players within the extended limbic system that create the subjective feeling of sadness. Many psychiatrist-neuroscientists are actively dissecting how these various distributed regions link together past memories, present experiences, and the attribution of feeling tones to them. This complex set of interconnections permits us to subjectively experience emotions such as sadness and also cognitively attribute these feelings to ourselves by saying, "Gosh, I sure feel depressed today."

How Are Mood Disorders Treated?

The treatment of mood disorders is one of the biggest success stories in modern psychiatry, or for that matter in modern medicine. Effective treatments are available for both mania and depression. Four general classes of treatment are available: mood stabilizers, antidepressants, electroconvulsive therapy, and psychotherapy.

Mood Stabilizers

Mood stabilizers are primarily used to treat bipolar disorder. As the name suggests, they abort the activity of the emotional roller coaster and even out the mood swings so that the emotional temperature fluctuates mildly around the zero point. Because bipolar disorder is a very debilitating illness, mood stabilizers are a godsend to people who previously used to swing between mania and depression. Because being "high" can be subjectively fun for a person with mania, people with this condition sometimes dislike mood stabilizers because they get rid of the highs. Usually, however, people with mania eventually develop insight about the destructive effects of manic episodes. For example, the gifted poet Robert Lowell, who suffered from severe bipolar illness, spoke gratefully about the benefits of taking lithium after it became available in the early 1970s. One of his friends described Lowell's remarkable improvement on lithium:

> He showed me the bottle of lithium capsules. Another medical gift from Copenhagen. Had I heard what his trouble was? "Salt deficiency." This has been the first year in eighteen he hadn't had an attack. There'd been fourteen or fifteen of them over the past eighteen years. Frightful humiliation and waste. He'd been all set to taxi up to Riverdale five times a week at $50 a session, plus (of course) taxi fare. Now it was a capsule a day and once-a-week therapy. His face seemed smoother, the weight of distress-attacks gone.

Lithium is in fact one of the earliest psychoactive drugs to be discovered that is still in use. Lithium is a naturally occurring salt, which is adjacent to sodium on the chemical periodic table. Lithium chloride, chemically very similar to the sodium chloride that we use to salt our food, was first used medically as a salt substitute for people who had high blood pressure and needed to consume a low-sodium diet. High doses of lithium chloride turned out to make them sick, however, and so lithium was abandoned as a salt substitute. In the late 1940s the Australian psychiatrist John Cade noticed that lithium chloride also produced sedating effects and experimented with using it on agitated psychotic patients. He observed that they responded well.

People did not follow up on Cade's observation immediately, however, and it was not until the 1960s that Mogens Schou, a Danish psychiatrist, completed systematic studies of the therapeutic effects of lithium in mania. He demonstrated that when treated with lithium, manic patients recovered almost completely from their manic episodes within a few weeks. He went on to explore whether lithium might also be helpful in normalizing mood over the long run by conducting careful longitudinal studies of a relatively large series of bipolar patients. People who had previously been prone to relapse into manic or depressive episodes were found to have a stabilized mood, with their recurrence rates markedly attenuated. During the 1960s and early 1970s, techniques were developed to monitor blood levels of lithium so that it could be administered safely. It has been in use in the United States since 1970, and it continues to be one of the most important mood stabilizers currently available.

While many patients with bipolar illness are stabilized with lithium, some do not respond and continue to cycle up and down. For these patients, another class of drugs, which were originally used in neurology as anticonvulsants, has been added to the armamentarium. They include carbamazepine, valproate, and clonazepam. These medications appear to be particularly useful for patients who are prone to rapid cycling.

Antidepressant Medications

The first of the antidepressants, imipramine, was introduced in the early 1950s, at just about the same time that chlorpromazine was being developed. In fact, it was produced by making a slight change in the chlorpromazine molecule with the hope of finding a newer and better antipsychotic. The Swiss company that synthesized the compound, CIBA Geigy, performed what were then standard procedures for developing new treatment agents. They gave samples of the compound to a Swiss psychiatrist, Roland Kuhn, who tried it out in various doses with several different patients. Kuhn observed that this compound had little effect on delusions and hallucinations, despite its similarity to chlorpromazine. On the other hand, it seemed to help with the symptoms of depression, and so he

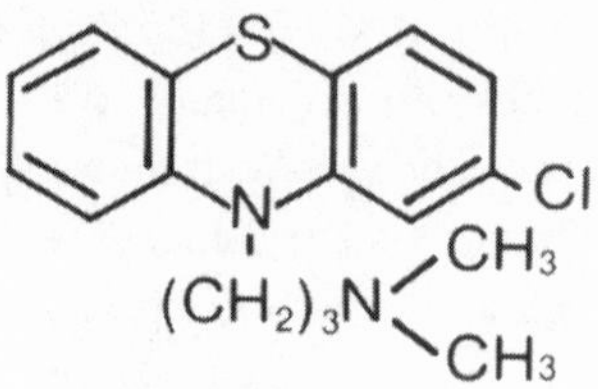

Chlorpromazine
(the first neuroleptic)

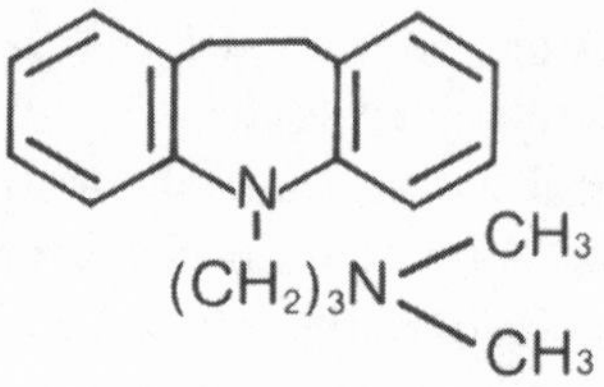

Imipramine
(the first antidepressant)

Figure 9–4: Similarity of Chemical Structure of Chlorpromazine and Imipramine

explored its effects in people who had mood syndromes. They responded remarkably. Soon the drug crossed the Atlantic and was used by pioneering psychopharmacologists in the United States.

Imipramine, also known as Tofranil, is an "ancient" drug by modern standards, but it continues to be one of the best treatments available for severe depression. It is the first of the group of drugs that came to be called the tricyclic antidepressants because their chemical structure consisted of three rings. The structure of the chlorpromazine molecule and its successor, the imipramine molecule, are shown in Figure 9–4.

Other tricyclic antidepressants were subsequently developed, as competing drug companies sought to capture a share of the growing antidepressant market with other new compounds. Soon clinicians had four or five to choose from. As more and more drugs were available, it became clear that most of the tricyclics were equally effective in reducing depressive symptoms, but they differed in side effects. Some were more sedating than others. For example, despramine (Norpramin) is the least sedating, while amitriptyline (Elavil) is among the most sedating. Clinicians have learned to exploit the side effects of these drugs by giving the more sedating ones to people who have severe insomnia and anxiety and using the less sedating ones for people who are more anxious or agitated. It also became clear that most of the tricyclic antidepressants took some time to achieve their therapeutic effects—usually two to three weeks. Some of them had side effects that patients found uncomfortable, such as dry mouth or constipation. Sometimes patients had to be warned that they needed to "live through" the unpleasant initial period of side effects in order to achieve the pleasant result of a good therapeutic effect in two to three weeks.

The most widely used tricyclics include imipramine, amitriptyline, nortriptylene (Pamelor), protriptyline (Vivactil), doxepin (Sinequan), and clomipramine (Anafranil). These tricyclics have been joined by other agents that are sometimes referred to as atypical, because of their chemical structure and mixed neurochemical profile. They include amoxapine (Asenden) trazodone (Desyrel), maprotiline (Ludiomil), mirtazapine (Remeron), and bupropion (Wellbutrin).

The monoamine oxidase inhibitors (MAOIs) soon became available as alternatives. Although they did not have as many unpleasant side effects, they were somewhat less effective as antidepressants. Therefore, they were usually tried only after a patient had failed to respond to one or two of the tricyclics. They became what physicians call "second-line treatments." The MAOIs include isocarboxazid (Marplan), phenelzine (Nardil), and tranylcypromine (Parnate). The major problem with the MAOIs is their interaction with a variety of foods, requiring people to avoid red wine, aged cheese, and chocolate. The combination of MAOIs with these popular foods can produce a dangerous and sometimes life-threatening rise in blood pressure. Many people do not want to bother to monitor their diet closely enough to take the MAOIs.

The recognition that some of these antidepressants affected serotonin transmission led the American company Eli Lilly to focus on developing a "selective" agent that would work primarily on serotonin, with minimal effects on norepinephrine. This innovative and forward-looking concept led to the development of an entirely new class of antidepressant agents, the selective serotonin reuptake inhibitors (SSRIs). The first of these was Lilly's fluoxetine, which was marketed as Prozac and subsequently made famous by Peter Kramer's book *Listening to Prozac*. Kramer made the interesting point that these new medications are "mood brighteners" that can be used to cosmetically alter one's view of the world, much as cosmetic surgery is used to make people more physically attractive. This provocative idea has led to widespread discussions among philosophers and ethicists about the pros and cons of psychoactive drugs. It also has led many people to turn up in a doctor's office requesting a prescription for Prozac to make them feel "better than normal."

I first learned about fluoxetine before it was even given the name Prozac. A group of other prominent psychiatrists and I (referred to by drug companies as "opinion leaders") were invited to Eli Lilly for input and suggestions about their new compound. When I heard the description of its side effects, I silently predicted that it would never take off. This just shows how wrong an intelligent "opinion leader" can be! In contrast to the tricyclic antidepressants, the side effects of fluoxetine included insomnia, anorexia, weight loss, and restlessness. Since those were the key target symptoms of melancholia, which provided my own model for "classic depression," I thought both psychiatrists and their patients would quickly reject the drug. What I failed to appreciate was the fact that many people would be willing to deal with those side effects in exchange for the energizing effects of Prozac, and that indeed people who were moderately overweight or had hypersomnia would welcome the side effects.

Prozac took off like a rocket and has been going strong ever since. It was joined by other SSRIs as competitors: paroxetine (Paxil), sertraline (Zoloft), fluvoxamine (Luvox) and citalopram (Celexa). In addition, other companies developed a combined serotonin-norepinephrine reuptake inhibitor, venlafaxine (Effexor), and a serotonin transport blocker antagonist, nefazodone (Serzone). These drugs act more quickly than the tricyclics, primarily because they rapidly increase energy levels and make people feel more spontaneous and confident.

No drugs are without their side effects, however. For some people with depression, the insomnia and anorexia *are* unpleasant, just as I originally hypothesized. In addition, some people experience the energizing effects of the SSRIs as anxiety and become more tense and anxious. For some, the capacity to achieve sexual orgasm is reduced. Finally, some people become more impulsive and disinhibited on the SSRIs, although this is rarely a serious problem.

Electroconvulsive Therapy (ECT)

ECT can be an extremely effective treatment for depression, and sometimes for mania, although it has been given bad press and tends to carry a very negative stereotype. Films like *One Flew Over the Cuckoo's Nest* have portrayed ECT as a vindictive treatment used to punish patients for bad behavior and turn them into living vegetables. This portrayal makes good theater, but it is a very inaccurate representation of ECT in contemporary psychiatry.

Modern ECT is a useful treatment for patients suffering from severe depression who do not respond to treatment with medications or psychotherapy. The administration of ECT involves applying a small electrical current that activates the brain and produces seizure-like activity, which is measured on an EEG monitor. Prior to the administration of the current, the patient is given a short-acting barbiturate that puts him to sleep, followed by a muscle relaxant that prevents a physical seizure from occurring. (Prior to the development of such "modified" seizures, patients risked having violent muscle contractions because physical convulsions did occur, just as they do during an ordinary epileptic seizure.)

Typically, ECT is administered in a series of four to eight treatments, which are given three times a week. The primary side effect is memory loss, particularly for events right around the time of the ECT treatment. The memory problems may worsen as the number of treatments increases, but they usually clear up completely within a few weeks after the ECT is stopped. There is no evidence of any significant long-term memory impairment as a consequence of ECT, and it remains the treatment of choice for people with severe depression that does not respond to medications, a high potential for suicide, cardiovascular disease that precludes the use of antidepressants, and pregnancy. It produces a rapid remission of depressive symptoms, with a good response occurring in approximately 80% of patients, in comparison with the 50–70% response rate seen for any single antidepressant medication.

Psychotherapy

For some patients, psychotherapy may be the treatment of choice, either because the person prefers not to take a medication, or because there are personal problems that would benefit from psychotherapeutic intervention. Psychotherapeutic treatments of depression focus on helping people identify negative cognitive responses and to retrain these patterns so that they are replaced by more positive or affirmative schemas. Psychodynamic psychotherapies may also assist people in exploring the ways that prior experiences lead to the tendency to respond with depressive emotions.

Many people who choose to be treated with medication benefit from psychotherapy in addition to the medication. There is an unfortunate tendency, promulgated primarily by the economics of the health care delivery system, to minimize the importance of psychotherapy and to attempt to replace it with medications alone. Too often, people suffering from depression are assessed briefly in order to make a quick diagnosis, and then simply given a prescription for some type of antidepressant medication. Experiencing an episode of mood disorder is often a blow to a person's confidence and self-esteem. Most people who have had an episode of depression would welcome a sympathetic ear and some helpful advice as their depressive "wound" begins to heal. Ideally, this should be accomplished by a single physician who is trained in both psychotherapy and psychopharmacology, who can provide continuity of care, who can see the patient as a person, and who can help the person sort out the social and emotional factors that are causing distress or that may have worsened as a consequence of the episode of mood disorder. Unfortunately, this ideal type of integrated treatment is becoming increasingly rare as health care is fragmented, and as reducing the bottom line is considered more important than reducing human suffering.

CHAPTER **10**

DEMENTIAS
A Death in Life

Though wise men at their end know dark is right,

Because their words had forked no lightning they

Do not go gentle into that good night

. . . .

And you, my father, there on the sad height,

Curse, bless, me now with your fierce tears, I pray.

Do not go gentle into that good night.

Rage, rage against the dying of the light.

—Dylan Thomas

"Do Not Go Gentle into That Good Night"

Wayne adjusts his rearview mirror as he turns onto the highway. His eyes rest briefly on a handkerchief, shoved into the armrest on the door of his old but immaculate Lincoln Towncar.

She was never without a handkerchief, always presentable, confident, content.

There was a subtle groove worn into her passenger seat from the many miles of their lives together. Every day she found an adventure. She would lean forward, pouring over the roadmap, plotting a course to the next town with an antique store or the fastest route to the impending graduation, wedding, or birthday. He wasn't much for weddings and the like, and never could figure out the attraction of antiques. But her joy spilled over the driver's side and he would find himself eager to reach the next stop for no other reason than to see her smile, pluck up her handkerchief and bound from the car like a kid on the first day of school.

Don't know when she stopped looking at the map.

A drizzling rain is smearing the windshield. He hits the wipers as he pulls off the road, tears rising in his eyes and smearing his vision of the world even more.

I can't do this, can't do it. I'd rather be visiting her grave. God help me, I AM visiting her grave.

Pearl is 64. She now lives in a nursing home, Fairview Manor, having been moved there by her reluctant husband, Wayne, six months ago.

Wayne, who is 65 and still running his business as a plumbing supply dealer, visits her on Saturday and Sunday, as well as at least one day during the week. He usually finds her seated in a chair with a tray to restrain her from getting up, since she will otherwise wander into other rooms. If allowed to roam freely, she sometimes hits other people, calls them names such as "bitch," or slips and falls. Each time he comes, Wayne is overcome with guilt and a psychological pain that is gut-wrenching. He was no longer able to care for her himself at home, but he still wonders if he has done "the right thing." When he leaves, he is depressed. He carefully follows the latest scientific news about research on Alzheimer's disease, hoping that a breakthrough will occur that will help Pearl before it is too late. He is filled with terror that he will develop the same problem, or—much worse—that it will afflict one of their three children or seven grandchildren eventually.

Pearl and Wayne were married 45 years ago. They grew up together in the same small Texas town and were childhood sweethearts. By high school they had decided that they wanted to marry one another. Both their parents wanted them to go to college. Wayne, one year older, headed off reluctantly to Galveston, making the 60–mile trek back home each weekend so that he could see Pearl. Occasionally she would visit the campus and attend a fraternity party, spending the night with an older friend in one of the local sororities. Pearl started to attend Galveston the next year. Neither could wait until they graduated so that they could get married. Wayne thought Pearl was the most beautiful girl in the world, and she thought him the most handsome boy. As their love intensified, their physical contact became more passionate, and they found they could not wait (years!) until they graduated to unite their bodies to one another. During the summer after Pearl's freshman year, she discovered that she was pregnant. In terror and contrition they confessed their plight to both sets of parents, accepted the appropriate mixture of disapproval and blessing, and were rapidly and quietly married. They moved to Houston to begin their new life together.

Despite the unpropitious beginning, it was a happy marriage. Neither had been much of a student, and neither was that disappointed about having to drop out of college. Wayne began to work for a plumbing supply company that was owned by a distant cousin. A good worker, he soon became invaluable. Eventually he became the owner. The company caught the rushing tide of the postwar building boom and prospered. Poor and frugal initially, Pearl and Wayne also gradually became prosperous. Their first child Bill was followed quickly by two others. Their first tiny apartment was followed by a tiny house. The early years together were not rich in material goods, but they were rich in mutual love.

By their mid-thirties Wayne and Pearl were leading the very life that they had dreamed of in high school: tending a trio of healthy children who starred in Little League and swim club, making plans to build their "dream house" eventually, playing tennis together as a strong pair in mixed doubles tournaments, and vacationing each year in places like Disneyland or Hilton Head. By their forties they actually owned their dream home, were skiing blue and black runs in Park City each winter, and were sending the kids off to college. By their early fifties they had begun their "second honeymoon on a full-time basis." All the kids were gone, and so they finally had the whole house to themselves. Each still believed the other to be "the most beautiful" or "the most handsome." They were still very much in love. It was almost too good to be true.

And it was. By her early sixties Pearl's personality began to change. Always very much a lady, she began to swear occasionally. She never drank much, but now she seemed to become intoxicated when she consumed a single drink. Disinhibited, she would pick fights with close friends. She even began to accuse Wayne of having an affair with the wife of his best friend. She also began to misplace things and to have trouble remembering names. After spending a perplexed year or two struggling with these changes, Wayne decided that they should see a psychiatrist. After several hours of interviews, a magnetic resonance scan, and some psychological tests, the bad news was announced. Pearl appeared to have a type of dementia known as Alzheimer's disease. The future was uncertain, since the course of the illness could vary. Some people decline slowly, others more rapidly. The long-term outcome would be a progression of the changes in personality and memory, leading eventually to complete impairment.

Pearl turned out to have a bad case of the disease. By her mid-sixties she was already confused most of the time. It was nearly impossible to take her anywhere, since she was irritable and disruptive. She was frequently incontinent, and therefore she had to wear diapers all the time. The only resemblance to the joyous and beautiful Pearl in the wedding picture that still sat on the living room desk was in facial features. The person herself was nearly gone. Wayne tended her dutifully nonetheless. He hired a nurse to remain with her during the day, while he remained with her in the evenings, changing her diapers and keeping her clean and well groomed. He was grateful that he was still strong and in good health, so that he could lift her. But when he paused to think (which he rarely dared to do), he had to admit that he felt as if he had been catapulted into the center of hell.

As Pearl became more deteriorated, irritable, and verbally and physically aggressive, it became very difficult to find anyone who would care for her during the day. Wayne often stayed home from work to tend to her needs, but he could not do this on a regular basis unless he chose to retire, which he did not feel ready to do. Their children all lived in distant states—California, Georgia, Massachusetts. (He was grateful about that because he did not like having them see their mother in such a transformed state. It was hard enough for him.) Their family doctor began to urge him to seek nursing home placement. Reluctantly, he finally did. Now he found that he had simply traded the pain of chronic worry for the pain of chronic guilt. He could hardly bring himself to walk into Fairview Manor. He could still see the Pearl that once was, hiding behind the vacant eyes that no longer recognized who he was.

What Are the Dementias?

The dementias are diseases of aging. They are like a ticking time bomb: As the elderly population accumulates during the next several decades, the dementias threaten to overwhelm all of us. The development of a dementia in an older person brings despair to the person's loved ones, particularly the husband, wife, or children who must care for their progressively fading family member. The dementias are relatively uncommon in people who are younger than age 65, but the risk increases steadily with increasing age. About 5% of people over 65 have severe dementia, while 10% have mild problems. By age 80 the figures rise to 20% for severe dementia and by age 90 the figure reaches 30%. Among people who have other serious illnesses, such as heart disease, the rates are even higher.

Since the number of people over 65 is outstripping the growth of the general population, the dementias will become a steadily greater problem in the future. As the case of Pearl illustrates, the costs are high. In addition to experiencing the pain of watching a beloved wife, husband, or parent slowly fade away mentally, family members may also be financially devastated as they watch their retirement savings be drained away by the costs of nursing home care. Other medical costs to families and society as a whole will rise as well, as our ability to keep people's bodies alive exceeds the pace of our ability to give them a relatively normal mental life. People with dementia slowly and inexorably "lose their minds," while their bodies often remain robustly healthy. Many older people actually hope that they may die quickly from a heart attack rather than confront the possibility of the steady mental decline that occurs with the dementias.

The dementias are not a single illness. They are divided into several groups, which have slightly different causes and courses. The major ones include Alzheimer's disease and vascular dementia (also known as multi-infarct dementia). Other less common forms also occur, such as Huntington's disease or the dementia that is sometimes associated with Parkinson's disease. While these various kinds of dementia have different causes, their clinical presentation is remarkably similar.

The hallmark of all of the dementias is impairment in memory and cognition, which is accompanied by a general decline in the ability to relate or to function at work or in the home and in social settings. At least in the early stages, the person is relatively alert or conscious. As they pass into their fifties, nearly all people begin to notice some decrease in their ability to remember names or details from recent experience (e.g., "Now where did I leave my car keys?"). The memory loss associated with the dementias is much more severe than this normal forgetfulness, and it actually interferes with the ability to function normally. Likewise, normal aging tends to be associated with a mild slowing in the ability to solve intellectual problems or puzzles, such as doing calculations in one's head, but people with dementia have a severe slowing that prevents them from functioning normally. In addition to the problems with cognition and memory, people with dementia also have a variety of "noncognitive" symptoms, such as delusional thoughts and inappropriate suspiciousness, hallucinations, agitation, or depression.

Dementia usually develops insidiously. Early in the game, it may be hard for either the victim or for family members to decide whether Mom or Dad is just experiencing normal aging. The diagnosis is even more difficult to make when physical problems are present, such as failing eyesight or hearing, since these can slow down functioning and put extra stresses on the ability to think efficiently. When both partners in a couple are alive, they often help compensate for each other's weaknesses. Sometimes dementia is first recognized after the serious illness or death of one person in a pair. The remaining spouse is then unable to function on his or her own. Even in a situation like this, it can be difficult to disentangle depression and grief from early dementia. If the problem is a true dementia, however, the severity of the cognitive impairment will eventually become evident, and it is usually accompanied by a decreasing interest in things, changes in personality, and lability in emotions.

The types of cognitive impairment that occur in dementia are sometimes divided into four general groups, all of which start with the letter "a," which means "absence of" or "loss of" in Greek. These are listed in Table 10–1 and might be called the "the four A's of dementia."

Amnesia, the loss of the ability to remember information or experiences, is the most common and is usually the earliest sign of the illness. Agnosia, the inability to recognize and identify objects even when the person can see clearly, is a sign of the later stages of dementia. Apraxia, problems in remembering how to perform relatively simple motor tasks such as brushing one's teeth or driving a car, is another later sign, as is aphasia, loss of the ability to speak. After amnesia, problems with apraxia are probably the most likely to call attention to the occurrence of a dementia. Sometimes the first sign will be the fact that Mom has gotten into a car accident or has gotten lost while trying to drive home from the supermarket. In people who are still working, intellectual skills may decline, so that the person has problems doing tasks that were once routine. The personality or emotional changes that occur in dementia may also be among the earliest to call attention to the disorder. A person who managed to deal patiently with customers for many years may become irritable or even rude, much to the surprise of coworkers who have been around him for many years.

As the dementia advances, memory impairment becomes more severe. The memory impairment of dementia initially affects the ability to remember recent events or perform tasks with "working memory," the short-term memory bank that we use to do simple calculations in our head or remember a phone number long enough to write it down. Early on, people with insidiously developing dementia may still have relatively intact memories of the past. This sometimes manifests itself in their preference to talk about things that happened long ago, particularly events from their childhood or young adult life, since they can still remember these perfectly well. In fact, people with developing dementia may cover over their memory loss by selectively discussing the things that they still know well. Eventually, however, even experiences or events from the

TABLE 10–1

The Four "A's" of Dementia

Amnesia	Loss of memory
Agnosia	Loss of ability to recognize objects
Apraxia	Loss of knowledge about how to do things
Aphasia	Loss of speech

remote past are remembered poorly. The person can no longer remember the names of children or grandchildren, birth dates, or other well-learned information. As dementia advances, the changes in personality and emotional responsiveness also become more severe, and the person no longer seems to be "the same." Social skills that are almost natural reflexes, such as "having good manners," may cover symptoms of dementia in earlier stages, but they also gradually are lost and no longer help to preserve the person's image of good mental health.

The noncognitive symptoms of dementia can sometimes be even more troublesome than the cognitive ones. Nearly two-thirds of those with Alzheimer's disease develop psychotic symptoms, such as delusions and hallucinations. A parent who was once trusting and welcomed neighborhood children or pets into the yard may become fearful and concerned, suspecting that they or their parents are playing tricks on her. Marital relationships may become strained because the insidiously developing dementia leads to suspicions of trickery, infidelity, or outright abuse. Pearl might have begun to tell the neighbors that Wayne was hiding kitchen utensils or clothing from her, secretly having an affair with the checker at the supermarket to whom he gave an especially cheery smile, or performing sexually abusive acts with her. Sometimes delusions like this may have some connection with the memory problems, but sometimes they also seem to "come out of the blue." Whatever their source, they can be very embarrassing to family members. If the person is in a care facility, delusions about thievery or intrusions may develop around the other people who live in nearby rooms. Restlessness, agitation, and irritability are also troubling symptoms. A person who was once quite sweet and gentle may become physically aggressive, striking out and hitting those nearby unexpectedly. A declining mind in an otherwise strong body can become quite dangerous, particularly if the body belongs to a relatively large man. This symptom can make people with dementia especially hard to care for, either at home or in a care facility.

Alternately, people with a developing dementia may become steadily more withdrawn and apathetic. In this instance, it is especially important to decide whether the person suffers from real dementia or from *pseudo-dementia*, a "false dementia" that sometimes accompanies depressive illness. As described in chapter 9, the symptoms of depression have some overlap with the cognitive symptoms of dementia. Depressed patients may have trouble remembering and may have a generalized slowing of their mental skills, in addition to losing interest in things and becoming apathetic and despondent. The distinction between dementia and pseudodementia is especially difficult to make during what could be the early stages of dementia. Making the distinction is very important, however, since one is literally walking down a road that forks in two very distinct directions. If Mom is apathetic and withdrawn and having trouble think-ing because she is feeling depressed, perhaps because Dad died six months ago, she should be treated with medications and perhaps even electrocon-vulsive therapy (ECT), since ECT is particularly effective in older people who are clinically depressed. On the other hand, if she is suffering from dementia, ECT is a very inappropriate treatment, since it will at least temporarily worsen her failing memory.

It is important to realize that people with dementia may also develop a secondary depression that may respond to treatment. As many as 40% of people with Alzheimer's disease develop mild to severe depressive symp-toms, and depression is even more common in people with multi-infarct dementia. Clues indicating that the person with dementia has a superim-posed or secondary depression that may respond to treatment with anti-depressants include weight loss, self-deprecating comments ("I can't do anything!"), frequent crying spells, or recent changes in sleep patterns leading to troubling insomnia.

Late-life psychosis may also occur in the elder years and be confused with irreversible dementia, since people with this problem have a mixture of cognitive and psychotic symptoms. This disorder is relatively uncom-mon, but it is an important consideration because people with this syn-drome usually improve substantially if treated with antipsychotic medica-tions. Typically, the dementia associated with late-onset psychosis arises in people who have had a hearing loss or severe visual problems such as cataracts or retinal disease. Confronted by failing sensory input, they mis-interpret what is going on around them and become suspicious, with-drawn, and outright delusional. They may also become neglectful of their appearance and lose interest in what is going on around them, becoming withdrawn and apathetic. Typically, they will improve markedly if given appropriate treatment for their primary hearing or visual problem, assuming successful treatment is available. They may also benefit from very low doses of antipsychotic medications.

Given that dementia has such confusing boundaries—with normal aging on one side and with other mental illnesses such as depression on the other side—how can one be sure that a person does or does not have dementia? Early on, a definitive diagnosis can be quite difficult, and it is usually best to err on the side of assuming the better rather than the worse. That is, Dad may be simply aging normally, or Mom may in fact be feeling depressed. Most doctors tend to take this conservative approach. Sometimes family members are even faced with the frustrating task of convincing the doctor that something is really wrong, since they can recognize that the personality changes or mental disabilities are truly abnormal in the context of what Mom or Dad is typically like or typically able to do. This situation arises most frequently when Mom or Dad is especially bright or socially skilled and can be quite good at covering up subtle impairments, particularly for short periods of time.

Several different kinds of laboratory testing can be helpful, however. The family doctor can do some simple laboratory tests in order to determine whether general medical problems such as liver disease, thyroid disease, or vitamin deficiencies are present, since all of these can cause mental slowing and cognitive impairments and are treatable causes of a dementia syndrome. Typically, however, such tests are negative, and more complicated tests are necessary. An MR scan of the brain (or occasionally the older CT scan) may be quite helpful.

An MR scan will tell whether the symptoms are due to a brain tumor, a relatively rare cause of dementia that is readily seen by this test. A scan will also show whether or not the person has had one or more "small strokes," which lead to the syndrome of vascular dementia (described below). It may also show whether there is shrinking of the caudate nucleus or changes in the substantia nigra, brain regions that are affected by Huntington's disease and Parkinson's disease; typically, however, these illnesses must be fairly far advanced to show up clearly on an MR scan. Finally, the MR scan may show a generalized loss of brain tissue, perhaps with a more selective decrease in either frontal cortex or temporal lobe regions. Interpreting MR scans in older people who have a generalized shrinkage of the brain can be somewhat tricky, however, since generalized tissue loss also occurs with normal aging.

Figure 10–1 shows scans from four individuals: a healthy person in his twenties, a healthy person in his forties, a healthy person in his sixties, and a person also in his sixties who has Alzheimer's disease. As these MR scans show, tissue is slowly lost in the normal brain due to aging changes, and these changes become more pronounced even in healthy older people. The difference between the tissue loss in Alzheimer's disease (the most common cause of dementia) and in normal aging can be subtle. When the brain changes are severe, however, the diagnosis is more obvious. In some instances a PET or SPECT scan may also be helpful, since people with dementia typically have characteristic decreases in cerebral blood flow in temporal and parietal regions.

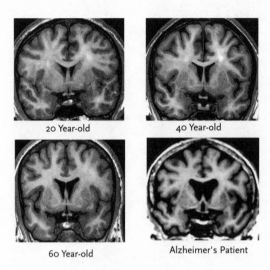

20 Year-old

40 Year-old

60 Year-old

Alzheimer's Patient

Figure 10–1: MR Scans Illustrating Changes as a Consequence of Aging and Disease

Neuropsychological testing may also be helpful because people developing dementia may have a characteristic pattern of impairments. Usually the neuropsychologist administers multiple tests that assess different aspects of cognitive function, such as memory, attention, and language skills, as well as tests of general information and motor dexterity and speed. People who are developing a dementia may have relatively normal performance on some tests that tap into knowledge stores they learned some time ago (e.g., tests that assess general information or vocabulary), but have relatively worse performance on memory tests that require them to exercise their short-term memory. As is the case with brain scans, however, testing early in the illness may still be equivocal and require careful interpretation, often tempered by knowledge of the person's baseline function and insights gleaned from family members about changes in functioning. Neuropsychological testing can also be useful in tracking the course of cognitive impairments if a person has been treated for depression or a late-onset psychosis. If the dementia is secondary to these other mental disorders, then neuropsychological testing typically improves as treatments reverse the primary problem.

Dementia is a syndrome—a group of signs and symptoms that occur together in a recognizable pattern. In fact, however, there are several different forms of dementia, each of which can be considered to be a somewhat different disease, since each has different causes, subtle differences in symptoms, and subtle differences in course and outcome. Once dementia secondary to depression or late-life psychosis has been excluded, the vast majority of people with dementia suffer either from Alzheimer's disease or vascular dementia, which together probably account for 80–90% of all cases. Other important, but less common, kinds of dementia include Huntington's disease, Parkinsonian dementia, Pick's disease, dementia associated with AIDS, and several others. Let's look at each of them in a bit more detail.

The Many Faces of Dementia
Alzheimer's Disease (AD)

Alzheimer's disease was first described earlier in the twentieth century by the German psychiatrist Alois Alzheimer, a member of the historic team of investigators in Munich who laid much of the groundwork for the scientific study of mental illnesses: Emil Kraepelin, Franz Nissl, and Korbinian Brodmann. This was a group of clinician-scientists who cared for patients by day and retreated to their labs by night to try to identify the brain mechanisms behind the painful symptoms that their patients displayed: loss of memory, loss of the ability to think clearly, tormenting voices, black despair. Alzheimer noticed that those patients who developed a cognitive decline late in life had clear and consistent brain abnormalities that could be seen with Nissl's stains. Nerve cells looked as if they had exploded and turned into messy tangles of string, and there were also areas that looked as if puddles of sludge had accumulated. These "tangles" and "plaques" became the pathological markers for the illness that Kraepelin began to call Alzheimer's disease, in honor of his friend's discovery.

Alzheimer's disease continues to be a scourge of those who are aging, a disease with a recognizable brain pathology for which the mechanism is not known but which seems to be neurodegenerative. It is now clear that it is at least partially familial and that its age of onset varies from the forties through the eighties or nineties. Approximately fifteen years ago several different investigators made an interesting observation that gave a new clue. They noted that patients who suffer from a neurodevelopmental disorder, Down's syndrome or trisomy 21, were also predisposed to develop Alzheimer's disease prematurely (in their twenties and thirties) and to display its characteristic plaques and tangles. This was a disease for which the mechanism *was* known, and it led molecular biologists to begin the exploration of Alzheimer's disease. Patients with trisomy 21, a multisystem disorder with mild mental retardation as its core feature, are born with an extra copy of Chromosome 21. The mechanism causing Alzheimer's disease therefore might be some function performed by a gene on that chromosome.

The story of the search for the Alzheimer's gene illustrates both the highs and the lows of the search for the molecular mechanisms of mental illnesses. The clue provided by the association between Down's syndrome and Alzheimer's disease led scientists to turn their attention to Chromosome 21. A site on this chromosome was identified that produces a substance called amyloid precursor protein (APP). The plaques of Alzheimer's disease consist of accumulated masses of a protein called amyloid, which acts like "brain sludge," destroying the capacity of neurons to communicate with one another. For a short time in 1987, it appeared that a complete explanation of the mechanism of Alzheimer's had been identified, when investigators found a mutation in the gene for APP in fifteen patients suffering from Alzheimer's disease. Finding the gene created the hope that the accumulation of the deadly amyloid plaques could be blocked and the disease prevented.

It soon became apparent, however, that many cases of Alzheimer's disease did not have this mutation, and hopes for a quick cure had to be abandoned. As investigators returned to the search using new variants of conventional linkage analysis, Chromosome 19 was found to be implicated, and another substance was identified as a potential additional villain in the disorder: the ε4 allele of apolipoprotein E (APOε4 for short). Individuals who carry two copies of the APOε4 allele have a dramatically increased risk for developing AD—90% by the age of 80. The APOε findings raise the possibility of screening individuals to determine whether they are predisposed to develop the disease and of designing a treatment that will strengthen the concentration of a healthy variant of APOε known as APOε2, which could protect potential victims from developing the disease. Functional imaging studies have indicated that carriers of the ε4 allele have measurable differences in brain blood flow—before the illness has begun or signs of cognitive impairment are present. This has increased hopes for finding a premorbid marker that can be used to identify people who might benefit from preventive interventions.

Further linkage studies have implicated genes on two additional chromosomes: 1 and 14. At the moment it is not clear whether amyloid accumulation is a *cause* of neuronal death in AD or, conversely, a *result* of cell death. Another factor, presenilin, has been added to the story as well. It only accounts for rare cases (about 5%) and has two variants, presenilin-1 (Chromosome 14) and presenilin-2. A laboratory test for the presenilin-1 mutation is currently available. It is not clear whether the many chromosomes identified represent causative factors or susceptibility factors, or how many would need to be attacked therapeutically to treat or prevent the disease. Finally, it is not clear whether the essence of the illness is an accumulation of beta amyloid protein (BAP), whether it is a consequence of an inflammatory immune response provoked by amyloid accumulation, whether it is a problem in clearing amyloid using APOE, or whether yet another protein known as Tau is the major villain. Tau appears to be more closely related to the formation of tangles. Sorting out the answers to these questions will be crucial in developing a treatment for Alzheimer's disease.

Apart from genes, other risk factors have been identified for Alzheimer's disease. Women have a somewhat higher rate of Alzheimer's disease, although this may simply reflect the fact that women tend to live longer and therefore have a greater "age of risk." Other risk factors include a history of head injury, low educational and occupational level, and having a first-degree relative with Alzheimer's. If a relative of a person with Alzheimer's disease lives to age 90, he or she has a 50% chance of developing the illness.

Alzheimer's disease cannot be diagnosed definitively while a person is still living, since the diagnosis requires identification of its characteristic brain pathology at autopsy. This characteristic pathology consists of the senile plaques and neurofibrillary tangles originally identified by Alzheimer. Early in the illness, the plaques and tangles usually form in the temporal lobes, and particularly in the hippocampus. This localization explains why mild memory impairment is typically the earliest symptom of the illness. As the disease progresses, the plaques and tangles become more widely dispersed throughout the brain. In severe end-stage cases they are seen throughout the cerebral cortex.

Neurochemically, the primary abnormality is in the cholinergic system. As described in chapter 4, cholinergic neurons in the brain arise in a sub-cortical region called the nucleus basalis of Meynert. These cholinergic neurons then send connections throughout the brain and provide a modulatory function. Studies of people suffering from Alzheimer's disease indicate that the cholinergic neurons have degenerated. Clinically, people have observed for many years that drugs that block the cholinergic system, such as antihistamines, can produce memory impairments or trigger confusion, particularly in predisposed individuals. As described below, treatment strategies have tried to exploit this knowledge and identify ways to improve cholinergic activity in people suffering from Alzheimer's disease.

Alzheimer's disease is the most common of the degenerative dementias. It accounts for approximately 50% of all cases of dementia. It is unfortunately very common, affecting approximately 2.5 million Americans, and not overlooking the rich or famous, such as Ronald Reagan. It usually begins insidiously in the fifties, sixties, or beyond. Some people decline very slowly and live on for ten to twenty years after the initial diagnosis. Others go downhill more quickly and may die within eight to ten years after onset. Investigators are intensively exploring preventive and treatment measures, and we are hopeful that the tools of molecular biology and genetics will produce major breakthroughs in treatment and prevention of Alzheimer's disease over the next several decades. The nature of these possible breakthroughs is discussed in chapter 12.

Vascular Dementia (Multi-infarct Dementia)

Vascular dementia, also known as multi-infarct dementia, is the second most common cause of dementia. Approximately 20–30% of people with dementia suffer from this form of the illness.

Classically, people with vascular dementia develop their symptoms because they have had a series of "small strokes." A typical "large stroke" produces motor symptoms such as paralysis (which occurs on the opposite side from the brain region injured by the stroke), and sometimes it may also cause difficulties with speech or perception. If a stroke is relatively small (i.e., due to the occlusion of a smaller blood vessel, affecting a relatively small region of the brain), or if it occurs in a brain region that does not have motor functions, it may pass unnoticed. Strokes of this type are called silent. People who develop multi-infarct dementia usually have had a series of strokes, some of which may be large but some of which may be small and not even diagnosed at the time they occurred. If the initial strokes are silent, then intellectual decline is typically the first thing to be noted. Even if the strokes are large, mental functioning may still be intact, and the mental decline may only occur slowly if subsequent strokes occur. Not all people who have strokes go on to have dementia. This syndrome arises as a consequence of repeated strokes.

The strokes occur in people who have a vascular disease of some type. During a stroke, the blood vessels supplying the brain become clogged by clots, which shut down the blood supply in a specific brain region. (This is called an infarction by physicians—hence the name "multi-infarct dementia.") The manifestations of a stroke vary depending on the location and size of the blood vessel that develops the clot. If it occurs in a large vessel supplying the motor and sensory cortex, then paralysis ensues. This is the most typical and familiar form of stroke.

The onset, symptoms, and course of vascular dementia are somewhat different from Alzheimer's disease. The first signs may occur relatively suddenly, after a moderate or major stroke has occurred. Thereafter, the person may stabilize and have no or minimal cognitive impairment. However, if subsequent large and small strokes occur, there is a stepwise deterioration, with increasing cognitive impairment after each stroke. People who have had strokes may have one or two of the "four a's" mentioned earlier. Aphasia, or impairment in the ability to speak clearly, is especially common in people who have had left hemisphere strokes affecting the language centers of the brain. People who have had right hemisphere strokes may have apraxia because their visuospatial centers have been damaged. However, they would not be diagnosed with dementia unless they also had other associated cognitive symptoms, such as impairment in memory functions (amnesia). If a sufficient number of strokes occurs, however, the person with multi-infarct dementia develops both multiple and severe cognitive impairments, and usually some of the noncognitive symptoms such as psychosis or depression as well.

Depression is a particularly common symptom both in stroke and in vascular dementia. While the occurrence of depression was once attributed to the psychological impact of paralysis and perhaps aphasia, Bob Robinson, an expert in vascular dementia at the University of Iowa, has demonstrated through research using animal models and pharmacologic interventions that post-stroke depression is primarily a consequence of the disruption of norepinephrine tracts in the brain. People with strokes affecting the left hemisphere are particularly vulnerable to depression because connections with emotional centers are often disrupted.

Anything that increases the risk for cerebrovascular or cardiovascular disease is a risk factor for multi-infarct dementia. High blood pressure is perhaps the most important risk factor, along with high cholesterol and diabetes. If hypertension can be controlled, the risk of developing vascular dementia is substantially reduced. People who are at risk for stroke from the development of clots also benefit from taking a small amount of aspirin, which reduces the tendency to clot and may prevent the formation of clots in blood vessels in the brain. In the case of vascular dementia, the course and outcome are closely linked to the overall course of the underlying vascular disease, and so that is the primary emphasis of treatment. People with a depression associated with vascular dementia often benefit from taking antidepressants, however.

Although vascular dementia and Alzheimer's disease have been described as if they are two different things, and indeed they are at an abstract level, they may in fact overlap in individual people. There is no reason that a person cannot develop both. The major difference between these two forms of dementia is in the course of the illness, with Alzheimer's disease having a slow downhill deterioration and vascular dementia having a stepwise course. However, the symptoms that people have are more or less the same, including the cognitive symptoms (the four a's) and the noncognitive symptoms (psychosis, depression, agitation, and apathy).

In real life, doctors and family members may have difficulty deciding which type of dementia is actually present. Diagnostic tests may be helpful, but are not always decisive. The pattern of abnormality measured with neuropsychological testing may be nearly identical in the two disorders.

In patients with vascular dementia, an MR or CT scan will usually show evidence for multiple strokes, and this is perhaps the most decisive test. Multiple strokes also produce the brain shrinkage seen in Alzheimer's disease, however, with visible cortical atrophy and ventricular enlargement. So MR scans may even be confusing as well. The possibility of Alzheimer's disease co-occurring with vascular dementia, particularly in people who have a relatively severe downhill course, can only be determined by searching the brain at autopsy for plaques and tangles. Therefore, families that are concerned about carrying an increased risk for Alzheimer's disease may choose to request a postmortem examination of the brain for any family member who has had a dementia.

Huntington's Disease

Huntington's disease is a relatively rare form of dementia. It is interesting because it illustrates how many different levels of an illness we need to understand in order to develop improved treatments or prevent mental illnesses. Most of this story has already been told in chapter 5.

Unlike Alzheimer's disease, which is associated with a complex mixture of genes and diverse nongenetic risk factors, Huntington's disease is very simple. It is probably the most genetic of all the mental illnesses. It is due to a single genetic mutation, which has been located on the short arm of Chromosome 4. This genetic mutation is transmitted according to classical Mendelian patterns. It is autosomal-dominant and fully penetrant, which means that carrying one copy of the gene virtually guarantees that a person will develop the illness. That gene was passed on from a parent who carried one copy of the gene. Therefore, the child of a person with Huntington's disease has a 50/50 chance of developing the disorder. We now have highly accurate clinical tests that permit people who have a parent with Huntington's disease to determine whether or not they have inherited the gene. If they are in the unlucky 50%, they are guaranteed to become ill, and if they have already married and have children, each of their children has a 50/50 chance of developing the illness. If they did not inherit the gene, however, they have escaped the family curse, and they are also free to conceive children without risk of passing the illness on. The genetics of Huntington's disease are pretty melodramatic. It is a story in which a particular site on the chromosome can wear a black hat and cause a menacing illness, or it can wear a white hat and confer normal mental health.

Against this stark background of black-hat and white-hat genes, a complex clinical drama unfolds. The person destined to develop Huntington's disease receives the single gene at the time of conception. This gene, located on Chromosome 4, has an abnormality known as multiple trinucleotide repeats, repeated strands of nonsense code that throw sand in the wheels of genetic communication, causing harmful information to slowly and steadily accumulate. Despite knowing where the gene is and how it is abnormal, we still do not know what its product is, though we can be sure that it is something that affects the development and functional integrity of the brain, and even that it affects a specific brain region, the caudate nucleus. However, the developing baby who carries the black-hat gene is perfectly normal at birth and continues to be perfectly normal for the next forty or fifty years. Whatever the gene does, its effects are delayed for a very long time. If the man or woman carrying the gene does not know about the presence of this dreadful taint, as was fairly likely before the availability of genetic testing, he or she will have children who may themselves receive the same black-hat gene. Since the gene causes the illness, it has probably been doing its mischief somehow and somewhere in the brain for a number of years and perhaps since early in fetal development. In a sense, the person could be said to have Huntington's disease since birth, even though the overt signs of the illness do not manifest themselves until much later. We do not yet know enough about the brain processes that lead to Huntington's disease to point to a specific time when we can say that the disease has begun.

Although we do not know exactly what the gene does, we do know that it affects the caudate nucleus in the brain. This nucleus, described more fully in chapter 4, is one of the subcortical nuclei in the brain. It has usually been considered to modulate motor function. Since the symptoms of Huntington's disease are cognitive and emotional, however, it is clear that the caudate must do much more than moderate motor functions. Exactly how the caudate is damaged is still unknown. It is very puzzling that a single gene could have such a specific effect on a single subcortical nucleus of the brain, and that a single subcortical nucleus could in turn affect so many aspects of mental function.

Typically, the disease first becomes manifest as a slight change in personality. A person who had a friendly and sunny temperament begins to be moody and irritable. Outbursts of anger may occur, or there may be prolonged periods of despondency. Many people with Huntington's disease present first as having clinical depression. Alternately, mild problems with concentration, attention, or memory may be the first complaints. These can be difficult to disentangle from depressive symptoms.

The very first Huntington's patient whom I saw, for example, was a

40–year-old woman who had fallen into a profound despondency, murdered her son, and then made a failed suicide attempt. I made a tentative diagnosis of Huntington's disease rather than depression, based on the fact that her father had suffered from Huntington's disease. This was in the era before the gene had been discovered or genetic testing was available, but the autosomal dominant transmission of the disease was already well known, having been described by George Huntington way back in 1872. She did not yet exhibit the telltale motor abnormalities of the disorder, which she began to display several years later. Eventually, everyone with Huntington's disease develops the motor symptoms of chorea and athetosis—writhing and squirming movements with facial grimacing. As the disease progresses, speech becomes impaired, the cognitive symptoms become steadily more severe, and the personality changes continue. Its victims typically die between five and twenty years after onset.

The earlier the onset, the more severe the symptoms and course. At present no preventative measures are available for those who carry the gene and will subsequently develop the manifest illness, and no treatments are available. As with other genetic disorders, we hope that eventually knowing about the gene and how it affects the brain will help us figure out how to prevent the expression of the gene that causes the illness to manifest itself. However, Huntington's disease illustrates how difficult and frustrating this task may be. We know so much, and yet so little.

Other Kinds of Dementia

Alzheimer's disease, Huntington's disease, and vascular dementia are the "big three" of the dementia syndromes, since they are the most extensively studied or the most common. There are several other kinds of dementia, however. Some of these may be much more prevalent than we currently realize. Furthermore, since the majority of people who are diagnosed as having Alzheimer's disease do not have the diagnosis checked through autopsy, some of them may be misdiagnosed and in fact have some other form of dementia. Therefore, it is important for families to recognize that there are other types of dementia, some of which may be hereditary like Alzheimer's disease and Huntington's disease, and some of which may be multifactorial or even purely environmentally caused.

Parkinson's disease may be associated with a dementia that is very similar to Alzheimer's disease. People with this illness may have the motor symptoms of Parkinson's, which include tremor, a relatively blank and immobile facial expression, and a shuffling gait. Some people who develop a mild dementia may also have very mild Parkinsonian symptoms, and this is a clue that the dementia is not due to Alzheimer's disease. The picture can become quite confused, however, because people with early dementia may be treated with antipsychotic drugs, and the extrapyramidal side effects of these drugs closely mimic the motor symptoms of Parkinson's disease. The dementia syndrome itself is similar to the one seen in Alzheimer's disease, with mild cognitive and noncognitive symptoms that may progressively worsen. A definitive diagnosis of Parkinson's dementia, as distinguished from Alzheimer's disease, can only be made at autopsy, which reveals Lewy bodies and loss of nerve cells in the substantia nigra.

Parkinson's disease, like Alzheimer's disease, appears to be a disorder that has multiple causes. It is modestly familial, suggesting a genetic component. It occurs in purely nongenetic forms as well. For example, the influenza epidemic of 1917 produced a subsequent epidemic of Parkinson's disease in a subgroup of people who came down with that flu. In this instance, the causative agent was a virus. People who have sustained head injuries are also prone to develop Parkinson's disease. Muhammad Ali is perhaps the most famous example.

A variation on the dementia due to Parkinsonism is a syndrome called *Lewy body dementia*. This syndrome shares features with Parkinson's disease, in that people have problems with motor rigidity and loss of facial expression. However, this variant of Parkinson's disease is characterized by a more prominent loss of memory, frequently with confusion, that occurs early in the course of illness and is excessive in relation to the motor difficulties. This disorder, like Parkinson's disease, is known to have a certain type of finding on the postmortem evaluation called a Lewy body. Lewy bodies are spheric inclusion bodies that contain neurofilaments and other proteins. Lewy body dementia may be distinguished from Parkinson's dementia by its lack of response to L-dopa treatment, the typical treatment used in Parkinson's disease. In fact these patients may become much worse with this medication, even experiencing visual hallucinations.

Pick's disease (also called frontotemporal dementia) accounts for approximately 5% of the dementias. Like Parkinsonian dementia, it is clinically indistinguishable from Alzheimer's disease and can only be definitively diagnosed after death. The first sign of Pick's disease may be a noticeable decline in interpersonal social conduct. Socially odd or disinhibited behaviors are sometimes very evident before cognitive impairment is discernable. A hint of this diagnosis can be obtained from an MR scan, however, because a telltale sign of Pick's disease is selective frontotemporal atrophy. Because of this distribution, the clinical symptoms may be tipped somewhat more in the direction of personality change and inappropriate behavior, in addition to disturbances of memory. At autopsy the disease is differentiated from Alzheimer's disease by the absence of plaques and tangles and the presence of Pick's bodies, which contain remnants of nerve cells and stain in a characteristic way. The causes of Pick's disease are unknown, but it does have a mild tendency to run in families.

Creutzfeldt-Jakob disease was a relatively obscure disorder until the 1990s, when a few cases developed around the world, particularly in Europe, and were traced to infected meat. Thereafter, it became widely known as "mad cow disease." Creutzfeldt-Jakob disease is a particularly interesting form of dementia, since it is transmitted like an infection. The earliest described cases occurred in cannibals who resided in the South Pacific and consumed human brains, through which the illness was transmitted. More recently Stan Prusiner has demonstrated that the infection is in fact transmitted through prions (protein-based infectious agents), rather than through classic infectious agents such as viruses—the first example of this form of infectious transmission. Prusiner was awarded the Nobel Prize for this discovery in 1997. People who develop Creutzfeldt-Jakob disease have the classic symptoms of dementia, such as problems with memory and changes in personality and behavior. They also have motor symptoms such as weakness and incoordination. Once they have become infected, their disease progresses rapidly, and they lapse into coma and die within a few months to a year.

HIV dementia is another example of dementia occurring secondary to a nongenetic cause. In this case, the human immunodeficiency virus (HIV) invades and damages nerve cells. In fact, cognitive and emotional symptoms may be the first clue that infection with HIV or AIDS is present. The mental symptoms associated with AIDS include depression, delusions, apathy, anhedonia, and hallucinations, in addition to more cognitive symptoms such as memory impairment, confusion, or problems with attention. In fact, mental symptoms are often the first ones for which people seek treatment when they develop AIDS. People with AIDS have a variety of neural symptoms. This is because brain tissue has very high concentrations of a group of receptors called CD4 receptors, which are targeted by the HIV virus. In addition to cognitive and emotional problems, people with AIDS may have a variety of motor symptoms, such as poor coordination, weakness, or even paralysis. Fortunately, for this form of dementia, treatments have been developed that have markedly reduced the progression of the illness, so that the many people infected with HIV are spared the full-blown AIDS syndrome, and those with AIDS are spared a progressive cognitive dysfunction.

Alcohol-related dementia is another example of a nongenetic form of dementia. A well-recognized disorder called Wernicke-Korsakoff syndrome results from a deficiency of the vitamin thiamine. It occurs in some people who suffer from chronic alcoholism. People who develop this problem often have a rapid occurrence of a characteristic mixture of symptoms: abnormalities in eye movements, motor incoordination, disorganized speech, and severe impairments in memory. Some of these symptoms may clear up quite rapidly when the thiamine deficiency is treated.

How Can Dementias Be Prevented or Treated?

While changing demographics have ballooned the number of people over age seventy, my generation is in the unenviable position of being "the Sandwich Generation." We are stuck in the middle of two important responsibilities: caring for our growing children and caring for our aging parents. We will soon be passing this responsibility on to baby boomers. The aging population for whom they will be responsible will be even larger, as the human life span becomes longer and longer. Simultaneously caring for both parents and children is no small challenge. Yet that is precisely the challenge that the baby boomers will be facing as their children pass through school and their parents begin to develop dementia and require either home care or nursing home care.

The baby boomers themselves are just beginning to approach their sixties. Looking down the road, they can see that the painful possibility of developing a dementia themselves is not far away. One might say, in fact, that for all of us at the present time, the risk of developing a dementia is about as real as death and taxes. We are all very interested in figuring out what the future holds in the way of treatments for these illnesses. Even more, we are interested in figuring out whether they can be prevented. All of us would prefer to age gracefully with a sound mind in a failing body, rather than having a failing mind in a sound body. Increasingly, the fear of developing a dementia will no longer be simply an abstraction for people in the forty-to-sixty age range. Instead, like Wayne, they will be reminded of this risk every time they go to visit a spouse or parent who must now reside in a care facility, or when they struggle to cope with maintaining a loved one in a protected environment in the community. As families are scattered over the country, and as children and parents become more and more geographically dispersed, worries about aging parents become even greater.

Neuroscientists and pharmaceutical companies are well aware of the problem and are actively studying the mechanisms that cause nerve cells to become sick and die and the ways that injured cells can repair themselves. The announcement that neuroscientists had discovered that new nerve cells could grow in one region of the brain, the hippocampus, was greeted with great interest. Up to that time, the inability of neurons to divide and replicate themselves like the other cells of the body had been a central dogma of neuroscience. If hippocampal cells could duplicate themselves, perhaps other brain cells could as well, and we could figure out ways to prevent or treat neurodegenerative disorders and replace cells injured by trauma. As often happens, however, this development was given excessive "hype" in the popular press, perhaps raising false hopes. The hippocampus is an unusual and relatively primitive part of the brain, and therefore the cellular reduplication observed there may not generalize to other brain regions. Nonetheless, the discovery still represents a possibility that the fate of those destined to develop dementia may become less hopeless.

Since the dementias usually afflict people who have been mentally healthy and normal up to that point, figuring out how to prevent such people from experiencing a mental illness in late life is perhaps an even more important goal than figuring out how to develop a treatment after the dementia has begun. Therefore, much of the emphasis in research has been to identify risk factors in predisposed individuals. The recognition of the role of APOε4 and its applicability as a genetic screening test is an example. Of course, this and other screening tests raise a variety of ethical and economic questions. Putting those aside, however, let's just ask whether screening families with dementia for this gene is medically useful and should be done routinely. There is no simple answer, since checking for the APOε4 allele does not provide a definitive diagnostic test. Unlike Huntington's disease, for which the Chromosome 4 screen provides a definitive yes or no, the APOε4 screen can only indicate a maybe. At best, carrying the APOε4 allele raises the probability that a person will develop dementia to about 30 or 40%. If the gene is present and the test is positive, the person is faced with a 30–40% possibility that an illness may develop for which there is as yet no preventive treatment. So in this instance the positive genetic test is not only inconclusive, but it does not suggest any useful intervention.

The purpose of identifying the genes involved and developing a screening test for mental illnesses that are hereditary or partially hereditary is to identify people who carry the genes so that preventive measures can be put in place early on. Physicians and neuroscientists would like to design a preemptive strike that will arrest the invasion of these scourges before they attack. Identifying the genes is the first step in building the arsenal of preventive measures, but only the first. The second step is to figure out how the genes work their mischief in the brain. Do they cause the energy power packs (the mitochondria) to wear out? Do they attack specific neurochemicals in crucial brain regions such as the dopamine system of the substantia nigra in the case of Parkinson's disease or the nucleus basalis of Meynert in Alzheimer's disease? (Since dopamine and acetylcholine seem to be involved in these illnesses, that is a plausible scenario.) Do they attack the protective skin of nerve cells, their membranes, and prevent them from regulating the relationship between the internal cell environment and the cell's external world? Do they disrupt the delicate signaling machinery of the cells, the receptors? These are all good questions.

The causative genetic mutation has been nailed down only for one relatively uncommon form of dementia, Huntington's disease. Multiple genes have been identified for Alzheimer's disease, but their specific effects and their utility for designing treatments and preventions is still in a controversial and exploratory phase. During the next decade or two, the remainder of the story is likely to unfold. Figuring out how the genes work their mischief will take some time. As this story continues, the final step will be to figure out medications and other interventions that can march in and attack the damage created by the destructive messages promulgated by the black-hat genes.

While we wait to hear the good news about the successful identification of genes, their mechanisms, and new drugs, is there anything else we can do? The answer is yes.

Whatever type of dementia we may be predisposed to develop, exercising good physical hygiene can help. The brain is the most important factory in our body, using a substantial portion of the energy consumed each day. It needs a healthy supply line through our blood vessels and an adequate supply of chemical resources to assist in the breakdown and elimination of toxic byproducts and wastes. It needs its owner to engage in an exercise program to keep the heart pumping well and delivering an adequate blood supply, while also reducing the accumulation of sludge in the pipes. It needs a healthy diet containing an adequate balance of fats, proteins, and carbohydrates that will give the brain the food it needs. Unfortunately, diet fads come and go, and we need to learn more about the nutritional balance that is correct for the brain. Since the brain is very rich in fat, which it uses as the insulation for axons and other parts of the cell membrane, the low-fat diet that is designed to reduce sludge in arteries may also be depriving the brain of a critical nutrient.

In the absence of definite information, a diet that includes moderate amounts of a variety of foods is probably best. For women, the protective effects of estrogen supplements for the brain are relatively well established, although each individual woman must weigh the related risks, such as cancer of the breast or endometrium. Evidence is also good that intake of a modest amount of vitamins can help the brain manufacture proteins and get rid of the wastes, as its nerve cells work assiduously, and sometimes frantically, to get the right messages sent to the right locations. Other dietary supplements, such as ginkgo biloba, have also been proposed as helpful additives to the diet. Ginkgo appears to increase the blood supply to the brain. While theory suggests that this may be helpful, the experimental tests to demonstrate this are still in progress. Vitamin E, taken in modest amounts, helps cells all over the body break down the dangerous free radicals that are released as a by-product when our cells burn dietary fuel to produce energy so the cells can run.

Can we supplement good physical hygiene with good mental hygiene in order to decrease our risk for developing a dementia? Again, the answer is almost certainly yes. As for sexual function or muscle strength, the advice here too is "use it or lose it." It has been solidly demonstrated that a higher educational level substantially reduces risk for Alzheimer's disease. Approximately 40% of people whose educational level is low will develop dementia by age seventy, while the risk is only 10% for those with a higher educational level (i.e., college education). These numbers pull apart even more sharply by age eighty, reaching 60% in contrast to 20%. Of course, educational level may not tell the whole story, since people who are more highly educated also enjoy other benefits, such as more financial resources to eat a good diet or more medical care and more opportunities to exercise regularly.

Whatever our educational level in the past, we can all do something to reduce our risk for dementia in the present and future. Not only can all of us improve our physical lifestyle, but we can also potentially improve mental hygiene by exercising our brains. Sitting in front of a TV and watching soap operas or football games while sipping a beer or munching on potato chips is probably much less healthy for the brain than working crossword puzzles, reading books, learning new skills such as using computers, speaking a foreign language, or taking on a new hobby that exercises the mind and makes it solve new problems or see things in new ways. As we age, we all find it is easier to do less, but our brains will be much happier and healthier if we do more—enjoying our old friends, making new ones, thinking, feeling, exercising—keeping our minds and bodies active in every way possible.

As each older individual does his best to maintain good physical and mental health, scientists are hopeful that the efforts of the Human Genome Project and Venter's Celera will speed up the process of finding the genes that produce Alzheimer's disease and other mental illnesses. As these genes are identified, the task of figuring out how they work their mischief will be passed on to neuroscientists, psychiatrists, and neurologists, who will make the connection between genes, gene products, and human thought and behavior. The task of figuring out how the mischief can be reversed or prevented will fall into the hands of molecular and cell biologists, who will design chemical strategies to repair the damaged or dying nerve cells. In the meantime, the victims of the dementias and other mental illnesses, as well as their loved ones, await hopefully the time when the genes can be found and understood . . . and ultimately the time when the diseases can be prevented.

The discoveries probably will not come in time to help Pearl, but perhaps Wayne will be able to put them to good use. If all goes well, Wayne and Pearl's three children will never have to suffer the pain confronted by their mother and father.

CHAPTER 11

ANXIETY DISORDERS
The Stress Regulator
Goes Wild

The world is too much with us; late and soon,

Getting and spending, we lay waste our powers:

Little we see in Nature that is ours;

We have given our hearts away, a sordid boon!

—William Wordsworth

"The World Is Too Much with Us"

I'll never get caught up. I need a break! I feel so overwhelmed. I just can't cope with it all. I need a chance to kick back, but I'll never find the time. I'm so stressed out!

Those are the refrains of the twenty-first century.

As we look back to the world of one hundred years ago, we see a time of innocence and simplicity that we view with nostalgia. Traveling by horse and buggy and communicating by written letters has a quaint appeal. Would we trade what we have now for what we had then? Maybe yes, or maybe no, depending on how much we value "modern conveniences" in comparison with the complexity, confusion, and pressures of modern life. But anyway we don't have a choice. The twenty-first century has already happened. We can't turn back the clock.

It is a strange paradox that we have more and more things to make our lives easier, and yet the daily routine seems to be getting harder and harder. We have lots of labor-saving devices—clothes washers, dishwashers, fancy electric or gas stoves, cuisinarts, vacuum sweepers, coffeemakers, juicers—an endless array of household appliances. We have fast food, delis with great take-outs, pizza delivery. We have telephones in our homes and cell phones in our cars so we can stay in constant contact, and some of us even have pagers as well. We have cars (and often multiple cars), airplanes, and subways. We have television, the Internet, CDs, DVDs, and Walkmans. We have computers—on our desks, in our briefcases, and in our hands.

Quite simply, we have a lot. We may have more, and gotten it faster, than our human brains were designed to deal with. Consequently, many of us feel anxious and out of control. Some of us have such severe anxiety that we are sometimes disabled by it. We have an anxiety disorder.

✼   ✼   ✼

With her eyes still closed, Michelle could sense that the gray light of day had not yet begun to glint through the window shades. Here we go again, she thought. *Some day I'd like to just be able to sleep late, or even to be awakened by my alarm clock instead of my internal clock.* She lay there thinking, beginning to feel the rising flood of panic inside her. Her heart started to pound and even to skip a beat from time to time.

God I can't stand this. It's starting to drive me crazy. I used to be fine, but now I wonder if I'm going to have a heart attack. Granddaddy Fred dropped dead unexpectedly in his late fifties when he went out to shovel snow. Grammy Jo just found him lying in the driveway. She was nearly scared to death herself. Dad is only 55, and he has had a couple of heart attacks and triple bypass surgery. Mom, Jerry, and I are so afraid that something is going to happen to him. There it goes again. My heart is running a mile a minute. I need to check my pulse.

Groping down to her wrist, Michelle put her forefinger on the pulsating artery. Since it was dark she couldn't count using the second hand on her watch, but she could tell that it was going very fast and that it was sometimes missing a beat.

She took several deep breaths. She could barely breathe.

I'm suffocating. I can't get my breath. I wonder if I am going to have a heart attack and die myself. Maybe I should call 911.

She jumped out of bed and stood up. She felt dizzy and lightheaded. A wave of nausea started in her stomach and rose to her throat. She ran to the bathroom and gagged in the toilet. She could feel her hands trembling as she bent over and gripped the seat.

I wish I could go back to bed. I wish I could just go to sleep. I don't think I can face another day of gulping down a cup of coffee, jumping on the subway, and rushing in to work. When I get there, there will be a stack of papers on my desk. Unanswered letters. Unanswered phone calls. Unanswered e-mail. People will be coming into my cubicle and wondering why I'm so behind on my assignments. I am supposed to be preparing that presentation on our new product for next week, and I'll never get it done in time. I'll probably have two or three more attacks like this off and on during the day. I never even know when they are going to happen. I'll have to run off into the bathroom and puke. People will think I'm pregnant. I am only twenty-seven years old, and I am probably going to die from a heart attack before I hit thirty. I promise myself—I'll make an appointment to see my doctor today and find out what has gone wrong.

Michelle's problems began only two or three months ago. It all started while she was commuting into work. She had just moved to New York from Michigan in August, and she was thrilled that she had gotten a job in marketing at Pfizer. What an opportunity! Just the corporate headquarters, located in one of the many skyscrapers creating the canyons of Manhattan, seemed to occupy an entire block. When she had done her job interview a few months earlier, she had surveyed the occupants of the elevator, feeling frumpy in her Midwestern clothes and vowing that she would become as sleek, haughty, and indifferent as these harried but elegant New Yorkers who stood beside her, grasping their bags of bagels and cups of coffee.

Apparently her 3.8 GPA, her letters of reference, and her interviews had been persuasive. They gave her a three-hour "minitest," which required that she read about a new drug for treating sexual dysfunction in women, decide what remaining clinical trials should be done, choose a possible name for the new drug, and design a preliminary marketing plan. Although she didn't know much about sexual dysfunction, she had a good time with the assignment. She chose Elegra as her name, thinking it would be a good match for Viagra. For her promotional theme in the marketing plan, she came up with "Mrs. Field's cookies in the kitchen and Marilyn Monroe in the bedroom." Everyone seemed to think it was pretty impressive that she could come up with so many ideas on such short notice.

The salary they offered her seemed astronomical to a girl from Michigan, whose parents had struggled through layoffs and rehirings during the ups and downs of the auto industry and who preached economic security as the highest goal to which she should aspire. A couple of her friends from her undergrad days at the University of Michigan had warned her that the salary would not really be that much money in New York, but she just couldn't believe it. Luckily, she was able to stay with them while she hunted for an apartment. She was stunned by the prices. Rent alone was going to eat up half of her take-home salary, and prices for food were astronomical. It would be a while before her Michigan wardrobe would be replaced by garb from Bloomies or even Banana Republic or the Gap. She did treat herself to a handsome Coach briefcase and a Mont Blanc pen—conspicuous symbols of her new professional status. As for her apartment, the notion of a classy West Side pad had to be replaced by a few rooms in a Brooklyn brownstone and a half-hour subway commute.

The first problem began one morning as she descended into the gray, subterranean subway world. Although she had never been in the sewers of Paris, the subway stations made her imagine what they must be like. As she settled into her seat (feeling lucky to get one), she began to think about how hard it must be to keep all the trains from running into one another as they hurtled from various directions through those narrow, dark underground tunnels. If her train were to crash, she could be injured, mutilated, crippled for life, or killed. It would be a horrible death because she would be trapped underground. It would be like being buried alive.

As she thought about these possibilities, she started feeling more tense and anxious. She glanced around at the other people on the subway, indifferently reading their newspapers and magazines, and wondered why they weren't more concerned. She herself suddenly experienced an overwhelming rush of fear and panic. Her heart started to pound, and she felt as if she was going to suffocate. She began to take deeper and deeper breaths, but this only made her feel more short of breath. That was the first time she worried about suddenly dying of a heart attack. She wanted to jump up and run away or ask someone for help, but she knew this was impossible. So she struggled to get a grip on herself, and after five minutes or so the feelings of panic slowly subsided. She climbed out of the subway, walked a few blocks to work in the midst of the bustling mass of humanity, and stepped into the elevator to ride up to her cubicle on the 28th floor. As the elevator door closed and she felt the intangible pressure of ten other bodies around her, she began to experience the same sense of being trapped and closed in, but she was able to brush it off and settle into work that morning.

Things were frantic that day. She could never tell what kind of mood her boss, Debbie, was going to be in. Debbie was ten years older and was juggling the demands of a workaholic husband employed on Wall Street, a three-year-old in day care, and a golden retriever that chewed its toenails. The tone of the day was often set by whether her son, Tony, was going to need tubes in his ears or whether Shelby the retriever was going to need to have his paws bandaged. Whatever Debbie was unable to finish seemed to end up on Michelle's desk, and it was always Michelle's fault if things got behind schedule. The schedules themselves often seemed totally unrealistic, with two-day turnarounds for tasks that ought to take two weeks. What had she gotten into? Michelle wondered. Maybe she should have stayed back home in Michigan and gone to work for General Motors. Less glamorous. Less exciting. But certainly less stressful. Well, on with the day. And she did get through it.

But the next day, on the subway, she had another attack. It was the same as the first one—terror, pounding heart, suffocating, feeling trapped, thinking she might die. She had another one that came out of the blue while she was just sitting at her desk at work, reading and thinking about a project. Over the next few weeks, they occurred more and more frequently. She even started to get them in the middle of the night or early in the morning. One day she called in sick, simply because she couldn't stand the thought of getting on that f——ing subway and probably having another attack.

After two or three months of this, she knew she had to see a doctor. One of her friends had had a heart problem called paroxysmal atrial tachycardia (PAT) and eventually had surgery for it. Michelle had by now convinced herself that she probably also had PAT. She described all of her symptoms to the doctor, as well as her worrisome family history of heart disease. He listened to the story carefully and then obtained multiple laboratory tests, including a chest film, an electrocardiogram, an echocardiogram, and a lot of blood work that included tests for cholesterol, lipids, and thyroid.

When she came back for a debriefing one week later, he explained that all her tests were normal. She did not know whether to feel pleased, relieved, or disappointed. What could be causing these terrible symptoms? There was obviously something wrong with her. The doctor explained that her problem was probably related to stress—moving to a new city, taking on a new lifestyle, and having a high-pressure job. He thought a tranquilizer might help and gave her a prescription for Valium. If this did not help, however, he would probably need to refer her to a psychiatrist. That was about the worst news Michelle had ever heard. It was bad enough to have these frightening attacks. It was still worse to have the doctor imply that it was "all in her head." These problems were real physical problems, not mental ones! She was determined to improve and make the attacks go away, and so she dutifully took her first Valium when she went to bed that night.

It only helped a little. During the next few weeks the attacks lessened slightly, but she was still having two or three a day. It was interfering with her work. She was beginning to wonder if she might lose her job. So she called the doctor again and asked for a referral. She knew that half the people in New York City were probably seeing psychiatrists, but she wasn't neurotic like them. She was a stable Midwesterner who rarely got sick and could cope with just about any physical or mental challenge that was thrown her way.

As she sat in the waiting room of the psychiatrist's office one week later, she picked up *The New Yorker* and leafed through it, furtively glancing over at the secretary, feeling curious about who else might be coming and going through the ominous doors, and wondering how deeply the psychiatrist might try to probe her inner psyche. This was ten times worse than when she had to have her first pelvic exam in order to get birth control pills.

The psychiatrist, Dr. Schein, turned out to be pretty nice. The degrees on the wall indicated that he had gone to medical school in Wisconsin and done a residency in psychiatry at Columbia. He was certified by the American Board of Psychiatry and Neurology, and he was also a Phi Beta Kappa and a member of some other honorary medical fraternity called Alpha Omega Alpha. Those all looked like pretty good signs. He looked to be about 45, with a bit of gray at his temples and a warm smile. Instead of beginning by asking her about her problems, he spent about five minutes making small talk about where she worked, where she was from, and how she liked the Big Apple. That made her feel much more comfortable, and it helped that he seemed to understand the mild culture shock that she had been experiencing. As he began to ask more specific questions about her symptoms, her family history of heart disease, and her current lifestyle, she felt increasingly more comfortable. She was finally able to open up and for the first time explain to another person how she really felt:

"I'm scared to death, I feel like I've lost control of my life, sometimes it feels as if I really am going to die. I can't even predict when these attacks are going to occur, and I'm afraid they are going to ruin my life forever."

Dr. Schein smiled gently and explained that she was having a very well-recognized problem. In fact, so much had been written about it in magazines and newspapers that many people who came into his office with these problems already knew their diagnosis and were requesting a specific treatment. The problem was something called panic disorder. Her attacks were panic attacks, not attacks of cardiac disease. The basic problem was mental rather than physical, although this was obviously a rather false distinction, because the symptoms were really *both* mental and physical. Sometimes panic attacks occur because a person is put in a situation that reminds them of a bad experience in the past. Sometimes they seem to come on out of the blue without apparent explanation. Sometimes, as was the case with Michelle, they come on when a person is in a new and stressful situation and feels somewhat overwhelmed or out of control. At this point, he didn't think they should worry too much about the "whys." They would make cutting down the number and frequency of the panic attacks their main goal, hoping to eliminate them altogether eventually. Lots of new kinds of medications were available that helped most people with panic attacks.

"Does this mean I'm a mental patient?" Michelle asked.

Dr. Schein chuckled.

"Well, yes and no. I *am* a psychiatrist, and my job is to treat people who have *mental* illness, and panic disorder *is* a mental illness. So in that sense, I guess you are a mental patient. But you sure have lots of company. Prominent people like Mike Wallace and Betty Ford and William Styron are 'mental patients.' In the same sense lots and lots of less famous people see psychiatrists and are 'mental patients' who receive therapy or medications for their problems every day. Two or three percent of the people in the United States have your same problem, panic disorder. So there ought to be no stigma attached to having a 'mental illness.' In a sense, the distinction between mental and physical illnesses is arbitrary, as your symptoms are already telling you. They may arise from your mind, but they are expressing themselves in your body, and you and I both know that they are 'totally real' and not 'all in your mind.' Now, try to stop being so hard on yourself. Let's just get you back to normal."

Dr. Schein explained that he could choose from many options for medication treatment of panic disorder. He chose a particular medication with the goal of matching the drug to the person's personality and lifestyle. He thought that a drug called Paxil might be the best bet. This drug was tried-and-true for reducing the frequency of panic attacks, although people with less need to be "in control" sometimes developed a physiological dependency if they took it for a long time or in high doses. He was not very worried that Michelle would have this problem. He would provide a relatively low dose and would titrate it up as needed to prevent or eliminate the attacks. After about six months, if the attacks were gone or very infrequent, he would start tapering it back down again and eventually discontinue it. If this medication did not work, however, he had lots of other options.

When he saw her in another week, they would spend another hour together discussing things she could do in her work setting and other aspects of her life to reduce the risk of precipitating panic attacks or to bring them under control. For openers, right now, he suggested she stop drinking coffee and other drinks that contain caffeine, since they tend to precipitate or worsen panic attacks. But he could also help her figure out some ways to change her responses to stress or ways of thinking about herself that might help control the panic as well. He conveyed a sense of confidence that if they worked together, the panic attacks could either be eliminated or made so infrequent that they would become completely manageable. When Michelle left the office, she already felt so much better that she barely minded having become a "mental patient."

As they discussed at the first appointment, Dr. Schein and Michelle worked together over the next six months, combining medication with learning techniques for adjusting to the many changes that were occurring in her life. The Paxil helped rather dramatically, and so Dr. Schein did not need to try any of the other possibilities available to him.

During those six months and the years thereafter (as she rose higher and higher in the corporate ranks), Michelle figured out that she really did not want to become a sophisticated (and stressed-out) New Yorker after all. She would make a successful career out of being a girl from Michigan who could masquerade as a New Yorker when she chose. It worked well.

What Are Anxiety Disorders?

The anxiety disorders are a group of conditions that share a single theme. In all of them a response that is normal and adaptive has turned into a monster inside us. The monster is "pathological anxiety," which gives us symptoms such as tenseness and fear that are unwanted, excessive, and not right for the situation we are in. This pathological anxiety jumps in and attacks just when we don't want it, or when it makes no sense at all. It may hit us when we have to give an important presentation, crippling our ability to communicate. It may wake us up in the middle of the night, when there is no obvious trigger and we'd prefer to be getting our ZZZs.

Our capacity to feel anxiety and fear is a blessing that can also become a curse.

Anxiety and fear are emotional responses that are built into our brains to help us survive. They kick in at the right time to help us get the juices of excitement and attention going. Michelle did better at her Pfizer job interview because she was mildly anxious, just the right amount to make her brain switch into a higher gear and spin out all those good ideas that eventually helped her land the Pfizer job. A college freshman, free from parental supervision for the first time, may have a ball partying until two weeks before exam time. Then anxiety and fear set in (at least if he and his parents are lucky), and he grabs his books and starts cramming enough information into his revved up brain to make it through finals. If his brain is good enough and his anxiety is set at exactly the right level, he may achieve peak performance and ace all his courses. If he gets too anxious and panics, either while studying or in the middle of an exam, he is in deep trouble. The trick is finding just the right level of anxiety to do well. Too much or too little are both bad. Anxiety disorders occur when our anxiety regulator gets reset at too high a level, and so we experience pathological anxiety rather than adaptive anxiety.

A great deal of research during the past century has illuminated the mechanisms of this pathological resetting.

The story begins early in the twentieth century with the work of Ivan Pavlov. Pavlov, a great Russian psychologist, discovered the process that we now call conditioning. His work founded the behavioral school in psychology, introduced objectivity and experimental rigor into the study of why animals and human beings behave as they do, and laid the foundations for thousands (perhaps even millions) of subsequent experiments, Ph.D. dissertations, and scientific articles.

The essence of his work was to observe how learning occurs and how pairing together two unrelated stimuli could change behavior. His earliest work showed how dogs could "learn" to salivate and secrete digestive juices in their stomachs when they heard a tone instead of when they saw food. He achieved this result by pairing a tone and the presentation of a bowl of food at the same time, so that the dogs learned to associate the two stimuli. Somehow their brains wired them together and saw them as associated. Later, the dogs could then be exposed only to the tone, without any food around, and their stomachs would still secrete gastric juices.

Pavlov didn't know how the brain rewired itself to produce this rather strange response, and he and many subsequent generations of behaviorists didn't much care. They were intrigued with the fact that behavior could actually be changed through the manipulation of the environment.

This process is called conditioning. In a typical conditioning experiment, the presentation of food is called the conditioned stimulus (CS), while the tone is the unconditioned stimulus (UCS). The food is called conditioned because the dog already has previous associations and reactions to food, while the tone is unconditioned because at least initially he doesn't have these preprogrammed responses. The presentation of food (the CS) normally produces a conditioned response (CR). Because learning has occurred, however, a UCS now produces the same CR as the CS. (If this terminology is confusing, don't bother to learn it, as long as you get the point that "unnatural triggers" can produce the same result as "natural triggers" when they are paired together.)

Although Pavlov didn't much care about the whys and wherefores, recent generations of neuroscientists *have* worried about them. We now explain conditioning in terms of brain plasticity, invoking the Hebbian principle that "neurons that fire together wire together." As described in the section on plasticity in chapter 4, the process of learning and conditioning causes new connections to be formed in the brain that link together new pieces of information, create associations, and change the way we think and respond. Our modern understanding of brain plasticity explains how conditioning and learning occur. It helps us understand both how "irrational" pathological anxiety can arise, and also how pathological thoughts and feelings can be reduced or eliminated through psychotherapeutic and cognitive treatments.

Despite the Russian revolution and the First World War, the contributions of Pavlov did not pass unnoticed in the world of psychology and psychiatry. John Watson, widely regarded as the founder of American behaviorism, applied Pavlov's ideas to the study of human anxiety. He demonstrated that the same principles of pairing stimuli to produce learning could be used in human beings and could provoke "mental" responses such as fear, in addition to more "physical" responses such as salivation or gastric secretion. In 1920 Watson and Rayner reported on their work in the famous case of "Little Albert," using the techniques of Pavlovian conditioning. Little Albert, a young boy, was taught to have a pathological fear of rats by experimental conditioning. When Little Albert first saw a rat, he found it interesting and amusing. However, he was frightened and showed an exaggerated startle response when he was exposed to an unexpected loud noise created behind him when Watson and Rayner hit an iron bar. Watson and Rayner then paired together the two stimuli, making the frightening noise at the same time that Little Albert reached out to pet the rat. Soon, they found that they were able to remove the loud noise and observe that Little Albert now responded to the rat with fear and an exaggerated startle response. Essentially, Little Albert had "learned" to have an "animal phobia" that he did not previously experience. Conditioned learning could therefore be used to explain the development of pathological anxiety and fear.

The classic work of Pavlov and Watson has now been extended by the subsequent work of many neuroscientists. We now understand many of the mechanisms that explain how our brains get "broken" so that we develop pathological anxiety and fear.

Walter B. Cannon was perhaps the earliest scientist to give a plausible and lasting explanation for the physiology of fear. Cannon is famous for describing the "fight or flight" response, which is evoked when we are exposed to a frightening or dangerous stimulus. Cannon pioneered the study of the autonomic nervous system, which is the neural regulator of all of our basic bodily functions, such as hunger, thirst, breathing, heart output, or sexual drives. The autonomic nervous system is divided into two parts, called the sympathetic and parasympathetic systems, which work with one another to balance the regulation of our basic survival drives. The sympathetic nervous system works primarily through epinephrine and norepinephrine, more familiar to lay people under the generic term "adrenalin."

Cannon proposed that our bodies respond to fearful or dangerous situations by an outpouring of adrenalin from the autonomic nervous system. This outpouring prepares us to respond in a way that will help us survive. Neuroscience textbooks usually illustrate Cannon's point by using examples of dangerous stimuli that few of us ever see, locked as we are in urban environments, such as a snarling bear or a rattlesnake.

Imagine instead how your body would respond if the relative tranquility of your Sunday outing to the local shopping mall was interrupted by the appearance of a masked terrorist, who was shouting, waving an automatic weapon, and obviously about to begin firing. Your eyes would dilate, and your heart would begin to pound. You would be preparing to run away and hide, or (if unusually brave or well-trained) to formulate a plan to disarm the terrorist. The outpouring of adrenalin tunes up our bodies to readiness. Our eyes dilate so that we can see more, and our hearts beat faster so that more blood can be shunted to the right parts of our body for reacting appropriately. Specifically, noradrenalin sends blood to muscles and increases blood flow in the brain. Seeing the terrorist sets up a series of reactions. It begins with the recognition of danger, which leads to feelings of fear and anxiety, which are accompanied by neurochemical outpourings that help us run away or fight back.

Cannon also conducted research demonstrating that the regulation of the autonomic nervous system occurred in the hypothalamus, the small brain region just below the

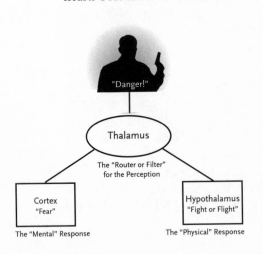

Figure 11–1: Walter B. Cannon: The Emergency Reaction to Danger

thalamus. He described it as a master regulator of the various bodily and appetitive functions, which controlled the production of norepinephrine. He was also aware that the thalamus, the filter in our brains, functions as a router that sends the sensory perception of the terrorist both to the cerebral cortex (where the feeling of fear or anxiety was experienced) and to the hypothalamus (where the autonomic arousal was begun). For the first time, a brain circuit was described that could explain the physical and psychological components of anxiety. This circuit is shown in Figure 11–1.

The next major contribution to the understanding of anxiety and fear came from James Papez, a professor at Cornell University who reintroduced the concept of the limbic system, which had been previously named and identified in the nineteenth century by the French physician Paul Broca. The concept of the limbic system, briefly described in chapter 4, has been pivotal in the understanding of mental functions and mental illnesses. In a striking imaginative leap, Papez proposed the idea of a reverberating circuit in the brain, where different components of emotions interacted and communicated with one another. Papez went beyond Cannon's formulation that stimulus perception runs through the thalamus to the cortex or the hypothalamus. He added other crucial modules to the overall process of emotional response, such as the cingulate gyrus, the hippocampus, the fornix, and the mammillary bodies. In the Papez model, stimuli perceived by the sensory cortex were funneled back down to the more primitive cortex of the cingulate gyrus and the hippocampus. Papez knew that these structures were connected to the mammillary bodies (a subregion of the hypothalamus that looks like a

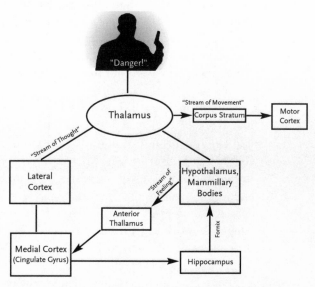

Figure 11–2: James Papez: Emotion as a Product of the Limbic Circuit. In the "Papez limbic circuit," the hypothalamus, anterior thalamus, cingulate gyrus, and hippocampus interact to create a reverberating circuit in medial portions of the brain that process and produce emotions.

pair of breasts) by way of the hypothalamus, which then communicated back to the thalamus, and in turn back to the cingulate gyrus. Long before the functions of the hippocampus and its adjacent region the amygdala were well understood, Papez hypothesized that emotion tended to build up in these regions, while the cingulate gyrus helped form the connections that created awareness of the emotions. He described three streams that were involved in emotion, which are shown in Figure 11–2, and which were embodied in the interconnections between the multiple brain regions of the limbic system.

The seminal work of Papez was subsequently supplemented by the ideas of Paul MacLean, Walle Nauta, and Mortimer Mishkin. Nauta, a Boston neuroanatomist who also contributed to our understanding of the prefrontal cortex, proposed that the limbic system should be defined based on the various structures that connect it to the hypothalamus. He gave anatomic reality to the Papez circuit by pointing out that the inter-connections between the hypothalamus, the amygdala, the hippocampus, and the cingulate gyrus are all reciprocal and therefore can, in fact, pro-duce the ongoing feedback originally implied by Papez. In Nauta's view, the hypothalamus collects the more primitive "visceral" sensory informa-tion, but input also comes to the limbic circuit through "higher" cortical associations such as the prefrontal cortex or the inferior temporal lobes. Mishkin added the pivotal observation that the amygdala played a key role in the perception of danger and the experience of fear, and that the hippocampus played a separate but related role by connecting meaning with perception. With this integration the modern concepts of the limbic system were nearly complete.

The final touch in understanding the limbic system has returned to the work of Pavlov and explored how the autonomic nervous system "learns" to run out of control. This work, to which Joseph LeDoux of New York University and Michael Davis of Yale have been major con-tributors, has laid down foundations for understanding how pathological anxiety might arise in human beings and produce the broad range of anxiety disorders. They have figured out how the hippocampus and the amygdala work together to produce built-in memories that may either help us or come back to haunt us, sometimes without our even being aware that this is happening.

They have used a variety of different techniques from neuroscience, which include the lesion technique (i.e., finding out what deficits occur when specific brain regions are injured), mapping neuronal pathways using tracers, stimulating particular brain regions and observing the behavioral responses, and examining how different kinds of stimuli affect learning.

The central theme that runs through all of these experiments is a paradigm known as fear conditioning. Fear conditioning harks back to Pavlov and Watson. Animals, particularly rats, are studied in a controlled experimental setting. In a typical experiment, the rat would first be exposed to a sound and then later receive a paired stimulus: the sound and an electrical shock are given at the same time. The shock produces a typical fear response in the rat, which is not unlike what many of us might do when the terrorist waives his gun at the shopping mall. If a rat is running through a field and sees a dangerous predator, such as a large bird flying overhead, its natural response is to dive for cover and freeze. The experimental rat in the cage is doing essentially the same thing. Measurements of sympathetic arousal, such as elevations in blood pressure, indicate that the freezing response in the rat is mediated by sympathetic autonomic arousal. Like Pavlov's dogs or Little Albert, the rats rather quickly "learn" that the tone signals danger. Soon they learn to freeze in response to the tone alone. In fact, rats do not even need to hear the tone. They quickly learn to freeze as soon as they are put in the conditioning box. This is called contextual conditioning, because the context alone evokes the fear response.

The fear conditioning paradigm permits neuroscientists to examine brain functions and responses that are common to most species. Rats, dogs, monkeys, and human beings all learn to avoid danger through fear conditioning. For us human beings, fear conditioning is probably the mechanism that explains anxiety disorders. Contextual conditioning may explain why locations, smells, and other cues can evoke fear, even when they are not obviously dangerous. At an unconscious level the memory associations in our hippocampus/amygdala recognize that something unpleasant or dangerous previously occurred in a particular setting, even though we are not consciously aware of the connection between the context and our response.

The fear conditioning experiments have made it clear that the "almond" of the brain, the amygdala, plays a critical role in helping us respond with fear and anxiety. In fact, LeDoux calls the amygdala the "hub in the wheel of fear." His model for the role of the amygdala in fear is shown in Figure 11–3.

The key evidence for this model is the fact that damage to the amygdala interferes with the ability to be conditioned to learn a fear response, based on lesion studies in many different species. In this model, the thala-

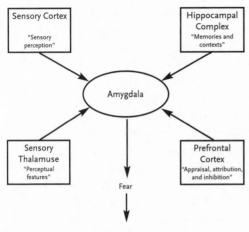

Figure 11–3: Joseph LeDoux: The Amygdala as "The Hub in the Wheel of Fear." The "emergency reaction" to fear includes dilated eyes, increased heart rate and blood pressure, outpouring of adrenalin and cortisol, startle reflex, freezing, fleeing, and fighting.

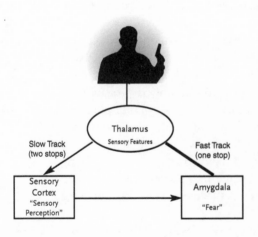

Figure 11–4: Conditioned Fear: Fast and Slow Tracks to the Amygdala. The fast track permits a "quick and dirty" response that may turn out to be an overreaction. The slow track permits the cortex to assess the stimulus more accurately and to revise the response if appropriate.

mus and the sensory cortex are also important, much as in the earlier models of Cannon and Papez. The perceptions flow in through the thalamus, which has a direct pathway to the amygdala. This pathway bypasses the cortex and is used to respond very rapidly to dangerous stimuli. The sensory thalamus also has input to the sensory cortex, which forwards information on to the amygdala by "snail mail," allowing for more refined and complex assessment of the nature of the stimulus. These two tracks, the "fast track" and the "slow track," are shown in Figure 11–4.

An example of these two mechanisms in human beings is when we round a corner and nearly bump into someone, pulling back in a startle reaction (the fast route). We then recognize that we have bumped into an old friend after we look at the face, leading us to smile and hug the person (the slow but more accurate pathway). The amygdala also makes its appraisal of a situation by referring to other brain regions. It must reference stimuli against memories, using several regions that are conveniently located very nearby: the hippocampus, the parahippocampus, and the entorhinal cortex. These regions contain information about the associations of the stimulus. For our rat who was placed in the conditioning cage, the hippocampus contributes the contextual information that says, "Oh oh, it's that cage again. That's the place where I used to get those shocks, and where I now hear those tones, and where I could get shocked again." Of course, the rat probably does not have this conscious awareness of what is going on, nor does the rat have language to articulate such formalized ideas. But the rat hippocampus does contain the information about the context in which the danger occurred. In human beings, it appears that the amygdala and hippocampus are used for different kinds of emotional memory. The amygdala is used for "fast reaction" unconscious responses, such as the startle response, while the hippocampus is what kicks in when we recognize our friend's face after rounding the corner and say, "Gosh, its Bill. It's been weeks since I saw him."

Most of the basic research on the neural basis of fear and anxiety has been with laboratory animals. The fear/anxiety pathways are among the phylogenetically oldest and appear to be present across all vertebrate species. Nevertheless, readers are probably saying to themselves, "What does this have to do with me? There *is* a difference between mice and men." And of course there is.

Fortunately, because of the power of modern neuroimaging techniques, we can also study the human brain and measure its blood flow, thereby mapping the circuitry of fear and anxiety in the human brain in vivo. In PET studies conducted in Iowa, we have visualized how the normal brain responds to unpleasant fear-evoking stimuli. In these PET studies, a group of healthy volunteers were shown a standardized set of pictures, at the rate of one every two seconds, which portrayed images that most of us would find frightening, anxiety-provoking, or unpleasant. Examples are mutilated human bodies, a deceased and severely emaciated African child who had starved to death, and a fly crawling on an open and bloody wound. We compared the brain's response to these unpleasant images to the response produced by viewing pleasant images, such as a couple looking at each other with a loving expression. This comparison produced a very interesting result, which is shown in Figure 11–5 (placed in the color section in the middle of the book).

The unpleasant fear-engendering stimuli produced increases in blood flow in multiple brain regions, most of which are components of the well-recognized fear pathways in animals. Activations were seen in the amygdala, hippocampal complex, thalamus, anterior cingulate, and inferior frontal regions. These regions represent the classic primitive "limbic brain." Interestingly, the pleasant images activated very different brain regions, nearly all of which were in the cerebral cortex—that brain area that is substantially enlarged in human beings as compared to other creatures. This seems to imply that it takes the "higher" parts of our brain to recognize pleasure, joy, and happiness. Unfortunately, fear and anxiety seem to come easier to us, having been built into the deepest and most basic parts of our brains.

Other scientists have examined the neurochemical basis of fear and anxiety. Cannon's early work on the "emergency reaction" suggested that adrenalin is the primary neurotransmitter that produces the "fight or flight" response. Adrenalin is present only in the peripheral nervous system (i.e., the nerves that are located outside our brain and communicate with our muscles and visceral organs such as our heart, lungs, and digestive system). Adrenalin is secreted from the central part of the adrenal glands in response to stress, and it is one source of the various manifestations of the emergency reaction, such as a pounding heart. But if adrenalin does not reach the brain, what explains the "mental" components of fear and anxiety? Adrenalin affects the brain by stimulating a small island of gray matter in the base of the brain (called the brain stem), which forwards a message on to the locus ceruleus, which is the site where the brain cells that produce the "twin" of adrenalin reside. This brain-based twin is called noradrenalin (also called norepinephrine). (Locus ceruleus means "skyblue area," because the cells in this region contain a beautiful pale blue pigment.) The locus ceruleus releases noradrenalin by direct pathways to both the amygdala and the hippocampus. In this way, the limbic brain and the "physical body" are coactivated at the same time during the stress response.

Adrenalin and noradrenalin are not the only neurotransmitters involved in the stress response, however. While the inner part of the adrenals is sending out adrenalin, other parts are also activated by stress and send out the neuroendocrine transmitter, cortisol, which has been discussed in more detail in chapter 9 ("Mood Disorders"). Produced in proper amounts and at the right time, cortisol also helps us adapt to stressful situations. Cortisol has effects in many parts of the body. In order to facilitate response to stress it regulates the sleep/wake cycle, mental arousal, and the immune system. The regulation of cortisol production is mediated through feedback loops that have already been shown in chapter 9 and that try to create exactly the right balance between "too little" and "too much" cortisol.

Not discussed in chapter 9, however, is the fact that both the hippocampus and the amygdala are filled with cortisol receptors that participate in the feedback regulation of stress. These two adjacent and complementary brain regions play different roles in anxiety and survival mechanisms. The amygdala is our "fast track" responder, which causes us to react without "thinking." As discussed in the memory section of chapter 4, this type of memory has been called "implicit" or "unconscious" memory. The hippocampus, although also a relatively "primitive" region, is somewhat more sophisticated and might be considered another piece of the "slow track" system. As we have seen, the hippocampus is responsible for associating contexts and perceived stimuli. It plays a role in what has been called explicit or conscious memory—the kind that we are aware of when we intentionally retrieve or recall memories. As discussed in chapter 4, it also plays a role in long-term consolidation of memories through the mechanism of long-term potentiation. These two complementary brain regions also have somewhat different roles in the regulation of cortisol. The amygdala stimulates cortisol release, while the hippocampus suppresses it.

Bruce McEwen, a neuroscientist at Rockefeller University, has contributed pivotal work to our understanding of the effects of cortisol on the hippocampus and its relationship to stress and memory. The amygdala is relatively "tough." It does not seem to be damaged when chronic cortisol stimulation occurs under stressful conditions. But the hippocampus is more sensitive and vulnerable. McEwen has shown that chronic stress eventually causes the neurons in the hippocampus to lose dendrites and spines, so that the hippocampus actually shrinks. This was originally observed in animal studies, but it has now also been confirmed in the human brain using structural MR imaging. An example of a normal hippocampus and a very small hippocampus, shrunken by stress, are shown in Figure 11–6.

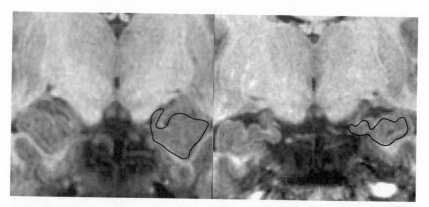

Figure 11–6: Examples of Normal and Abnormal Hippocampal Size
The hippocampus on the left is normal, while the one on the right is smaller, as a consequence of repeated stress.

These observations have important implications for our understanding of the interrelationships between memory, stress, and anxiety. The precise relationships are still being worked out, as we learn more and more about unconscious memory, conscious memory, and the effects of cortisol on memory encoding, consolidation, and storage. Ultimately, as we understand how this network of brain regions works, we will achieve a synthesis between psychoanalytic theories of the unconscious and the conditioning theories of behaviorism, which are probably not as far apart as people once believed. Our questions about the nature and role of unconscious memories, forgotten memories, false memories, flashbulb memories, and intrusive unwanted memories will probably all be explained during the next several decades, as we understand more and more about how the hippocampus and amygdala work with one another and with other sites distributed throughout the brain.

What does all this have to do with the development of anxiety disorders? How does it explain the fact that Michelle began to develop her unbearable attacks of panic soon after she began her job at Pfizer? Is there really a connection between fear conditioning in rats, the conditioning of Little Albert, and the millions of people who suffer from anxiety disorders?

Very likely, there is a strong connection.

Michelle was already stressed out on that morning when she descended into the subway and had her first panic attack. No doubt the process of moving to New York "tuned up" her sympathetic nervous system a bit and made her more anxious than she would have been back home in Michigan. For some reason, going deep into the bowels of the subway system that particular day evoked associations that made her brain say "Danger!" It had to be some association specific to her, since lots of other people were in the subway along with her and were not at all bothered. As a "newbie" to the subway system in New York, and a Midwesterner accustomed to wide open spaces, she was no doubt more vulnerable to being frightened by the thoughts that occurred to her, such as "being trapped in the sewers of Paris like Jean Valjean," or the natural fear of being buried alive. Perhaps she had recently read about a subway accident or a train crash. Perhaps she had recently seen *Les Miserables*. Perhaps, sometime in her early childhood, she had had a frightening experience because she was closed in somewhere—shut in a closet, trapped in an elevator, or hemmed in on an overcrowded bus.

The individual contextual triggers that Michelle was carrying in her hippocampus could probably be recovered if a therapist searched long and hard enough. Whatever they were, they combined with her tuned-up autonomic nervous system and the unpleasant stimulus of the crowded subway to produce the same basic fear or panic attack as the field mouse confronting a predator, the rat in the conditioning cage, or one of us staring at the gun-waving terrorist in the shopping mall. Once the panic attack occurred, she was conditioned. Having a second attack in the same situation the next day was not inevitable, but it was somewhere between possible and probable. If she had been a "mellow Melissa" or a "cool Christine," maybe she would have escaped the recurrent cycle. But she was who she was, with her network of memories and her intense, aspiring, upwardly mobile style. So she was a prime candidate for both the success that permitted her to land a job to begin with and the development of recurrent panic attacks. Although Michelle did not know of any family history of panic disorder, the tendency to be anxious does run in families, and so she might also have had a genetic predisposition lurking in the background that she did not know about. Perhaps she increased her vulnerability on that particular morning by drinking a second cup of coffee instead of her usual one. Very likely, Michelle's particular response on that particular morning was the unfortunate result of the convergence of multiple factors that "tipped her over the edge."

Fortunately, Michelle managed to get her life back together very quickly, helped by both medications and psychotherapy. How do the medications used to treat anxiety disorders actually work? And how and why does psychotherapy also work? Those are very interesting questions.

The medications used to treat anxiety disorders are described at the end of this chapter. The most striking thing about them is that many different kinds of medication are helpful. The common theme to all of them is that they all affect neurochemical systems in the brain—a fact that is hardly surprising. What is surprising is the diversity of the neurochemical systems. The older antianxiety agents, the benzodiazapines, largely affect the inhibitory GABA system of the brain. Presumably they tone down the effects of the overactive noradrenergic system, which arises in the locus ceruleus and projects diffusely all over the brain, as was shown in chapter 4. There are direct noradrenergic projections to both the amygdala and hippocampus. Reducing input to those regions may have a specific effect on the conditioned fear response produced by the amygdala or on the evocation of contextual memories in the hippocampus. Alternately, GABA may reduce anticipatory anxiety through direct effects on noradrenergic activity throughout the cortex.

The selective serotonin reuptake inhibitors (SSRIs) have joined the therapeutic armamentarium against anxiety during the past several years. These medications target the serotonin system and increase serotonergic tone. Since they are frequently used as antidepressants, it may appear unlikely that they would also be effective for anxiety. However, evidence from multiple sources suggests that treatment with SSRIs over several weeks to months blocks noradrenergic activity in both direct and indirect ways. Noradrenalin metabolites have been measured in patients who have received twelve weeks of treatment and have been found to be reduced. Long-term treatment with SSRIs also appears to reduce adrenal production of cortisol, blocking it at the level of the hypothalamus. Further, the serotonergic neurons also have direct projections to the amygdala, where they probably block the excitatory inputs from both the sensory cortex and the thalamus; this may be the primary mechanism by which they reduce anxiety.

More interesting, however, may be the question of how various kinds of psychotherapy, behavioral therapy, and cognitive therapy also are effective for treating anxiety. Here too, our growing knowledge of the neurobiology of anxiety mechanisms provides useful clues. We now know that the production of anxiety has both subcortical (amygdala) and cortical components. Very likely, the medications work by targeting the subcortical components or neurochemical aspects of the cortical components.

The various "mental" therapies probably target the cortical components and the memory systems of the hippocampus. Behavioral therapies, such as systematic desensitization, produce what Pavlov would call a "deconditioning." Our improved understanding of learning and memory suggests that this deconditioning occurs by retraining hippocampal neurons to reorganize the contextual cues that they store, so that they are no longer associated with a danger signal and no longer produce the fear conditioning response. Psychotherapy, in which the contextual memories are explored, may attack the same problem from the cortical level. Multiple regions in the prefrontal cortex have projections to the hippocampus, the amygdala, and the thalamus. Although we do not, as yet, fully understand how these "higher cortical" projections are manipulated through psychotherapy to modify the fear response and reduce excessive anxiety, research over the next decade is likely to clarify these mechanisms.

Kinds of Anxiety Disorders

Several different disorders share the common feature of having the adaptive anxiety response lose its control over the accelerator so that people feel as if they are in a vehicle that is careening down the street out of control. Each of these differs somewhat in triggering mechanisms, symptoms, and subjective experiences, but all share the trait of pathological anxiety. These disorders include panic disorder, phobic disorder, posttraumatic stress disorder, generalized anxiety disorder, and obsessive-compulsive disorder.

Panic Disorder

Michelle's mental illness was a textbook case of panic disorder. The essence of panic disorder is having "panic attacks." A panic attack is an abnormal version of the intense fear reaction that we all feel when we encounter something dangerous, such as a person waving a gun nearby, a predatory wild animal, or a car ride down a mountain with a reckless driver. Although people with panic attacks may have some kind of unconscious trigger lurking in the background, the actual panic attacks usually come on without exposure to a conscious trigger.

Panic attacks have both physical and psychological elements, and the relative balance of these two kinds of elements may differ from one person to another. The first time a panic attack occurs, the person immediately recognizes that something is wrong and becomes alarmed. The physical components of a panic attack include shortness of breath, a feeling of suffocation, pounding heart, chest pain, queasiness in the stomach, and tremulousness. When these physical feelings predominate, the person may think she is having a heart attack and may go see a doctor or visit the emergency room. Various laboratory tests may be obtained, such as an electrocardiogram, and these are usually found to be normal. Frequently, the person is then reassured that it was just "an attack of nerves," with the implication that the problem "amounts to nothing." If the panic attacks recur, however, the victim gradually concludes that the problem definitely "amounts to something" and eventually seeks psychiatric help.

If the mental components are more prominent at the outset, then psychiatric help is often sought immediately or after an obvious pattern of recurrent attacks. For people who have prominent mental manifestations, the core symptom of psychological panic is particularly intense, and they subjectively feel stark terror. In addition, they may have a sense that they are losing control or going crazy, that they are somehow cut off from reality, or even that they are about to die. Table 11–1 lists the various kinds of symptoms that people experience during a panic attack, divided according to whether they are more physical or more mental.

Almost everyone reading this book has had one or two of these symptoms or perhaps even one or two experiences of an actual panic attack.

TABLE 11–1

Predominantly "Mental" Symptoms	Predominantly "Physical" Symptoms
General sense of intense fear or discomfort	Chest pain or discomfort
Feeling disconnected from reality or detached from one's self	Pounding heart
	Shortness of breath or smothering sensation
Fear of losing control or going crazy	Feeling of choking
Fear of dying	Sweating
	Tremulousness
	Dizziness or faintness
	Nausea or stomach distress
	Numbness or tingling
	Chills or hot flashes

We all "get nervous" sometimes and feel our heart pound or perhaps become short of breath. Some of us can recall an experience when we had an episode of panic that was self-contained and never recurred—a time when we were unexpectedly called on in class and could not think of the answer, when we had to get up and give a talk, or when an airplane made a particularly turbulent landing. These normal blips of nervousness turn into panic disorder when enough symptoms are present and when the panic attacks recur on multiple occasions and become troubling to the person who is experiencing them.

The definition in DSM requires the occurrence of the subjective experience of intense fear in combination with at least four other symptoms listed in Table 11–1. As a rule of thumb, these panic attacks must also occur more than once or twice and must be subjectively troubling. Sometimes people do not want to recognize that their problem is "mental," and they go from one specialist to another, expecting to eventually get to the root of the problem. They may first see a cardiologist for their chest pain, going on to a pulmonologist for their shortness of breath and smothering sensation, then turning to a gynecologist for help with hot flashes or visiting a gastroenterologist for their nausea and abdominal pain. However, these specialists usually are not able to find the "butterflies" in the stomach or the heart attack in the chest. Fortunately, panic attacks often improve significantly when the correct diagnosis is made and treatment provided.

Sometimes people with panic attacks have a disabling complication known as agoraphobia. "Agoraphobia" literally means "fear of the marketplace" (agora = marketplace, phobia = fear, in Greek). This complication is an extreme human form of the freezing reaction in lower animals. Like the mouse who fears going out into an open field because this will make him the easy victim for a predator, the human being is fearful of losing his or her connection to a safe home base where security is ensured. People who have panic attacks with agoraphobia become fearful of having a panic attack in a public place where they will be out of control. Therefore, they tend to avoid crowded places, such as shopping malls, movie theaters, or even church. They may be afraid to go on airplane trips, to take long journeys in a car, or to travel through tunnels or across bridges. If agoraphobia becomes severe, a person can become totally housebound.

Phobic Disorder

Everyone has heard of "having a phobia," and nearly everyone has one. In fact, as I am fond of saying, everyone is entitled to have one or two phobias and to still consider oneself "normal." Table 11–2 lists some common

TABLE 11–2
Common Specific Phobias by Type

Phobia	Focus of Fear	Phobia	Focus of Fear
Animal type		**Blood, injection, injury type**	
Ailurophobia	Cats	Hemophobia	Blood
Arachnophobia	Spiders	Odynephobia	Pain
Cynophobia	Dogs	Poinephobia	Punishment
Entomophobia	Insects		
Ophidiophobia	Snakes	**Situation type**	
		Apeirophobia	Infinity
		Claustrophobia	Closed Spaces
Natural environment type		Topophobia	Stage fright
Acrophobia	Heights		
Amathophobia	Dust	**Other**	
Frigophobia	Cold weather	Gynephobia	Women
Keraunophobia	Thunder	Homophobia	Homosexuals
Nyctophobia	Night	Kakorrhaphiophobia	Failure
Phonophobia	Loud noises	Logophobia	Words
Photophobia	Light	Theophobia	God
Pyrophobia	Fire	Triskaidekaphobia	Number 13

phobias. Some people find it amusing to learn these names and impress their friends with the extent of their psychiatric vocabulary and (implied) psychiatric knowledge. We all have heard of claustrophobia, and perhaps even have it, and we may even know about acrophobia or homophobia, but how many of us know that our fear of spiders is called arachnophobia or our fear of the number thirteen triskaidekaphobia? People who build hotels and office buildings know all about triskaidekaphobia, even though they do not know its name, since one almost never encounters an elevator that goes to the thirteenth floor!

Apart from agoraphobia, which has already been described, phobic disorders are usually divided into two broad categories: social phobias and specific phobias. Social phobias are fears of doing things in front of other people that might lead to humiliation or embarrassment. Specific phobias are fears of specific things, such as cats or dogs. People with social phobias are afraid of various kinds of social situations, such as speaking in public, using public restrooms, making phone calls, eating in front of other people, or writing or performing calculations in front of others. Sometimes the social phobias are very extensive, so that a person avoids almost all social situations and becomes a recluse because of his or her various fears.

Phobias are very common. It has been estimated that 3–5% of the population have social phobias, while the various specific phobias may affect up to 25% of us. Since some fears, such as fear of snakes or spiders, are socially adaptive, at what point should having a phobia be considered to be a mental illness? The answer is pretty obvious: when it gets in the way of something that the person wants or needs to do. Being afraid of snakes is rarely going to interfere with a person's life. At worst, people who fear snakes may avoid going to zoos, or simply avoid that part of the zoo, perhaps to the disappointment of their children. On the other hand, cat lovers or dog lovers may be somewhat unsympathetic if a friend refuses an invitation to dinner out of fear of Fluffy or Fido.

People need and seek psychiatric treatment for their phobias when their phobias interfere with their daily lives. A travel agent can hardly afford to have a fear of flying or staying on the twentieth floor in a hotel. An aspiring teacher or executive needs to be able to speak in public, make telephone calls, and go out to lunch. Public restrooms can be pretty hard to avoid. People who have phobias so severe that they need to seek treatment for them comprise about 2% of the population. These people are often truly handicapped. Their fear is excessive or unreasonable, and it interferes significantly with their normal routine, their ability to function at work or school, or their social activities.

Posttraumatic Stress Disorder (PTSD)

Posttraumatic stress disorder (PTSD) was one of the earliest types of anxiety disorder to be medically recognized. A physician named J. M. DaCosta defined a condition that he called "irritable heart" in the *American Journal of Medical Sciences* in 1871. He described the syndrome in a soldier who developed it during the Civil War and came to him because he was experiencing chest pain, rapid and irregular heartbeats, and dizziness. DaCosta astutely recognized that the symptoms were not due to a "real" cardiac problem, but instead were characterized by hypersensitivity and overreactivity in the general nervous system. In literary form, Steven Crane provided a similar description in the *Red Badge of Courage*. Because this condition can be triggered by the stress of combat, it has also been referred to as soldier's heart. The terrible trauma of trench warfare during the First World War led to its description under other names, such as shell shock, battle fatigue, or traumatic war neurosis. The Second World War introduced new kinds of inhuman behavior to which people could be exposed, such as POW camps, various forms of torture, Hiroshima, and Dachau. Physicians also began to recognize the commonality of reactions to stress encountered in war and in civilian disasters, such as earthquakes, plane crashes, large-scale fires in buildings, or collapsing hotels. The media has portrayed these frightening situations in films such as *Sophie's Choice*, *The Towering Inferno*, and *Titanic*.

The decision to create a handbook summarizing psychiatric diagnoses is closely linked to the occurrence of the Second World War. For the first time, that event brought together psychiatrists from all over the United States with varied training backgrounds, who had to evaluate and diagnose military troops who themselves came from all over, with varying ethnic and psychosocial backgrounds. The need for consensus was obvious, although it was not instantly implemented. Many veterans who were not career soldiers were exposed to months or years of combat. Some developed what came to be called combat neurosis or traumatic war neurosis. In fact, several books were written on this topic after the Second World War.

During the postwar era, the Veteran's Administration developed a handbook of diagnoses that was used to assess veterans in the newly developing and extensive VA hospital system. This led in turn to a decision by the American Psychiatric Association to create the *Diagnostic and Statistical Manual* (DSM), which provided a somewhat more comprehensive summary of diagnoses used by psychiatrists.

One of the diagnoses was gross stress reaction, which was used for people who had experienced either combat or civilian catastrophes. This diagnosis was defined as follows:

> Under conditions of great or unusual stress, a normal personality may utilize established patterns of reaction to deal with overwhelming fear. The patterns of such reactions differ from those of neurosis or psychosis chiefly with respect to clinical history, reversibility of reaction, and its transient character. When promptly and adequately treated, the condition may clear rapidly. It is also possible that the condition may progress to one of the neurotic reactions. . . . This diagnosis is justified only in situations in which the individual has been exposed to severe physical demands or extreme emotional stress, such as in combat or in civilian catastrophe (fire, earthquake, explosion, etc.). In many instances this diagnosis applies to previously more or less "normal" persons who have experienced intolerable stress.

Inexplicably, the concept of gross stress reaction was dropped in the second edition of DSM (DSM II), although many clinicians were still familiar with the term and continued to use it. The third edition (DSM III) was begun in the early 1970s, shortly after the Vietnam War. Like World War I and World War II, that war had taken its toll on many young men. Psychiatrists turned to their official manual to find a diagnosis that might apply and found none. Some veterans wondered if this might not reflect yet another failure on the part of the government and the public to acknowledge the trauma and sacrifice of Vietnam veterans. The diagnosis of "gross stress reaction" versus "post-Vietnam syndrome" was, at least temporarily, a political hot potato.

I was a member of the original group of twelve people who formed the Task Force that created DSM III. The chairman, Bob Spitzer, asked me if I would assume responsibility for deciding whether the diagnosis of gross stress reaction or post-Vietnam syndrome should be included in the new manual. This was a "natural" for me, since I devoted much of my early psychiatric career to studying this very topic. My patients, however, were not veterans, but rather people who had sustained one of the most terrible injuries possible—painful and disfiguring burns. At the invitation of a surgeon who was also a close friend, Ed Hartford, I had made daily rounds on our burn unit for two years, listening to and consoling patients and their families.

Although my work with burn patients occurred more than twenty years ago, many of the people are still as vivid as sunsets. There was the 48–year-old electrical lineman who lost both his right arm and leg and confronted this loss with stoic bravery over a hospitalization that lasted three months. When I was helping him prepare for discharge and asked how he would negotiate the stairs in his house, he replied with a wry smile, "Oh, I guess I'll just have to slither up and down." There was the 25–year-old beautician who, unaware of the danger, attempted to light her charcoal grill by using gasoline as a starter. The explosion burned away much of her face, which she had to keep covered with a scarf for the next three or four years as she underwent repeated plastic surgeries to loosen the scar tissue that made her face look as if it had melted. There was the 28–year-old woman who had just moved into her new home with her husband, two-year-old daughter, and five-year-old son. Unknown to anyone, the natural gas company that was providing service to the new subdivision had mistakenly permitted a leak to occur in the lines. When her husband struck a match to light a cigarette, the entire house exploded in flames, killing him and the son, and leaving mother and daughter with disfiguring facial scars and severe burns on the rest of their bodies, including damage to their hands that made them nearly unusable. As she recovered, the mother had to endure not only her own pain but also that of observing the pain of her little daughter. There was the mother who also sustained severe disfiguring burns during a house fire, when she ran upstairs to rescue her sleeping baby. The effort was fruitless, since the baby had already died of smoke inhalation. For the rest of her life, the mother wore the badge of her bravery in her facial burns.

Because burn injuries require prolonged hospitalization, usually lasting two to four months, I got to know all of my patients very well, and I learned a great deal about how people respond to emotional and physical stress. Burn injuries are excruciatingly painful. They require frequent dressing changes, debridement, and surgery. These physical stresses are combined with the psychological stress of disfigurement and loss of function, and many of my patients had also experienced the death of a loved one or the burden of observing his or her suffering as well.

My patients almost universally shared certain symptoms. They would frequently relive the catastrophe, suffer recurrent nightmares, have an exaggerated startle response, or retreat from their experience and show "psychic numbing." Such symptoms occurred in almost every patient on the burn unit during those months of hospitalization, usually beginning when the person emerged from delirium after the first week or two. The posttraumatic symptoms typically continued throughout most of the hospitalization.

Because I was curious about how people who experience such terrible trauma adapt over the long run, I also did a concurrent follow-up study of people who had experienced burn injuries between two and ten years earlier. Some, although not all, continued to have a similar pattern of symptoms that persisted for years after their initial injury. In general, people who had problems before injury (e.g., learning disabilities, alcohol abuse, depression) were more vulnerable to having a persistent "stress reaction." Many patients had adjusted remarkably well to their previous trauma and their current disfigurement, having received loving support from community and family. They were grateful just to be alive, and they learned to find pleasure in the small things in life or to use religious faith to give meaning to their existence. I was in my late twenties to early thirties when I did this research, which I subsequently published in the *New England Journal of Medicine* and *Annals of Surgery*. As a young doctor and a young psychiatrist, my experience caring for these burn patients provided unforgettable lessons about human suffering, as well as human dignity.

Needless to say, when the group lobbying to reinstate the diagnosis of gross stress reaction, perhaps to be renamed as "post-Vietnam syndrome," approached me, they found a sympathetic ear and a person who knew exactly what they were talking about. Although combat and burn injury are different kinds of trauma, both lead to a final common pathway that is nearly identical, no matter what the trauma. Further, a substantial medical literature had already described a similar response in victims of other traumatic experiences such as torture endured during the Holocaust or at POW camps, or civilian catastrophes such as the Coconut Grove fire. I pointed out that it did not make sense to tie the name of this syndrome to a single type of trauma, and certainly not to a single war. Instead, I suggested we use the simple descriptive name "posttraumatic stress disorder." Working with representatives of the Vietnam veterans, I developed the set of diagnostic criteria that defined what came to be known as PTSD. I wrote an extensive description of this "new" yet old disorder in the text of DSM III.

If we think about mental illnesses as being due to a range of genetic and environmental factors, then PTSD is one of the most extreme examples of the "environmental end" of the continuum. People who develop PTSD would not have this disorder if they had not been exposed to some type of terrible stress. One of the biggest issues that we faced back in the 1970s, when the disorder was being officially defined again with specific diagnostic criteria, was to determine how severe that stressor had to be. The DSM III definition required that the stressor be "outside the range of normal human experience." A later revision of DSM broadened the definition of the stressor to include experiences that involved "actual or threatened death or serious injury, or a threat to the physical integrity of self or others." The broadening of the definition has made PTSD a somewhat more common diagnosis. PTSD is defined by three general groups of symptoms: (1) re-experiencing the original traumatic event in various ways such as intrusive memories, nightmares, or flashbacks; (2) avoiding situations that might evoke memories of the trauma or numbing of responsiveness, such as being unable to remember the event or having a restricted range of emotion; and (3) persistent symptoms of increased arousal (due to autonomic overactivity, such as hypervigilance, exaggerated startle response, or trouble falling or staying asleep).

Why do some people develop PTSD, while others do not? Not everyone who went to Vietnam came back with PTSD. Although 100% of my burn patients had acute PTSD during their first three months of hospitalization, only about 30% of them had it two to five years after the initial trauma. What causes some people to have a persistent stress reaction, while others are able to somehow recover and forget?

The answer is not simple. Three different factors seem to be relevant: the severity of the stressor, the personal and emotional resources of the victim, and the amount of psychological support received after the injury. The original definition of gross stress reaction suggested that most people who develop this syndrome are "normal." My own experience with burn patients made it clear that they were not all so lucky. Some experienced burn injuries because they had epilepsy and accidentally caused a fire during a seizure. Some were alcohol abusers and sustained a burn injury because they fell asleep smoking in bed after drinking excessively. These less healthy people tended to be the ones who had more difficulty with PTSD during the years after their initial trauma. Young children had a harder time than adults, who had already developed a sense of who they were and a set of coping mechanisms. Other things being equal, however, the more severe the stressor, the more likely the chances of developing PTSD. Even the strongest bone will break if it is subjected to enough stress. People exposed to repeated torture will have PTSD, at least acutely, at a nearly 100% rate, just as my burn patients did. If the bone is weak to begin with—say, due to osteoporosis—then it will break more easily. People with "weaker psychological bones" will develop PTSD when faced with less severe stressors.

PTSD is a powerful example of the fact that psychological experiences have neurobiological consequences and that the mind-body dichotomy is a misleading oversimplification. Clinicians and neuroscientists have applied modern neuroimaging tools to the study of people who have experienced PTSD, as caused by a variety of stressors, such as combat, rape, or torture. Changes in the hippocampus, measured with MR, have been a particularly consistent finding. Exposure to a trauma apparently sets up a cascade of events, which follow the fear/alarm pathways of the brain through the amygdala. The chronic outpouring of adrenalin and cortisol may eventually produce damage to the hippocampus, which appears to be especially sensitive to high levels of cortisol. Multiple studies have shown measurable decreases in hippocampal size in the brains of people suffering from PTSD. Figure 11–6 showed a normal hippocampus visualized with MR, and one with a marked decrease in size.

The hippocampal changes may account for the symptoms of PTSD, such as intrusive memories or nightmares. The various "autonomic" symptoms of PTSD are a consequence of the related overactivity in the sympathetic nervous system. High levels of norepinephrine (noradrenalin) produce symptoms such as hypervigilance, exaggerated startle response, or tachycardia. This response is part of the well-recognized "fight or flight" mechanism of the sympathetic nervous system, which prepares people to defend themselves in response to perceived dangers. A person with PTSD lives chronically with intrusive memories of perceived dangers. The goal of treatment is to reduce the chronic hyperarousal through a variety of therapeutic mechanisms. We do not know as yet whether measurable changes in brain regions such as the hippocampus are reversible with treatment, but our increasing recognition that the brain makes plastic changes in response to both good and bad experiences certainly suggests this possibility.

Generalized Anxiety Disorder (GAD)

Generalized anxiety disorder (GAD) is the mildest but most common of the various anxiety disorders. This is the psychiatric disorder equivalent to what lay people refer to as "being chronically stressed out." People with generalized anxiety disorder feel anxious and worried most of the time and complain of things like being keyed up, on edge, restless, easily fatigued, irritable, or tense. They may have trouble falling asleep or staying asleep, or they may feel that they have slept fitfully or restlessly. Their worries and anxieties may interfere with their ability to concentrate. Somewhere between 4% and 7% of the general population have problems with generalized anxiety disorder.

Some people with generalized anxiety say, "I've been like this all my life. I was just born this way." This subjective observation is partly supported by family studies, which indicate that the disorder does tend to run in families, affecting about 25% of first-degree relatives. If something is inherited, it is probably a "generally tuned up" nervous system, with the "danger regulator" set at higher levels of noradrenergic arousal.

Since people with generalized anxiety disorder tend to be chronic worriers, one of their worries is that "this may get worse." Some people with GAD in fact do. Approximately 25% may eventually develop panic disorder. Others may turn to alcohol or drugs to relieve their chronic tension, while others may eventually become depressed. Psychiatrists are now talking increasingly about the tendency for patients to have other problems in addition to generalized anxiety. The buzz word is "comorbidity," which means that a person has several disorders at the same time, such as depression and anxiety. In general, the greater the number of comorbid problems, the greater are the challenges to successful treatment. Some people struggle along for years with chronic comorbid generalized anxiety and dysthymia, although they do receive some relief from both medications and psychotherapy.

Obsessive-Compulsive Disorder (OCD)

Obsessive-compulsive disorder (OCD) is grouped with the anxiety disorders because people with this disorder often become quite anxious if they try to resist a compulsion or ignore an obsession.

Compulsions are repetitive acts that people perform for no obvious reason or to an extreme degree. A person with a hand-washing compulsion may wash her hands twenty or thirty times a day, making them red and raw. Other common compulsions include repeatedly checking, counting things over and over, and other ritualistic behaviors such as having to walk without stepping on cracks. If the person violates the compulsion, she becomes quite fearful and anxious. For example, a person with a ritualistic compulsion to avoid stepping on cracks may also have a mental monologue saying, "Stepping on a crack will break your mother's back." She will have to check on her mother's health if she accidentally does step on a crack.

Obsessions are persistent and troubling thoughts that run through a person's mind over and over and that are often recognized to be senseless. For example, a mother of a newborn child may have the repeated obsessive fear that she may grab a knife and stab her child. Obsessions often involve themes of sex or violence. Ironically, most people with obsessions and compulsions tend to be gentle and conscientious, quite unlike the terrible thoughts that run through their minds. Sometimes obsessions or compulsions occur within the context of religious practices. Although the rules have been scrupulously followed, the person feels he has committed an unforgivable sin for failing to meet perceived ritualistic requirements adequately. Some people become preoccupied with collecting and hoarding, accumulating stacks of shopping bags, rubber bands, or paper clips. An extreme concern with orderliness is another form of obsessive–compulsive behavior, with the affected person becoming extremely stressed if books, clothing, or other articles are not arranged in a particular way.

Just as everyone is entitled to one or two phobias, so too "normal" people may have an occasional obsession or compulsion. These normal personality traits become a disorder when they are accompanied by intense and crippling anxiety if the obsessions or compulsions are resisted or not adequately satisfied. OCD can become extremely severe. A person who is severely affected may spend the entire day getting dressed or undressed in order to get things exactly right, may walk endlessly around the house in ritualistic patterns, or may sit all day repeatedly counting stacks of coins, cards, or other objects. When OCD becomes this severe, the preoccupations may seem almost delusional.

People with OCD usually come across to the average person as scrupulous and gentle chronic worriers. Because of the incongruity between the frequent sexual or violent content of the obsessions and compulsions and the manifestly mild-mannered and conscientious personality of the person who possesses them, psychodynamic theories have suggested that the obsessions or compulsions are mechanisms being used to control primitive drives that are unacceptable to the person's ego or superego. Although this psychodynamic explanation is plausible, obsessions and compulsions are notoriously difficult to treat with psychotherapy alone.

Complementary theories have proposed that obsessions and compulsions are related to the repetitive behaviors that occur in neurological disorders such as epilepsy or the tic-like movements that occasionally occur after a strep infection (known to physicians as Sydenham's chorea and lay people as Saint Vitus' dance). Studies conducted by Susan Swedo of NIMH have now demonstrated links between the occurrence of OCD, the vocal tics that occur in Tourette's disorder, and strep infection. These neurobiological theories of OCD are supported by imaging studies that have shown changes in blood flow and metabolism in both the orbitofrontal cortex and the caudate nuclei, regions involved in regulating complex motor behaviors, formulating long-term plans, and generating abstract thoughts.

Neurochemical theories of OCD have received support from psychopharmacology studies. Most of the medications currently used to treat OCD, many of which are quite effective, are blockers of serotonin reuptake. The pharmacologic studies have sometimes led to oversimplified "too much/too little" formulations such as "too much serotonin makes you impulsive and too little serotonin makes you compulsive." Many studies have linked increased serotonergic tone with impulsivity, aggression, violence, and suicidality, and decreases in serotonergic tone with compulsivity, perfectionism, and passivity.

Treatment of Anxiety Disorders

Twenty to thirty years ago the majority of patients with anxiety disorders were treated with various forms of psychotherapy, including behavioral therapy. Sometimes these approaches were very successful, but sometimes nothing seemed to help enough. Panic attacks and OCD were particularly difficult to treat and tended to be characterized by relapses and recurrences even after periods of successful remission. Therefore, creative psychopharmacologists began to conduct clinical trials of various new medications. These studies frequently indicated that the medications were quite effective for anxiety disorders. In the twenty-first century, medications have supplanted psychotherapy as first-line treatments for anxiety disorders in many clinical settings. Nonetheless, behavioral therapy and psychotherapy are important as well and often effective in combination with medications.

Pharmacotherapy

Donald Klein of Columbia University conducted some of the earliest studies of the use of medications for the treatment of anxiety disorders. Specifically, he explored the effectiveness of tricyclic antidepressants for reducing the frequency of panic attacks and found them to be highly effective. Others have followed his lead and explored many other drugs that have been developed subsequently. At present, a person who seeks treatment for one of the anxiety disorders may be given a prescription for any of several different groups of medications. The first choice will vary depending upon the particular type of anxiety disorder, as well as the person's age and the occurrence of any con-comitant medical problems.

Benzodiazapines are widely used and may be quite effective. Some of the earliest antianxiety drugs were Miltown, Librium, and Valium. Valium continues to be widely used because of its excellent anxiolytic and mus-cle-relaxing properties, although with some caution because people can develop a physiologic dependency if they take large doses. Newer "ben-zos" have also been developed. Alprazolam (Xanax) is effective for panic attacks, but lorazepam (Ativan) and oxazepam (Serax) are also used. These medications work by binding to benzodiazapine receptors, which are linked to GABA receptors. As described in chapter 4, GABA is an impor-tant inhibitory neurotransmitter. Increasing GABAergic tone is thought to have a direct anxiolytic effect throughout the brain, but particularly on limbic regions that are overactivated in the anxiety syndromes. The major downside of the benzodiazapines is that they can be habit-forming, and they may also be sedating.

All of the antidepressant medications have been used to treat anxiety disorders as well. The early work of Klein and others suggested the utility of tricyclic antidepressants and MAOIs. When the newer serotonin reup-take inhibitors (SSRIs) became available, they became first-line treat-ments for many anxiety syndromes, particularly obsessive-compulsive disorder and panic disorder. These medications are described in more detail in chapter 9 ("Mood Disorders"). These medications alter the tone of the noradrenergic system and the serotonin system.

Beta blockers, originally developed for the treatment of hypertension, are also sometimes used to treat anxiety disorders. Beta blockers diminish noradrenergic tone and therefore attack anxiety and panic within the context of that particular chemical system. They are particularly effective for reducing autonomic symptoms such as tremulousness. Sometimes patients find them helpful in combination with other medications such as the antidepressants, which can have the side effect of causing trembling hands.

Behavioral Therapies

Behavioral therapies can be useful for anxiety disorders, particularly pho-bias and panic disorder. Systematic desensitization is the most widely used behavioral technique. An individualized program is worked out for the patient, so that she gradually increases her exposure to the feared situation or object. For example, a person with a cat phobia may begin by looking at pictures of cats, progress on to being in the same room with a cat, eventually sitting close to the cat, and perhaps ultimately touching or petting the cat. In conjunction with increased exposure to the feared stimulus, the person is also taught anxiety-reduction techniques, such as regular breathing or the ability to call up tranquil "counter-thoughts" or "counter-feelings" to reduce the subjective sense of anxiety.

The mechanisms by which behavioral therapies work can be conceptualized in several different ways. A behaviorist would say that the person had been "deconditioned," so that the associations between the unconditioned stimulus and the conditioned response are gradually lost. A neurobiological explanation would suggest that new memory traces have been laid down in a plastic brain, which replace the earlier connections that produced the anxiety reactions. A psychodynamic explanation would say that the person was learning new psychological techniques to cope with or repress unwanted anxiety. These three explanations, although they may derive from different schools of thought, are not necessarily incompatible with one another.

Psychotherapies

Many different types of psychotherapy can be used for the treatment of anxiety syndromes, ranging from classic psychoanalysis, through insight-oriented psychotherapy, to supportive psychotherapy. Supportive psychotherapy is perhaps the most widely used, and it is sometimes used in conjunction with medications, as in the case of Michelle. During supportive psychotherapy, the patient and the doctor often work together in order to explore the triggers that produce the anxiety and to learn ways to adjust lifestyle in order to reduce anxiety. The patient is usually an active participant in identifying ways to make adjustments and changes, guided by friendly support from the clinician.

Insight-oriented psychotherapy probes more deeply into the role of interpersonal relationships and past experiences, in order to explore how the problems with anxiety might have arisen. During this type of therapy, patients are often encouraged to review and discuss relationships with others, attitudes about themselves, and early life experiences. As they do so, they begin to achieve insight about how the problems of anxiety arise, and the insight assists them in confronting the sources of their reactions and changing them as needed.

Psychodynamic or psychoanalytic psychotherapy is also used for the treatment of anxiety disorders. This type of treatment is more intense and long-term. It differs from insight-oriented psychotherapy primarily in that it aims to recover the emotional aspects of early experiences and memories, as well as emotional components of current relationships, and to achieve change in the anxiety response through processes such as reliving and catharsis. Intense probing may eventually lead to insights, which the patient may then use in order to achieve changes in behavior and response. This type of intensive psychotherapy may be viewed as a long-term rebuilding and restructuring of the memories and emotional responses that have been embedded in the limbic system. We do not yet understand precisely how this occurs at the neural level, but its bedrock is probably "Hebbian plasticity." Again, mind and brain are simply two words to refer to the same process.

PART IV

BRAVE
NEW
BRAIN

CHAPTER 12

O BRAVE NEW WORLD
Conquering Mental Illness
in the Era of the Genome

How beauteous mankind is! O brave new world,
That has such people in't!
—William Shakespeare
The Tempest, v,i, 182–186

We also predestine and condition. We decant our
babies as socialized human beings, as Alphas or
Epsilons, as future sewage workers or . . . future
World controllers.
—Aldous Huxley
Brave New World

Concern for man himself and his fate must always
be the chief interest of all technical endeavors . . . in
order that the creations of our mind shall be a
blessing and not a curse to mankind.
—Albert Einstein
Lecture at the California Institute of Technology

During the coming century we will combine our knowledge of the human genome and our knowledge of the brain to develop new weapons with which to wage a war on mental illness that may eventually lead to a definitive victory. As Einstein reminds us, our minds can create powerful tools that can be used wisely, for good purposes. But these tools can also be used foolishly, and for evil ends. Aldous Huxley's futuristic novel *Brave New World* predicted in 1936 that scientists would develop methods to clone human beings, practice eugenics to create a society in which people were stratified into classes based on intelligence, and misuse knowledge in the service of selfish hedonism and a totalitarian worldview. His vision of the future is terrifying. Fortunately, it has not come true. But we have also never been so close to having the methods that could actually be used to achieve the sardonically evil brave new world that Huxley imagined. His cautionary tale reminds us that as we face our own brave new world that combines psychiatry with molecular neuroscience, we must make concern for mankind the guiding principle of our endeavors, so that "the creations of our mind shall be a blessing and not a curse."

What achievements can we expect? How will we be able to use science to create healthier minds and brains? Can we conquer other mental illnesses as we have already conquered neurosyphilis? Our goal should be nothing less.

What risks must we anticipate and avoid in this brave new world?

Waging War on Mental Illnesses: What Does the Future Hold?

We live in an era when biology and biomedical science have matured to a point where we can expect pivotal discoveries to occur. While we cannot anticipate exactly what they will be, we *can* anticipate that the nature of human life will change dramatically. Earlier medical achievements such as the discovery of insulin, which redeemed people with juvenile-onset diabetes from an inevitable death sentence, are likely to pale in comparison with future accomplishments in the treatment and prevention of mental illness and other major illnesses such as cancer.

We live in an era when two large knowledge bases will meet and mingle: the map of the human genome and the map of the human brain. The products of this union will be many. The synthesis of these two knowledge bases will give us the power to understand the mechanisms that cause major mental illnesses and to use this knowledge to relieve the pain of the millions of people who at present suffer from them. The time when we can realistically declare a war on mental illnesses, with some hope of eventually achieving a victory, has finally come.

In spite of the power of the many tools that we presently have to probe genes in the brain, however, we cannot expect that this war will be won either quickly or easily. Instead, progress will be slow and steady, with occasional setbacks but also occasional significant discoveries. Those who suffer, and the families of those who love them and who may fear that they too will develop a similar affliction, can face the future with guarded optimism. This optimism must be tempered by a recognition that major discoveries do not occur every week, or every month, or perhaps even every year. Nonetheless, three or four breakthroughs in our understanding of Alzheimer's disease, schizophrenia, mood disorders, or anxiety disorders will dramatically change millions of lives. Furthermore, such achievements will steadily build on one another as knowledge advances in the golden era of biomedicine that lies before us.

Defining the Target: Using Guidance from Maps of Brain Terrain

Winning the war requires that we know which weapons we should use and where we should be pointing our guns. For many years, our weapons have been primitive: shotguns that scatter the ammunition, cannons that produce a large crude shot, and arrows too weak to pierce the target. Our strategic plan as to where we should aim has been guided more by serendipity than science. We are now living in an era, however, when we can be guided by rational principles that derive from our knowledge of molecular and cellular biology and our understanding of how changes in one point of a complex dynamic biological system, the human brain and mind, can affect other sites in an elegant cascade that occurs each day in each individual human life. Table 12–1 illustrates the nature of this cascade.

As chapters 4–6 have shown, recent decades have not only added many weapons to our arsenal for attack on mental illnesses, but they have also given us a map of the terrain on many different levels in the cascade from genes through brain and mind. We no longer need to aim blindly and hope we might hit the enemy. At the most fine-grained level, we have

TABLE 12–1
Defining the Target: The Multiple Levels of Mind/Brain

Genes

↑↓

Gene expression

↑↓

Molecules (gene products, such as neurotransmitters or enzymes
that regulate brain development)

↑↓

Cells (including "chemical factories" inside cells and communication
systems in cell membranes)

↑↓

Chemical circuits (neurotransmitter systems that project throughout the brain,
such as dopamine or serotonin)

↑↓

Anatomical circuits (hardwiring between brain regions)

↑↓

Functional circuits (shifting metabolic activity in interconnected brain regions in
response to changes in stimulation both within the brain and in the outside world)

↑↓

The activity of the mind/brain ("normal" thoughts and feelings, or symptoms of
mental illness)

created a crude map of the human genome that will be steadily refined over the next few decades, until we reach a point when we have figured out exactly how many genes there are, what they do, and the mechanisms by which they are turned on and off. This fine-grained map of the genetic activity of our brains is complemented by large-scale maps of neurotransmitter systems, anatomical circuits, and functional circuits.

We now have maps of "normal brain terrain" that show us where to look when we want to understand how the mind/brain is able to learn, remember, or feel emotions. Ingenious techniques have also shown us how various brain regions are connected to one another, giving us increasingly refined wiring diagrams of neuronal connections and providing us with maps of anatomical circuits. We also have maps of chemical circuits, which have shown us the distribution of the multiple neurochemical systems of the brain and how they interact with one another to send specific messages or to fine-tune the level of activity within regions by exciting or inhibiting nearby nerve cells.

Now that we have relatively detailed maps of the chemical, anatomical, and functional circuitry of the normal brain, we have a basis for seeking out the sites of abnormality in the vast array of human mental illnesses. Using the in vivo tools of modern neuroscience, we can create comparison maps of brain terrain for diseases such as schizophrenia, bipolar disorder, major depression, Alzheimer's disease, panic disorder, autism, eating disorders, or attention deficit hyperactivity disorder (ADHD). As has been described in chapters 8–11, this process is already well underway. During the next several decades, we can expect to identify the abnormalities in brain geography and topography that define the various types of mental illnesses. Once this is accomplished, we will know where the enemy is. The techniques of molecular biology will give us the capacity to do precision bombing, while our maps of brain terrain will give us the targets at which to aim.

There will be interesting and unexpected surprises as this story unfolds. For most mental illnesses, we are still in the process of identifying the place or places in this cascade of events that a given illness is "caused." For most, we already know that, although genes will be a key factor, a single cause is not likely to be found. Therefore, as we define our targets for attack, we will work using two different insights, which may seem contradictory. First, our strategy for attack should be designed to search multiple points on this cascade. That is, finding the gene, or even repairing it completely, might not be enough to prevent a particular disease from occurring. We already know that possessing a gene explains, at most, less than 50% of the causality of most illnesses, including mental illnesses. Second, even before we find the cause or causes, we may be able to produce dramatic improvements by aiming treatment or prevention at one crucial site in the cascade. Since the brain is an interactive dynamic system, making an adjustment in one site can have reverberating effects in other places and produce a significant change.

The case of schizophrenia illustrates how some of the recent advances in neuroscience, based on the tools of neuroimaging, neurobiology, and molecular biology, have changed how we think. Our progress during the 50 years leading up to 2000 appears to be substantial, since "mental hospitals" were virtually emptied. The major achievement between 1953 and 2000 was the development of drugs, initially discovered by serendipity, that reduced psychotic symptoms and made patients more manageable. We later learned that these drugs were effective because they blocked dopamine receptors, which were located primarily in basal ganglia structures such as the caudate and putamen. Although emptying mental hospitals by reducing these symptoms was a significant achievement, the "clinical target" was nonetheless still insufficient. The earliest antipsychotics affected only psychotic symptoms and had little or no impact on negative symptoms or the more basic cognitive problems that define the essence of schizophrenia. At the anatomical level, the target was selected by accident rather than by rational knowledge of enemy location. By striking the basal ganglia, the older antipsychotics probably reduce symptoms through secondary effects on other interconnected brain regions. The newer "atypical neuroleptics" that became available in the 1990s and early 2000 have a better symptomatic target because they also attack negative symptoms and cognition (by mechanisms we are still learning to understand). But more dramatic and more rational approaches to designing treatments are now underway.

Using the tools of in vivo neuroimaging such as magnetic resonance and positron emission tomography, we have found new and sometimes unexpected "enemy sites." These neuroimaging studies suggest that we should explore alternative targets, both for the development of new treatments and in the search for causes. One of these new targets, for example, is the thalamus, shown to be smaller in schizophrenia by the anatomical structural tools of MR, to have abnormal cellular structure using the tools of postmortem tissue analysis, and to have abnormal functional connectivity using fMR and PET. The next stage of the story, likely to unfold over the next decade, will be to map the chemical systems and connections of the thalamus in more detail, to explore the neurodevelopmental processes that shape its connections at the genetic and cellular levels, to determine the neural functions performed by the thalamus, and to map the activities of gene expression and the proteins that genes produce. Somewhere in this small haystack we may find the quixotic needle that can be used to slay one of the biggest giants of mental illness.

Finding the Genes: The Long March to Finding Their Functions

Scientists once hoped that finding the genes for mental illnesses would be a simple process. A single gene, or perhaps two, would be identified, and the "cause" of a specific mental illness would then be discovered. We now know that the story is much more complicated and challenging.

First, most major mental illnesses are almost certainly caused by more than one gene. Scientists interested in the genetics of mental illness now murmur the mantra of "multiple genes of small effect." Sorting through the human genome to find the multiple genes for a specific illness will take time, but the detail on the map of the human genome is steadily accumulating under the leadership of Francis Collins and Craig Venter. Cleverly efficient computer technologies and statistical methods are now available to sift through this mass of information. Within a few years, we will probably be able to identify the genetic mutations that differentiate people who have developed manic-depressive illness (for example) from those who have not. These comparisons will help us determine both "bad alleles" that predispose a person to the development of illness, as well as "good alleles" that protect against it. We are already well on our way with Alzheimer's disease, which will probably serve as a model for the attack strategy for other diseases such as schizophrenia, mood disorders, and anxiety disorders.

Second, we now recognize that "finding the genes" is not enough. We must become increasingly nimble in moving up and down the dynamic and bidirectional cascade between genes and the activity of the mind/brain, as well as wise enough to recognize when it is time to pause on one level in order to survey the terrain in more detail. Discussions about "finding the genes" will ultimately switch to the topic of "functional genomics," the branch of genetics that focuses on figuring out what proteins the genes produce, what their functions are, and the medical mischief that occurs when the functions are abnormal.

We usually begin by figuring out "where the genes are." This information will not be simple. It will be something like: 50% of the people with manic-depressive illness have a mutation on Chromosome A, which has 4 alleles, one of which is "bad." Among those 50%, 10% have mutations on

Chromosome B, G, and X, while 30% have mutations on Chromosomes C, L, and Y, and the remainder have mutations on M and W. The remaining 50% of people with manic-depressive illness fall into smaller groups. Perhaps 5% will have a single major gene with a large effect that is incompletely penetrant, coupled with only one other gene. Others will be small groups of 5 or 10% (or even less) that have other polygenic patterns, containing some genes that overlap with those in group one. The story for schizophrenia, other mood disorders, and the various anxiety disorders (as well as the array of childhood disorders and other disorders not discussed in this book) will probably be quite similar.

This prediction of polygenic "causes" may seem pessimistic, but it is probably also fairly realistic. We will not win the war if we underestimate the power of the enemy. Nevertheless, the scientists leading the attack are likely to win many battles and ultimately achieve victory. The important question is not "Will it happen?" It is "How long will it take?"

As discussed earlier in this book, an important clue about the causes of mental illness lies in the fact that they are caused by other factors besides genes. That is, identical twins who share the same genes only have the same mental illness at most 50% of the time, indicating that the "ill twin" experienced some nongenetic influence that permitted his "bad alleles" to exert their toxic effect. Here, we may be looking at the first step in the cascade: the relationship between the presence of a gene and its expression. Much of the effort during the next few years will focus on the factors that cause disease genes to turn on, which will lead in turn to identifying preventive strategies that may turn them back off again.

Other efforts will move further up and down the cascade, attempting to determine how the expressed genes actually exert their effect. This process is the domain of functional genomics. It will explore how "bad alleles" produce an excess of "bad molecules" and lead to "bad functions." The molecules themselves may be fine, but there may be too many of them. Alternatively, the bad molecules produced could be truly "bad," in the sense that they have an improper chemical structure that has a toxic effect. The gene products are proteins: enzymes that facilitate chemical reactions or components of cell structure such as receptors. One smaller piece of the search process is called "proteomics," since it explores how defective genes can cause changes in the structure of the proteins that they produce, causing them to link or join improperly with other molecules. The long-term goal is to find the relationship between normal and abnormal genes, and then to identify their remotely distributed connection to normal and abnormal brain/mind functions.

Genetic Fingerprinting: The Ultimate Identity Card

Someday we all may be carrying the ultimate identity card: a small disc or chip that contains our genetic fingerprint. Each of us has a fingerprint that is totally unique to us: a form of identification far more precise than our actual fingerprints or a photograph of our faces. It is a profile of the individual genetic mutations that uniquely characterizes each of us, the single nucleotide polymorphisms (SNPs) referred to as "snips." This summary of personal genetic endowment is a quintessential definition of what each person actually is or is going to become. Right now, both the costly equipment and the technological tools that can scan a tissue sample from a single individual and summarize the information in a "snip profile" is not widely available. At the moment, the profiles are still crude, and our knowledge of what they mean is relatively vague. As the Human Genome Project matures, however, these blurry images will sharpen their focus and achieve amazing predictive power. The technology used to scan tissue samples will also become more widely available, and genetic fingerprints may then become widely used in both medical research and medical care.

A genetic fingerprint will tell us, for each gene, the specific allele that a person carries. For example, we already know that there are three different alleles for apolipoprotein E and that the ε4 allele increases the risk for Alzheimer's disease, while the ε2 allele appears to protect against it. We know that there is a similar pattern of protective and destructive effects for the BRCA1 and BRCA2 alleles that influence the development of breast cancer. As our armamentarium of information about the genes for specific diseases increases, and as we identify the patterns of alleles that predispose us to specific diseases, our genetic fingerprints will tell us which diseases we are at risk to develop. Inevitably, we are all predisposed to develop something. None of us will be free of disease alleles.

Importantly, we know that carrying a "bad" allele does not necessarily cause a specific disease to occur. It may only indicate that we are "at risk." A "positive snip" indicating a bad allele can serve as a "wake-up call" to warn us that we have the potential to develop a disease . . . and should perhaps initiate changes in our lifestyle that will reduce our risk. For example, if our genetic fingerprint tells us that we carry an allele that makes us prone to develop adult-onset diabetes, we can potentially forestall its effects or perhaps even prevent them by losing weight and adopting a better diet. If we know that we carry an allele that predisposes us to developing Alzheimer's disease, we can do more than review our will with our attorney. We can attempt to maintain healthy spines and synapses on our neurons by exercising brain cells and circuits through challenging mental activities—learning a new skill, doing arithmetic calculations in our heads, training ourselves to memorize names and faces, and of course reading books rather than watching TV or surfing the net. We can also maintain good general health by exercising regularly to ensure that our brain receives a good blood supply and eating a balanced diet to make sure that neurons are well nourished.

Using the information available through genetic fingerprints may eventually become as routine in medical care as electrocardiograms, X rays, or blood tests. They may be used not only to determine the predisposition to or presence of an illness, but also to make more intelligent decisions about how to treat it.

For example, one of the big puzzles that psychiatrists confront is why some patients respond to one drug and not another. At present, selection of a specific drug to treat depression, schizophrenia, or an anxiety disorder is done on a trial and error basis. Usually, a person is given a two- or three-week trial and then switched to another drug if the first one does not work. But nagging questions remain: Is the "treatment failure" due to choosing the wrong drug? Or is it due to choosing the wrong dose? Could it be that the dose was initially too high and that the person did not respond because he or she was already "overdosed?" Or could it be that the dose was too low because the drug was metabolized (chemically disassembled) too quickly? Did the enzymes in the liver, chemical scavengers programmed to break down and eliminate food and drugs and chemical byproducts so that our system runs using just the right juices, act too aggressively and kick the drug out before it even had a chance to act?

Consulting a patient's genetic fingerprint can help answer these questions. We already know that patients differ from one another in their rate of drug metabolism. Some people break drugs down quickly and are known as rapid metabolizers, while others do it more slowly and are prone to build up larger amounts of drug in the bloodstream on lower doses. One of the important liver enzymes that breaks down drugs is known as cytochrome P-450. The chain for this enzyme has already been characterized and its mutations identified. We know that the reason people vary in their rate of metabolism depends on their genetic endowment: whether they carry the wild-type allele or the mutated allele. If they have two wild-type alleles, they metabolize rapidly, while if they have one wild type and one mutation they metabolize more slowly. If they have two mutations, then they metabolize poorly. This genetic polymorphism can already be measured, but the laboratory procedures are too cumbersome and costly for routine use. When and if the genetic fingerprinting of point mutations is routinely done for everyone, however, we will have molecular explanations as to why a person may not be responding to a given drug. In fact, a psychiatrist can know in advance that a person is a slow or rapid metabolizer and prescribe a dose of medication based on scientific knowledge rather than trial and error, no longer delaying recovery because the dose was too high or too low.

As our knowledge of functional genomics increases, and as specific medications are developed that are aimed at the correct targets, genetic fingerprinting may be used to obtain other information that may be useful for diagnosis or treatment—for example, whether a person carries a polymorphism in a specific neurotransmitter or neuroreceptor gene (e.g., dopamine, D2 receptors) that is relevant to the occurrence of an illness. These applications are the wave of the future, and they are likely to be among the most useful. Someday people may carry their genetic fingerprints with them when they visit a psychiatrist and have them used to aid in the selection of treatment based on their specific pattern of alleles.

The Race to Create Newer and Better Drugs

People who suffer from debilitating mental illnesses, and their families, sometimes feel as if they are in a race against time. Will a new treatment become available that will halt or reverse the progression of Pearl's dementia before she dies? Will a new medication reverse the cognitive and emotional changes caused by Scott's schizophrenia in time for him to go to college, find a girlfriend, and lead a normal life? Will a medication be found that can completely stabilize Hal's mood swings, which can strike unexpectedly during a crucial moment in his business or family life? Creating new drugs is also a race for drug developers, who are forced to jump over a rigorous set of hurdles before a drug can be made available for public use. From both perspectives, the race sometimes seems to run in slow motion.

Drug development proceeds in a systematic manner through four phases. A drug cannot become available until it passes "phase 3 trials" and is approved by governmental agencies—the Food and Drug Administration (FDA) in the case of the United States. People interested in tracking drug development for a mental illness in which they have particular interest need to be aware of these phases, so that they can monitor progress realistically. The phases of drug development are summarized in Table 12–2.

TABLE 12–2
Steps in Drug Development

Phase 1	Phase 2	Phase 3	Phase 4
Preclinical trials, in which safety and efficacy are evaluated in animals; safety trials in healthy human beings	Open trials in patients to see if the drug is effective; studies to determine the correct dose	Double-blind trials to assess efficacy, comparing the new drug to either a placebo or an established drug	Further investigations of the long-term effects, side effects, or application to other related illnesses

New medications may be invented largely through good luck. The most difficult part of the process is to create and identify a useful compound. Based on their knowledge about the chemical structure of drugs that have worked in the past, pharmaceutical companies produce hundreds to thousands of compounds. Out of these compounds, they must identify the ones that might be helpful for treating specific diseases. This screening is typically done by testing in animals (usually rats). The rat brain does not have the extensive capacities for thought and emotion that the human brain does, and so scientists have developed a variety of so-called animal models of disease. An animal model is an effort to mimic the symptoms of a known mental illness through chemical or molecular manipulation. For many years, animal models were based on chemical manipulation. For example, the animal model that has been used to screen drugs for schizophrenia has involved giving rats high doses of amphetamines, which mimic some of the symptoms of schizophrenia, such as agitation and stereotyped behavior. A drug that reverses amphetamine-induced agitation and stereotyped behavior in rats without any adverse effects is moved on to the next step.

The molecular revolution has introduced a much more elegant way of conducting Phase 1 trials. As described in chapter 5, molecular engineers can now mimic the primary genetic damage of a disease, rather than the secondary consequences of the genetic damage as they are manifested in symptoms and behavior. Genes that produce an illness can be edited out (the "knock-out mouse") or spliced in (the "knock-in mouse"). Identifying drugs that will correct the effects of genetic injury can be very precise indeed, since they strike the enemy at its origins.

The development of genetically engineered mice that have the amyloid plaques of Alzheimer's disease exemplifies this type of Phase 1 strategy. These "Alzheimer's mice" were given a mutation in the gene that produces beta amyloid protein (BAP). This mutation caused them to have a steady buildup of large quantities of amyloid plaques in their brains, which could be verified through postmortem studies. They also showed "clinical" signs of Alzheimer's disease, because they could not remember how to make their way through a maze, based on previous experience passing through the maze.

These "Alzheimer's mice" are now used as a resource for Phase I drug treatments for Alzheimer's disease. So far several different strategies have been employed. One has involved vaccinating the mice with BAP. This strategy showed that vaccination both prevented plaque development if given early and even caused the plaques to melt away in year-old mice who had already developed severe plaque formation. The Alzheimer's mice can also be used to test drugs that work on other aspects of the disease. For example, one strategy for developing a drug to treat AD has been to block the enzymes (for example, gamma secretase) that are used to form beta amyloid protein. This approach, if applied early in genetically predisposed individuals, might permit a preemptive strike against the development of the illness.

Creating animal models of mental illness is a difficult task, since the models must mimic either the basic genetic defect that has been identified or the clinical features of the illness—a difficult challenge for diseases of the human mind. Nonetheless, the process is well underway for many major mental illnesses, and this will substantially speed up the process of drug development. However, people must be aware that the progression from Phase 1 through Phase 3 takes time. The current time frame for completing those three phases is approximately ten years. If given a high priority, some drugs can be developed at a faster rate, but this cannot be done for all drugs, and it is not easy to do.

While animal models are powerfully necessary tools for screening drugs, and while the mouse genome is surprisingly similar to the human genome (the overlap is estimated to be about 95%), the reach from mice to men is still long enough to make things difficult. Many drugs that appear to work well in animals fail in human beings, either because they do not have the desired effect of improving mental/brain function or because they cause damage in other organs of the body such as liver or kidneys. Phase 1 tests, conducted in small numbers of healthy human volunteers, determine if the drug has any toxic effects in human beings. Small changes in heart rhythm or changes in the enzymes used to monitor liver function are sufficient to trash the development of a drug during

Phase 1, and approximately 90% of drugs developed in preclinical trials fail Phase 1.

During Phase 2 the drug is subjected to a variety of tests. The first one may be an "open trial." In this case, the drug is given to people who have a particular illness in order to get an early read to see if it might actually work. A variety of different doses of the medication are usually tried in an attempt to figure out what the correct dose range might be. Open trials can be misleadingly optimistic, since both doctors and patients know that a new drug is being tried, and both doctors and patients are eager to see positive results. Therefore, patients and families should follow media reports closely, continually trying to figure out whether they are based on an open trial or the more rigorous "double-blind trial." In a double-blind trial, which is done during Phase 3, the new drug is matched against either a well-established drug or a placebo. (The word "placebo" means "I shall please." It is a compound that has no recognized therapeutic effectiveness, and it is used to evaluate the occurrence of "the placebo effect" that often makes open trials misleading.) In these rigorous double-blind trials, neither the doctor nor the patient knows which drug is being given. Both are blind. The new drug is considered to be valuable if it is at least superior to the placebo, and preferably superior to the recognized standard as well. In order to complete all three phases, drug developers usually test thousands of people. Coordinating and orchestrating these large clinical trials takes a very big effort, and only a few drugs manage to jump over all the hurdles and become available for treatment.

Replacing Damage Control with Preemptive Strikes: Early Intervention and Prevention

Our current approach to mental illness is usually to wait for a disease to develop and then treat the symptoms after they emerge. But wouldn't it be better to strike early, and perhaps preemptively? Instead of doing "damage control," wouldn't it be better to intervene before the damage actually occurs? This is the long-term goal of the psychiatric research community, but it is difficult to achieve.

The phrase "early intervention and prevention" has several meanings. One version of prevention is to identify those who are predisposed to developing a particular mental illness and to intervene early. The precise manner in which identification and early intervention are implemented will vary depending on the particular illness. As we define the multiple genes that lead to the various illnesses that we now call bipolar mood disorder, panic disorder, major depression, schizophrenia, and Alzheimer's disease, we will achieve increasingly precise information about the extent to which each individual carries alleles that predispose him to or protect against particular mental illnesses. Psychiatrists and other clinical and basic neuroscientists may work together to identify and promote basic public health programs for those who carry disease alleles. As described in the next section, psychiatrists have begun to think about ways to exploit the fact of brain plasticity and to identify and promote nonpharmacologic interventions that may arrest the onset or prevent the progression of mental illnesses.

Another strategy for arresting an illness before it becomes fully manifest is to use prophylactic medications that prevent its occurrence. However, no true prophylactic drug is as yet available for any mental illness, although this strategy is currently widely discussed. A few relatively controversial experimental studies that have used standard neuroleptic medications for schizophrenia in predisposed young people and have reported hopeful results. This effort is hampered, however, by three serious concerns. First, at present we do not have optimal tools for identifying high-risk individuals, since the predisposing alleles and the related nongenetic factors have not yet been mapped in detail. Second, the pharmacologic treatments used to intervene are also relatively crude, since they were designed primarily to work on symptoms rather than to do pinpoint bombing of abnormal processes that occur earlier in the gene-to-mind/brain cascade. Therefore, they may not work effectively in actually preventing the onset of the illness. Third, there is an ethical problem as well. Treating young people who are not yet ill may adversely affect their self-esteem and self-image, perhaps creating a self-fulfilling prophecy that may lead them eventually to become ill. Further, the medications may have adverse side effects, some of which occur in people who were misidentified due to our lack of optimal tools for diagnosis.

As improved treatments with minimal risk and a sharper focus on basic disease mechanisms are developed, however, such early interventions may become both common and appropriate, particularly when and if they can be coupled with precise identification of vulnerable individuals. Use of this strategy may be close for one mental illness. If a drug can be found that prevents beta amyloid protein buildup, and if it does not have serious side effects on other aspects of brain function or other organs within the body, then it might be used both to treat the symptoms of Alzheimer's disease and to prevent amyloid buildup in predisposed individuals.

Ultimately, we would like to prevent mental illnesses by identifying the mechanisms that cause them. This approach is, in effect, "total prevention," equivalent to developing a vaccine for polio or preventing malaria through the use of quinine. As these examples illustrate, so far this strategy has worked best for infectious diseases. Many feel that other biomedical illnesses, including those that affect the mind and brain, are too complex and multifactorial in their causes and are therefore likely to be forever refractory to total preventive measures.

Prevention may be possible for mental illnesses, however. Some, for example, Huntington's disease, Alzheimer's disease, and schizophrenia, are disorders in which the brain is apparently normal prior to onset of illness. In all three cases, these illnesses may have dormant seeds in place that only sprout when the appropriate climatic conditions occur in a particular season of life. Because it is a fully penetrant single-gene disease, Huntington's disease may be the "easiest" to prevent, although even it still has eluded our grasp. However, people with Huntington's disease are essentially well and normal for many years, and therefore blocking the effect of the Huntington's gene should permit us to prevent its expression. The key is to figure out what the gene actually does (i.e., its gene product and the effects of that product on brain development and degeneration). Schizophrenia and Alzheimer's disease are "less genetic," and very likely polygenic as well. This may not matter, however, if these diseases have a single common mechanism produced by multiple genes—whatever produces the plaques and tangles of Alzheimer's disease, or whatever produces the faulty wiring of schizophrenia. In these illnesses as well, the goal is to find the mechanism and to arrest it—perhaps by vaccinating with beta amyloid protein, as has already been tried with Alzheimer's disease.

Changing Minds: Basing Therapeutic Strategies on Brain Plasticity

When we think about improving treatment, our thoughts turn first to improving pharmacologic treatment. We forget the most important fact about the nature of the mind/brain and about human nature in general. Each of us is a unique person and has a unique brain primarily because each of us has had a different combination of life experiences that has shaped who we are. Furthermore, as we live each minute of each day and each day of each year, we make choices that change our brains and ultimately change who we are. Our brains are constantly rewiring themselves so that we very literally "change our minds."

The techniques of behavior therapy and psychotherapy have relied on the principles of brain plasticity, generally without realizing it, for nearly one hundred years. Just as our growing understanding of the gene-to-mind/brain cascade can permit us to find better targets at which we can aim pharmacological treatments, so too this knowledge can help us find better cognitive and emotional targets. While changing minds through nonpharmacologic therapy may be very helpful for the "serious casualties" (those with diseases such as dementia, schizophrenia, or autism), it is particularly applicable to the "walking wounded" (those with depressive and anxiety disorders and substance abuse problems).

As chapters 8–11 have shown, we now know substantial amounts about the brain systems that are affected in illnesses such as depression, panic disorder, or posttraumatic stress disorder. These disorders, which frequently appear to result from the response of a plastic brain to the cumulative insults of a painful or noxious environment, are well suited for targeted cognitive interventions. These interventions will be created by clever scientists who can synthesize what they know about human behavior and how to measure it with what they know about brain systems and how they can be modified. Such approaches have already been applied successfully in teaching children with dyslexia how to hear sounds and words more precisely so that they can rewire their brains and read and write more efficiently. Strategies for an attack on mood disorders might follow this lead and aim at such fundamental qualities of the illness as a lack of resilience in the face of adversity or an inability to regulate an intense response to it.

Historically, the development and use of psychotherapy has never been carefully monitored and subjected to the same rigorous blind testing that is required for drugs. Inevitably, this can lead to "junk therapies" or "quack therapies" that are difficult for the general public to identify and to distinguish from established and effective therapies. As new treatments that capitalize on the principles of brain plasticity are designed, it will become increasingly important to subject them to the same sort of rigorous trials that are used for pharmacologic therapies.

A Blessing and Not a Curse to Mankind

As biomedical tools grow more powerful, both scientists and private individuals will confront new and difficult choices. Great knowledge bestows great responsibility. Although we cannot anticipate all the risks and perils of our growing ability to modify human life as we seek to conquer human disease, we are obligated to ponder the ramifications of our new powers and to insure that they will be "a blessing and not a curse to mankind." We do not want to create Huxley's brave new world, in which we "predestine and condition," or in which "we decant our babies as socialized human beings, as Alphas and Epsilons." Instead, heeding Einstein's warning, we must do all we can to insure that "concern for man himself and his fate will be the chief interest of our endeavors."

Genetic Testing: Will We Know Too Much?

Only a few years ago, our primary worry about the misuse of genetic information was based on the recognition that some mental illnesses run in families. Because Uncle Will had schizophrenia or Mom suffered from manic-depressive illness, other family members worried that they might either develop these illnesses themselves or pass them on to their children. In many parts of the world, ill relatives are still treated like skeletons in the family closet, because of fears of a "hereditary taint." Children from these families may be seen as unworthy candidates for marriage and childbearing.

The technology of modern genetics far surpasses these old-style observations of transmitted hereditary taints in its potential for both use and abuse. As a multiplicity of disease genes and alleles are mapped, and as "snip and chip" technology becomes faster and more efficient, we can imagine many ways in which too much knowledge may be dangerous. Genetic fingerprints could eventually be created relatively easily. Who will decide when and if such fingerprints should be created? Who will have access to them?

The potential perils of genetic fingerprinting make our concerns about misuse of credit card information, banking records, or social security numbers look like child's play. Such records simply indicate what we have done. A genetic fingerprint indicates what we are.

Many frightening scenarios can be imagined. Employers, medical school and law school admission committees, or the military could require that we be genetically fingerprinted as a condition for hiring, admission, or acceptance. Despite strenuous efforts to maintain confidentiality and accuracy, some records are virtually certain to be lost, swapped, or leaked. The health insurance industry could band together and require that all applicants be fingerprinted prior to being insured, with the possibility that coverage might be denied if a person tests positive for some specified diseases, which might include mental illnesses. (Parity in the coverage of treatments for mental illness is still far from a reality in the United States and most of the rest of the world.) An investigative reporter might obtain fingerprint data on a politician or celebrity and use it as the basis for a tantalizing exposé. Presidential candidates might be required to submit their genetic fingerprints along with their income tax records before running for office. Feuding couples in bitter divorces might use fingerprint data as evidence against each other. Lawyers, detectives, or reporters might scavenge for tissue samples in order to create a genetic fingerprint without a person's knowledge, which could later be used as evidence against that person.

Fortunately, we are wise enough to anticipate these many risks. By the time meaningful and accurate genetic fingerprints become a reality, reasonable regulations will almost certainly be developed that will protect individual privacy and prevent the misuse of such information. The guiding principles are likely to be that the information will be kept strictly private, accessible only to physicians and to patients who wish to have the information. Obtaining unauthorized access to genetic information is likely to become a serious criminal offense.

Not that life will be easy, even with such safeguards in place. The possibility of having so much knowledge about our vulnerabilities to disease places a great burden on all of us. Even the simplest of cases is difficult. For example, right now any person from a family with Huntington's disease can determine whether he or she carries the abnormal gene. This information is very useful, since those who test negative know in advance that they are free from worry, and they can marry and have children without a concern about passing the trait on to their offspring. Those who test positive, however, know they face a grim future and may find the remainder of their healthy lives clouded by their ominous sense of foreboding.

Many intelligent people may decide they do not want to know their future, just as they may not want to know the sex of a future child when they have the opportunity. Nancy Wexler, one of the scientists involved in the discovery of the Huntington's gene and a potential carrier, chose not to know for many years. Since most genes are not fully penetrant like the Huntington's gene, and since most illnesses are polygenic and multifactorial, having information about the presence of a disease allele could be more harmful than helpful for some people. Worrywarts may walk under a sky filled with thunderclouds for the remainder of their life, not recognizing that there are many pockets of sunshine as well.

Although the choice about knowing will almost certainly be left up to each individual, life in many cases will not actually be that simple. Some spouses are likely to pressure one another to get the information. Parents may wish to know about their newborn children, or adult children about their aging parents.

Genetic Engineering: Will We Change Too Much?

Gene therapy, the capacity to replace a disease allele with a normal one, has been a goal of biomedical science for many years. The first instance of transplanting a gene to treat a disease was completed in 1990 by French Anderson on a young patient suffering from a disease called ADA (adenosine deaminase deficiency), characterized by an inadequate immune system that makes young children susceptible to many types of infections. Gene therapy, like all forms of transplant therapies, is difficult to execute, because our bodies are equipped with vigilant bodyguards within our immune systems that rise to combat dangerous foreign invaders. Ingenious techniques must be designed to get past these body-guards. This problem, coupled with the technical difficulties inherent in getting new genes into the right place, has slowed down the momentum of gene therapy for the moment. However, it is certain to remain on the biomedical agenda, along with other transplant and replacement thera-pies that can be facilitated by manipulation of gene expression.

How much can we, and should we, tinker with our genes and our other worn out and damaged body parts? As with genetic fingerprinting, we have many concerns. If DNA is the "blueprint of life," who has the right to play God by altering it? What guidelines can make such alter-ations permissible? Apart from actual gene transplants, what about the use of genetic knowledge to alter the gene pool in more subtle ways? How can we avoid the Huxleyan dystopia in which our knowledge of how to produce "better strains" of human beings could slowly slide down a slip-pery slope into a society composed of Alphas (who have the knowledge and wealth to use the new technology) and Epsilons (who lack these assets)? Although we have become comfortable with adults donating their organs or sperm for transplants or artificial insemination, how do we feel about the use of fetal tissue to replace diseased or missing cells in debilitating disorders such as Parkinson's disease?

We have many questions and fewer answers in this youthful era of the genome. In recognition of the problematic issues, an impartial Panel of Wise Men and Women, officially known as ELSI (Ethical, Legal, and Social Implications branch) was created under the auspices of the Human Genome Project, initially under the leadership of Nancy Wexler. They have made it clear that they will be guided by "concern for man himself and his fate" as they struggle to establish fair and reasonable guidelines that will govern the use of genetic engineering and the other ethical dilemmas that the new knowledge will create. The functions of this orig-inal committee (now disbanded) have now been dispersed through the various multiple components of the project, so that ongoing careful surveillance of all its aspects will occur.

Genetic Determinism: Will We Lose Our Sense of Autonomy?

"Our genes are our destiny." "There is no disease that does not have a genetic cause." Very famous scientists have made pronouncements like this. Therefore, it is little wonder that many people believe that our endowment of DNA will shape everything that we do and become. Dire warnings about humanity being governed by the iron hand of genetic determinism have filled the media. If genes rule all, then our future is predestined, just as Aldous Huxley predicted in *Brave New World*. We cannot choose which genes we inherit. They were given to us by random fate when our father's sperm united with our mother's egg—perhaps a consequence of their choice to conceive a child, or perhaps as an unanticipated "accident." In either case, if genes are destiny, we have no choice thereafter. Our future is predetermined by our genes.

The consequences of this line of reasoning are pernicious. Lacking free choice, we are absolved of responsibility. We simply live our lives pursuing whatever seems desirable at the moment, unguided by a sense of overall purpose or a moral compass.

Fortunately, the concept of genetic determinism is nonsense. For many reasons that have already been explained in many places in this book, genes alone are not destiny. They interact with a plethora of nongenetic factors over which we do have control. Since we do control those many nongenetic factors, we are in fact the arbiters of our own destiny and morally responsible for our decisions and actions. The only reason for even bringing the question of genetic determinism up again is to make sure that the answer is absolutely clear.

Medicating the Mind:
Humane Treatment, Cosmetic Surgery, or Self-Indulgence?

Aldous Huxley's prescient vision of a "brave new world" included not only genetic engineering, testing, and determinism, but also mind-altering drugs that were used for both control and self-indulgence. Is Huxley's description of "soma," a mind-altering drug, a prophetic forewarning of modern somatic therapies and their potential for misuse or abuse? Peter Kramer's widely read book *Listening to Prozac* also raised the possibility that in some cases our present ability to create more and better medications to treat the mind may be simply a form of cosmetic surgery, not a medical necessity.

In most instances, the psychoactive drugs available in our armamentarium are valued assets. We need more medications and better ones, not fewer. To think otherwise is to revert to the old tradition of stigmatizing mental illnesses and seeing them as moral failures caused by weakness of will or bad parenting. We should be grateful (and most of us are) that we do have medications that can "change minds." It is a blessing that we can lift a depression that might otherwise drive a person to suicide, reduce disabling panic attacks that interfere with daily function, or quiet the anguished thoughts and perceptions of schizophrenia or mania. We all believe that doctors have a moral obligation to reduce or relieve suffering. We do not expect people to have surgery without anesthesia, to be left to die after they experience a cardiac arrest, or to limp around on an unset broken leg. It would be strange indeed to apply a different standard to mental illnesses such as dementias, schizophrenia, mood disorders, or anxiety disorders.

Yet a nagging concern may remain in many people's minds. In some cases, are psychoactive medications used when they are not necessary and when a problem can be handled more appropriately in some other way? Some psychiatric disorders, such as depression or attention-deficit/hyperactivity disorder (ADHD), are diagnosed more frequently now than in the past, and psychoactive medications are often used to treat them. Are the medications being overused? Is this a form of "cosmetic surgery for the mind?"

This concern tends to arise for those disorders that have a continuum of severity. When such disorders are severe, there is usually no question that they represent an illness and little question about the need for medication. When the problem is milder, however, the situation is more complicated. When is an energetic and rambunctious child just too active and inquisitive? When is a period of sadness so severe that it becomes an illness best treated with medication? And when should treatment move from good behavioral management or psychotherapy to medication? These questions are more difficult to answer.

The power of pharmacology has perhaps given some people the sense that "popping a pill" is all that is needed to solve their problems. If responsibility for making changes is shifted completely from the individual to the medication being used, then moral responsibility is perhaps being inappropriately abrogated. Overactive little Johnny still needs consistent structure and discipline, both at home and at school, and this must be provided by the human beings around him. No pill can give him these crucial aids for calming his energy and focusing his attention. Miserable

Marilyn may need some pharmacologic help for dealing with her chronic blues and social anxieties, but she still has to take responsibility for getting to work on time, going out of her way to make new friends, or figuring out how to be a better friend herself.

Will the Growth of Biomedical Technology Dehumanize Psychiatry?

Some might argue that psychiatry already has been dehumanized!

Three different forces have swept through psychiatry during the past several decades. These include the "biological revolution," the emphasis on empirical description and objective diagnosis created by DSM III and DSM IV, and the "economic revolution" in health care. Each of these has brought good elements, but also carries the potential for abuse or misuse. Surveying the current psychiatric landscape with a jaundiced eye, one might observe that three forces converged on psychiatry at the same time like a wave of barbarian invaders. They ransacked its humanistic aspects and left a desolate landscape behind. A more sanguine view is that some of these changes have been good, that much humanism remains, and that we must work to preserve it.

A book that I wrote in 1983, *The Broken Brain: The Biological Revolution in Psychiatry,* described how psychiatry had been shaped by three different models: psychodynamic, behavioral, and biological. The psychodynamic or Freudian perspective was still dominant at that time. The book predicted a major paradigm shift:

> In America in the 1980s, the balance between these points of view has begun to shift. The emphasis is swinging, and swinging quite strongly, toward a biological model.

As *The Broken Brain* argued, the shift to a biological model occurred for many good reasons: the growth of a strong scientific base in neurobiology, the development of new and effective pharmacologic treatments, the value of reducing stigma by understanding mental illnesses within a medical framework as diseases of the brain, and the growing body of evidence that demonstrated brain changes and abnormalities in a variety of mental illnesses. As predicted, the biological model now prevails in most of psychiatry, and this shift has produced many benefits. The stigma against mental illness has decreased, and we have made enormous strides in understanding the neural basis of diseases of the mind. However, I now share with many the concern that psychiatry may have moved too far, and that it must make corrective adjustments to prevent losing its identity as the most humanistic of the medical specialties. After all, modern neuroscience also teaches us that the brain is plastic and that it can also be changed by psychotherapy—and should be.

An emphasis on empirical observation, coupled with the development of objective criteria for making diagnoses, has occurred at the same time. DSM III was published in 1980, and it had a major impact on psychiatric education and clinical practice over the ensuing decade. Again, introducing standardized approaches to diagnosis and assessment had many advantages, as discussed more fully in chapter 7. Essentially, psychiatry embraced a fundamental tenet of modern medicine: the importance of evidence-based approaches to diagnosis and treatment. This has been a significant achievement that has placed psychiatry on a sound clinical foundation. But again, it is time for reassessment and readjustment, particularly in the area of psychiatric education. In addition to learning and using diagnostic criteria, young psychiatrists must be taught to think first about the whole person and to appreciate that each one is interesting and unique, not simply a composite of symptoms that are used to make a DSM diagnosis and provide treatment according to a standard algorithm, making the erroneous assumption that "one size fits all."

The third force, the economic revolution in the provision of health care in the United States, also began in the 1980s and has had its greatest impact in the 1990s. Its impact on psychiatry has been especially bad, since the economic revolution follows principles that are unabashedly nonhumanistic. Spending time talking to patients, or listening to them, is now considered an expensive luxury, to be avoided whenever possible. But talking and especially listening are central to good psychiatric evaluation, and they form the basis for most psychotherapy.

The economic revolution has dramatically changed the philosophical framework that guides medicine. Essentially, medical care is now perceived and discussed primarily in economic terms, often to the dismay of both physicians and patients. The provision of medical care is now referred to as "the health care industry." Doctors are "providers," and patients are "consumers" or "clients." Large managed care organizations and health maintenance organizations (HMOs) formed during the 1980s, and they are now a powerful and even dominant force in American medicine.

This third force has changed the social contract between patient and doctor, which historically has been guided by the humanitarian principles of the Hippocratic oath, to an economic contract that is guided by the principles of free market competition. The principle that the health and welfare of each patient should come first has been replaced by the dictum that saving money and increasing the profits on the health company's bottom line should come first. Only too frequently, key decisions about patient care have been taken out of the doctors' hands and placed in the hands of "health care managers," who have minimal medical training and no direct experience in confronting human suffering. Health care managers may decide how long a doctor can see a patient, what medications she can prescribe, whether she can provide psychotherapy in addition to medications, how frequently the patient can be seen, and even the amount of information that can be included in a medical record. Many psychiatrists are being told that obtaining a comprehensive patient history, which includes personal information about family and social relationships and personal interests—those things that make the patient a unique individual—is a waste of time. Instead, the psychiatric history may consist only of a symptom checklist that is entered into a computer and serves as a basis for making a DSM diagnosis. Most managed care organizations and HMOs do not wish psychiatrists to do psychotherapy. They are considered to be "costly providers," whose time would be better spent just writing prescriptions.

In my experience, the majority of psychiatrists are discouraged and demoralized by these changes. They share the concern that they can no longer care for the patient as a unique and interesting person living in a specific social environment. They are dismayed that they can no longer provide the kind of humanistic care that they were trained to give. Too many have been reduced to minimal patient contact, brief interviews, and a primary emphasis on the provision of pharmacotherapy. The story of Jim and Mary, described in chapter 2, illustrates the quandary that has arisen in psychiatry because economic values have been given preeminence over humanistic values, due to changes in the structure of health care delivery.

There is no "easy fix" for the overemphasis on economics mandated by changes in our health care delivery system. The changes in education and training created through an overemphasis on DSM *are* the responsibility of psychiatry. They should and will be addressed by its governing boards and organizations, such as the American Psychiatric Association and the American Board of Psychiatry and Neurology. The economic revolution, and to a lesser extent excessive DSMism, have had a far greater impact on dehumanizing psychiatry than the biological revolution or the growth of biomedical technology.

Psychiatry is, and should remain, the most humanistic of the medical specialties. To create a polarity between science and humanism is to create yet another false dichotomy. The purpose of science is to advance knowledge, and knowledge can be used in turn to promote the health and welfare of human beings. The growth of biomedical technology in the era of the genome will give us unparalleled opportunities to reduce suffering by providing better diagnoses, better counseling, and better medications. If we live up to the opportunities that are our potential, the result will be a substantial improvement in the human condition by reducing the burden of mental illnesses throughout the world.

If the Mind Is the Brain, Where Is the Soul or Sense of Self?

What are the moral and religious implications of seeing mental illnesses as mind/brain diseases? If mind and brain are different facets of the same thing, where is the moral executor? Where is personal identity and the chance to make choices about what we might do or become? If we are the product of the activity of our brains, then where and who are WE? And where are WE after our brains cease to exist? Is there a difference between mind and soul?

This is a very good question, and primarily a philosophical or religious question rather than a scientific one. It is also such a large question that it is a topic for an entire (and different) book. But it is important enough that it cannot be totally ignored in this book.

One brief answer to this question comes from common sense. Although we cannot easily demonstrate its existence by scientific methods, we all have a sense of self. We rightly see ourselves as unique individuals, moral executors who are confronted with decisions that we freely choose to make, and linked by those choices to a bond with other people in a community that we call human society. As far as we know, only human beings have this sense of self that permits us to both act as free moral agents and also stand outside ourselves and appraise our thoughts and actions as "right" or "wrong." There are many different words for this sense of self: soul, spirit, conscience, consciousness.

Whatever we call it, as human beings we recognize the dual existence of both a sense of our individual identity and of an inexplicable force that transcends individuals and reflects a collective bond that we all have as living beings. This sense of individual self and a union with other human beings in our present, past, and future is the impetus behind selflessness, humility, compassion, and sacrifice. These are complex and abstract concepts for which we are unlikely to demonstrate a neural basis or mechanism. Although PET studies from our center have linked some of these concepts to neural circuits, especially the inferior frontal-cerebellar-thalamic-frontal circuit, such links are trivially reductionistic. If we want to prove the existence of the moral reality of an individual identity and a moral reality that also transcends individuals and links them to one another, we will learn more by looking at exemplary human lives. Mother Theresa will teach us more about the soul than a PET scan can. The recognition that each of us has an individual identity that we call a "self" or "soul," that our "self" is guided by a moral imperative, and that the moral imperative also transcends our individual "self" and links us to other human beings exists with indelible certainty across all cultures and continents. It is no coincidence that Jesus and Confucius independently came up with the Golden Rule.

The Role of Psychiatry: Curing People or Curing Society?

During recent decades one major social problem, the stigmatization of mental illness, has been successfully addressed, although not totally solved. Patients and families have taken the lead in creating this social change, while psychiatrists have stood beside them and fought hard as well.

Thirty years ago almost all mental illnesses were skeletons in the closet. A person diagnosed with depression would skulk guiltily into the pharmacy to pick up his prescription for one of the new antidepressant medications, barely willing to look the pharmacist in the eye. As recently as 1972, the Democratic candidate for the vice presidency, Tom Eagleton of Missouri, had to withdraw when the public and press learned that he had been treated for depression. Yet if we disqualified earlier candidates because of mental illness, we would never have had Teddy Roosevelt or Abraham Lincoln as presidents. Both also suffered from mood disorders.

The battle against public misunderstanding of mental illness has been led by three new national organizations that focus on diseases of the mind and brain. The National Alliance for the Mentally Ill (NAMI) was founded in 1979 and has grown into an enormous social and political force. Under the leadership of Laurie Flynn for its first two decades, NAMI is an organization composed primarily of people who have suffered from mental illnesses and their families. As its name indicates, it is an alliance of people who used to "hang separately" and now are much stronger because they "hang together" and advocate for funding for research and for improvements in treatment. They also have an ongoing antistigma campaign that repeatedly points out the many contributions that people with mental illnesses have made to society. The National

Alliance for Research in Schizophrenia and Affective Disorders (NARSAD), founded by Connie Lieber, has created a powerful research foundation that was able to award $2,154,000 in grants to 416 investigators in the year 2000. This level of funding has increased steadily since its foundation in the 1980s. NARSAD has enlisted help from many wealthy and prominent individuals, who have been candid about their own experiences with mental illness, thereby also helping to reduce stigma. Finally, the Charles Dana Foundation, led originally by David Mahoney and then by William Safire, has also built a powerful research foundation that seeks to aggressively attack, and ultimately prevent, the broad range of neurological and psychiatric illnesses using the tools of neuroscience.

As a support for these foundations, many prominent public figures have "opened the closet" and discussed their experiences with mental illnesses or those of their families. Examples include Ronald Reagan, Mike Wallace, Dick Cavett, William Styron, and Rod Steiger. Tipper Gore has described how psychotherapy helped her cope with the trauma of her son's serious injury. Many facets of the media have also taken a greater interest in neuroscience and mental illness. The *Wall Street Journal*, *New York Times*, *Washington Post*, and *New Yorker* have all run many articles on various aspects of schizophrenia, mood disorders, and new developments in genetics and treatments. National television programs, such as Bob Bazell's science series on NBC, include brief educational segments about mental illnesses. Hollywood and the film industry, which gave a negative tint to the public perception of mental illness with films such as *Snakepit* and *One Flew over the Cuckoo's Nest*, have introduced a positive spin with powerful and sympathetic films such as *Rainman* and *Shine*. Thirty years ago these subjects would have been dismissed with a "Who cares about autism?" or "No one wants to hear about schizophrenia." Now films about the mentally ill win academy awards.

The battle to improve public understanding and to reduce stigmatization of mental illnesses has made great progress. With the help of the "biological revolution," we have advanced beyond the misconception that mental illness results from a defect in a person's character that he or she can cure simply by "shaping up." We now recognize mental illnesses as biomedical diseases similar to heart disease or cancer. By and large, we have gotten past the vicious cycle of inappropriate blaming and failure to diagnose and treat with appropriate medications that prevailed 30 years ago. Should we do more? In particular, should psychiatrists do more?

At its best, psychiatry *is* a humanistic specialty. Because they care about people, psychiatrists often want to reach out and help as much as they can. Sometimes this has led psychiatrists to want to assume the role of helping more than individual patients. They would like to help society as a whole.

To aspire to this may, however, be aspiring to too much. Psychiatrists can help, but they must also recognize the limits of their power. The role of psychiatry in society certainly includes public education and the battle against misunderstanding. Psychiatrists also chip away modest fragments of collective social suffering by helping the individual people whom they see. However, psychiatry must recognize that its role is to treat diseases, not the social discontent of "unhappy people" or pervasive psychosocial malaise. We simply lack the knowledge to cure society as well as individuals.

Confronting this fact seems especially imperative at this time. Psychiatrists are frequently called on to prescribe quick treatments for a variety of social ills, such as the rising rates of crime and violence. Instead of appealing to the specialty of psychiatry to "fix" violence or reduce general unhappiness, all of us, as members of the human community, need to recognize that the sense of "self" in our post-turn-of-the-century world may be in need of repair. There has been a widespread move toward materialism, quick fixes, instant gratification, and a superficial sense of success, which is reinforced by the fast-paced cyberworld that we live in. The answer to our many current social problems must come from individual people, who must reappraise their sense of "self" and reach an appropriate perspective on what constitutes a sound moral compass and meaning in life. The need to search for a personal moral compass to guide our individual lives in the twenty-first century is a need that transcends medical intervention, but which has a very real impact on how we choose to employ medical science and what we expect from it. In the era of the genome, fraught as it is with a variety of crucial moral questions, we must all make an agonizing reappraisal of who we are, what life is, what life means, what we must do to help the other human beings who share our world with us, and what we can do to make it a brave new world.

REFERENCES AND SUGGESTED READINGS

Chapter 1

Murray, CJL, Lopez, AD, eds. *The Global Burden of Disease*. Geneva and Boston: World Health Organization and Harvard University Press, 1996.

Chapter 3

Kandel, ER. Biology and the future of psychoanalysis: A new intellectual framework for psychiatry. *American Journal of Psychiatry*, 156:505–524, 1999.

Chapter 4

Armstrong, E, Schleicher, A, Omran, H, Curtis, M, Zilles, K. The ontogeny of human gyrification. *Cerebral Cortex*, 1:56–63, 1995.

Cooper, JR, Bloom, FE, Roth, RA. *The Biochemical Basis of Neuropharmacology*. Seventh Edition. New York: Oxford University Press, 1996.

Baddeley, AD, Wilson, BA, Watts, FN, eds. *Handbook of Memory Disorders*. Chichester, UK: John Wiley and Sons, 1995.

Edelman, GM. *The Remembered Present: A Biological Theory of Consciousness*. New York: Basic Books, 1989.

Filipek, P, Richelme, C, Kennedy, D, Caviness, V. The young adult brain: An MRI-based morphometric analysis. *Cerebral Cortex*, 4:344–360, 1994.

Fuster, J. *The Prefrontal Cortex*. Philadelphia: Lippencott Raven, 1997.

Gazzaniga, MS, ed. *The New Cognitive Neurosciences*. Second Edition. Cambridge, Mass: MIT Press, 2000.

Heimer, L. *The Human Brain and Spinal Cord*. Second Edition. New York: Springer-Verlag, 1995.

Kandel, ER, Schwartz, JH, Jessell, TM. *Principles of Neural Science*. Fourth Edition. New York: McGraw Hill, 2000.

Mesulam, MM. *Principles of Behavioral and Cognitive Neurology*. Second Edition. New York: Oxford University Press, 2000.

Nauta, WJH, Feirtag, M. *Fundamental Neuroanatomy*. New York: W.H. Freeman, 1986.

Parent, A. *Carpenter's Human Neuroanatomy*. Ninth Edition. Baltimore: Williams and Wilkins, 1996.

Rakic, P, Sidman, RL. Histogenesis of cortical layers in human cerebellum, particularly the lamina dissecans. *Journal of Comparative Neurology*, 139:473–500, 1970.

Schatzberg, AF, Nemeroff, CB. *Textbook of Psychopharmacology*. Second Edition. Washington, DC: American Psychiatric Press, 1998.

Squire, LR. *Memory and Brain*. Oxford, UK: Oxford University Press, 1987.

Tulving, E, ed. *Memory, Consciousness, and the Brain*. Philadelphia: Psychology Press, 1999.

Zilles, K, Armstrong, E, Schleicher, A, Kretschmann, HJ. The human pattern of gyrification in the cerebral cortex. *Anatomy and Embryology*, 179:173–179, 1988.

Chapter 5

Cook-Degan, R. *The Gene Wars*. New York: W.W. Norton, 1994.

Faraone, SV, Tsuang, MT, Tsuang, DW. *Genetics of Mental Disorders*. New York: Guildford Press, 1999.

Frank-Kamenetskii, MD. *Unraveling DNA: The Most Important Molecule of Life*. Trans. Lev Liapin. Reading, Mass: Addison Wesley, 1997.

Gershon, ES, Cloninger, CR. *Genetic Approaches to Mental Disorders*. Washington, DC: American Psychiatric Press, 1994.

Griffiths, AJF, Miller, JH, Suzuki, DT, Lewontin, RC, Gelbart, WM. *An Introduction to Genetic Analysis*. Sixth Edition. New York: W.H. Freeman, 1996.

Hammond, C. *Cellular and Molecular Neurobiology*. San Diego: Academic Press, 1996.

Hyman, SE, Nestler, EJ. *The Molecular Foundations of Psychiatry*. Washington, DC: American Psychiatric Press, 1993.

Nurnberger, JI, Berrettini, W. *Psychiatric Genetics*. New York: Oxford University Press, 1997.

Twyman, RM. *Advanced Molecular Biology: A Concise Reference*. Oxford, UK: Bios Scientific Publishers, 1998.

Watson, JD, Hopkins, NH, Roberts, JW, Steitz, JA, Weiner, AM. *Molecular Biology of the Gene*. Fourth Edition. Menlo Park, Calif: Benjamin/Cummings Publishing Company, 1987.

Watson, JD, Crick, FHC. A structure for deoxyribonucleic acid. *Nature*, 171:737–738, 1953.

Chapter 6

Andreasen, NC, Nasrallah, HA, Dunn, V, Olson, S, Grove, W, Ehrhardt, J,

Coffman, J, Crossett, J. Structural abnormalities in the frontal system in schizophrenia: A magnetic resonance imaging study. *Archives of General Psychiatry*, 43:136–144, 1986.

Andreasen, NC, Arndt, S, Swayze, V, Cizadlo, T, Flaum, M, O'Leary, D, Ehrhardt, J, Yuh, WTC. Thalamic abnormalities in schizophrenia visualized through magnetic resonance image averaging, *Science*, 266:294–298, 1994.

Andreasen, NC, O'Leary, DS, Cizadlo, T, Arndt, S, Rezai, K, Ponto, LL, Watkins, GL, Hichwa, RD. Schizophrenia and cognitive dysmetria: A positron-emission tomography study of dysfunctional prefrontal-thalamic-cerebellar circuitry. *Proceedings of the National Academy of Sciences, U.S.A*, 93(18):9985–9990, 1996.

Andreasen, NC, Flaum, M, Swayze, V, O'Leary, DS, Alliger, R, Cohen, G, Ehrhardt, J, Yuh, WTC. Intelligence and brain structure in normal individuals. *American Journal of Psychiatry*, 150:130–134, 1993.

Belliveau, JW, Kennedy, DN, McKinstry, RC, Buchbindiner, BR, Weisskoff, RM, Cohen, MS, Vevea, J, Brady, T, Rosen, B. Functional mapping of the human visual cortex by magnetic resonance imaging. *Science*, 254:716–719, 1989.

Buchsbaum, MS, Ingvar, DH, Kessler, R, Waters, RN, Capelletti, J, Kammen, DP, King, C, Johnson, J, Manning, RG, Flynn, RW, Mann, LS, Bunney, WE, Sokoloff, L. Cerebral glucography with positron tomography. *Archives of General Psychiatry*, 39:251–259, 1982.

Frackowiak, RSJ, Friston, KJ, Frith, CD, Dolan, RJ, Mazziotta, JC. *Human Brain Function*. San Diego: Academic Press, 1997.

Gur, RC, Mozley, PD, Resnick, SM, Gottleib, GL, Kohn, M, Zimmerman, R, Herman, G, Atlas, S, Grossman, R, Beretta, D, Erwin, R, Gur, RE. Gender differences in age effect on brain atrophy measured by magnetic resonance imaging. *Proceedings of the National Academy of Sciences, USA*, 88:2845–2849, 1991.

Ingvar, DH, Franzen, G. Abnormalities of cerebral blood flow distribution in patients with chronic schizophrenia. *Acta Psychiatrica Scandinavica*, 50:425–462, 1974.

Johnstone, EC, Crow, TJ, Frith, CD, Husband, J, Kreel, L. Cerebral ventricular size and cognitive impairment in chronic schizophrenia. *Lancet*, 2:924–926, 1976.

Kapur, S, Zipursky, R, Jones, C, Remington, G, Houle, S. Relationship between dopamine D_2 occupancy, clinical response, and side effects: A double-blind PET study of first episode schizophrenia. *American Journal of Psychiatry*, 157: 514–520, 2000.

Kety, SS, Woodford, RB, Harmel, MH, Freyhan, F, Appel, K, Schmidt, C.

Cerebral blood flow and metabolism in schizophrenia: The effects of barbiturate semi-narcosis, insulin coma and electroshock. *American Journal of Psychiatry*, 104:765–770, 1948.

Krishnan, KRR, Doraiswamy, PM, eds. *Brain Imaging in Clinical Psychiatry.* New York: Marcel Dekker, Inc, 1997.

Latchaw, RE, Ugurbil, K, Hu, X. Functional MR imaging of perceptual and cognitive functions. *Neuroimaging Clinics of North America,* 5:2;193–205, 1995.

Mazziotta, JC, Toga, AW, Frackowiak, RSJ. *Brain Mapping: The Disorders.* San Diego: Academic Press, 2000.

Oldham, J, Riba, MB, Tasman, A, (eds.). *Review of Psychiatry,* Volume 12 Neuroimaging and Clinical Neurosciences. Washington, DC: American Psychiatric Press, Inc., 1993.

Ogawa, S, Lee, TM, Kay, AR, Tank, DW. Brain magnetic resonance imaging with contrast dependent on blood oxygenation. *Proceedings of the National Academy of Science* USA, 87:9898–9872, 1990.

Petersen, SE, Fox, PT, Posner, MI, Mintun, M, Raichle, ME. Positron emission tomographic studies of the processing of single words. *Journal of Cognitive Neuroscience,* 1:153–170, 1989.

Raz, N, Gunning, FM, Head, D, Dupuis, JH, McQuain, J, Briggs, SD, Loken, WJ, Thornton, AE, Acker, JD. Selective aging of the human cerebral cortex observed in vivo: Differential vulnerability of the prefrontal gray matter. *Cerebral Cortex,* 7:268–282, 1997.

Sedvall, G, Farde, L, Persson, A, Wiesel, FA. Imaging of neurotransmitter receptors in the living human brain. *Archives of General Psychiatry* 43:995–1005, 1986.

Toga, A, Mazziotta, J. *Brain Mapping: The Methods.* San Diego: Academic Press, 1996.

Weinberger, DR, Berman, KF, Zec, RF. Physiological dysfunction of dorsolateral prefrontal cortex in schizophrenia I: Regional cerebral blood flow (rCBF) evidence. *Archives of General Psychiatry,* 43:114–124, 1986.

Willerman, L, Schultz, R, Rutledge, JN, Bigler, ED. *In vivo* brain size and intelligence. *Intelligence,* 15,223–228, 1991.

Chapter 7

American Psychiatric Association. *Diagnostic and Statistical Manual of Mental Disorders.* Fourth Edition (DSM-IV). Washington, DC: American Psychiatric Press, Inc., 1994.

Andreasen, NC. Linking mind and brain in the study of mental illnesses: A project for a scientific psychopathology. *Science,* 275:1586–1593, 1997.

Charney, DS, Nestler, EJ, Bunney, BS, eds. *Neurobiology of Mental Illness.* New York: Oxford University Press, 1999.

Lishman, WA. *Organic Psychiatry: The Psychological Consequences of Cerebral Disorder.* Third Edition. Oxford: Blackwell Science, 1998.

Weissman, S, Sabshin, M, Eist, H, eds. *Psychiatry in the New Millenium.* Washington, DC: American Psychiatric Press, 1999.

Chapter 8

Andreasen, NC. Negative symptoms in schizophrenia: Definition and reliability. *Archives of General Psychiatry,* 39:784–788, 1982.

Andreasen, NC, Olson, S. Negative versus positive schizophrenia: Definition and validation. *Archives of General Psychiatry,* 39:789–794, 1982.

Andreasen, NC. Understanding the causes of schizophrenia. *The New England Journal of Medicine,* 340:645–647, 1999.

Andreasen, NC, O'Leary, DS, Cizadlo, T, Arndt, S, Rezai, K, Ponto, LL, Watkins, GL, Hichwa, RD. Schizophrenia and cognitive dysmetria: A positron-emission tomography study of dysfunctional prefrontal-thalamic-cerebellar circuitry. *Proceedings of the National Academy of Sciences, USA,* 93(18):9985–9990, 1996.

Andreasen, NC, ed. *Schizophrenia: From Mind to Molecule.* Washington DC: American Psychiatric Press, 1994.

Bleuler, E, translated by J. Zinkin. *Dementia Praecox or the Group of Schizophrenias (1911).* New York: International Universities Press, 1950.

Bloom, F. Advancing a neurodevelopmental origin for schizophrenia. *Archives of General Psychiatry,* 50:224–227, 1993.

Braff, DL. Information processing and attention dysfunctions in schizophrenia. *Schizophrenia Bulletin,* 19:233–259, 1993.

Carlsson, M, Carlsson, A. Schizophrenia: A subcortical neurotransmitter imbalance syndrome? *Schizophrenia Bulletin,* 16:425–432, 1990.

Creese, I, Burt, D, Snyder, S. Dopamine receptor binding predicts clinical and pharmacological potencies of antischizophrenic drugs. *Science,* 192:481–483, 1976.

Crow, TJ. Positive and negative schizophrenic symptoms and the role of dopamine. *British Journal of Psychiatry,* 137:383–386, 1980.

Davidson, M, Reichenberg, A, Rabinowitz, J, Weiser, M, Kaplan, Z, Mark, M. Behavioral and intellectual markers for schizophrenia in apparently healthy male adolescents. *American Journal of Psychiatry,* 156:1328–1335, 1999.

Davies, N, Russell, A, Jones, P, Murray, RM. Which characteristics of schizophrenia predate psychosis? *Journal of Psychiatric Research*, 32:121–131, 1998.

Frith, CD. *The Cognitive Neuropsychology of Schizophrenia*. East Sussex, UK: Lawrence Erlbaum, 1992.

Goldman-Rakic, PS. Working memory dysfunction in schizophrenia. *Journal of Neuropsychiatry & Clinical Neurosciences*, 6:348–357, 1994.

Hirsh, SR, Weinberger, DR. *Schizophrenia*. Oxford, UK: Blackwell Science, 1995.

Holzman, PS, Levy, DL, Proctor, LR. Smooth pursuit eye movements, attention, and schizophrenia. *Archives of General Psychiatry*, 45:641–647, 1976.

Jacobsen, LK, Rapoport, JL. Research update: Childhood-onset schizophrenia: implications of clinical and neurobiological research. *Journal of Child Psychology and Psychiatry*, 39:101–113, 1998.

Jones, P, Murray, RM. The genetics of schizophrenia is the genetics of neurodevelopment. *British Journal of Psychiatry*, 158:615–623, 1991.

Kane, J, Honigfeld, G, Singer, J, Meltzer, H. Clozapine for the treatment-resistant schizophrenic: A double-blind comparison with chlorpromazine. *Archives of General Psychiatry*, 45:789–796, 1988.

Kraepelin, E, Barclay, RM, Robertson, GM. *Dementia Praecox and Paraphrenia*. Edinburgh, UK: E&S Livingstone, 1919.

Moldin, SO, Gottesman, II. At issue: Genes, experience, and chance in schizophrenia—positioning for the 21st century. *Schizophrenia Bulletin*, 23(4):547–61, 1997.

Sedvall, G, Terenius, L. *Schizophrenia: Pathophysiological Mechanisms*. Amsterdam: Elsevier, 2000.

Seeman, P, Lee, T, Chang-Wong, M, Wong, K. Antipsychotic drug doses and neuroleptic/dopamine receptors. *Nature*, 261:717–719, 1976.

Walker E, Lewine R. Prediction of adult-onset schizophrenia from childhood home movies of the patients. *American Journal of Psychiatry*, 147(8):1052–1056, 1990.

Weinberger, D. Implications of normal brain development for the pathogenesis of schizophrenia. *Archives of General Psychiatry*, 44:660–669, 1987.

Chapter 9

Andreasen, NJC, Canter, A. The creative writer: Psychiatric symptoms and family history. *Comprehensive Psychiatry*, 15:123–131, 1974.

Andreasen, NC. Creativity and mental illness: Prevalence rate in writers and their first-degree relatives. *American Journal of Psychiatry*, 144:1288–1292, 1987.

Andreasen, NC, Rice, J, Endicott, J, Coryell, WH, Grove, WM, Reich, T. Familial rates of affective disorder: A report from the National Institute of Mental Health Collaborative Study. *Archives of General Psychiatry*, 44:461–469, 1987.

Cameron, OC, ed. *Adrenergic Dysfunction and Psychobiology*. Washington, DC: American Psychiatric Press, 1994.

Coryell, W, Endicott, J, Andreasen, NC, Keller, MB. Bipolar I, Bipolar II, and unipolar major depression among the relatives of affectively ill probands. *American Journal of Psychiatry*, 142:817–821, 1985.

Coryell, WH, Endicott, J, Keller, MB, Andreasen, NC, Grove, WM, Hirschfeld, RMA, Scheftner, W. Bipolar affective disorder and high achievement: A familial association. *American Journal of Psychiatry*, 146:983–988, 1989.

Goodwin, FK, Jamison, KR. *Manic-Depressive Illness*. New York: Oxford University Press, 1990.

Jamison, KR. *An Unquiet Mind*. New York: Alfred A. Knopf, 1995.

Kelsoe, JR, Ginns, EI, Egeland, JA, et al. Re-evaluation of the linkage relationship between chromosome 11p loci and the gene for bipolar affective disorder in the old order Amish. *Nature*, 342:238–243, 1989.

Klerman, GL, Lavori, PW, Rice, J, Reich, T, Endicott, J, Andreasen, NC, Keller, MB, Hirschfeld, RMA. Birth-cohort trends in rates of major depressive disorder among relatives of patients with affective disorder. *Archives of General Psychiatry*, 32:689–695, 1985.

Leonard, BE, Miller, K, eds. *Stress, the Immune System and Psychiatry*. Chichester, UK: John Wiley and Sons, 1995.

Rice, J, Reich, T, Andreasen, NC, Endicott, J, Van Eerdewegh, M, Fishman, R, Hirschfeld, RMA, Klerman, GL. The familial transmission of bipolar illness. *Archives of General Psychiatry*, 44:441–447, 1987.

Watson, SJ, ed. *Biology of Schizophrenia and Affective Disease*. Washington, DC: American Psychiatric Press, 1996.

Winokur, G, Clayton, PJ, Reich, T. *Manic Depressive Illness*. St. Louis: CV Mosby, 1969.

Chapter 10

Bottino, CM, Almeida, OP. Can neuroimaging techniques identify individuals at risk of developing Alzheimer's disease? *International Psychogeriatrics*, 9(4):389–403, (Dec) 1997.

Burns, A, Levy, R, eds. *Dementia*. London: Chapman and Hall, 1994.

Coffey, CE, Cummings, JL. *Textbook of Geriatric Psychiatry*. Washington, DC: American Psychiatric Press, 1994.

Folstein, MF, ed. *Neurobiology of Primary Dementia*. Washington, DC: American Psychiatric Press, 1998.

Haass, C, De Strooper, B. The presenilins in Alzheimer's disease: Proteolysis holds the key. *Science*, 286(5441):916–9, (Oct) 1999.

Haroutunian, V, Perl, DP, Purhoit, DP, Marin, D, Khan, K, Lantz, M, Davis, KL, Mohs, RC. Regional distribution of neuritic plaques in the non-demented elderly and subjects with very mild Alzheimer disease. *Archives of Neurology*, 55(0) 1185–91, (Sept) 1998.

Iqbal, K, Winblad, B, Nishimura, T, Takeda, M, Wisniewski, HM. *Alzheimer's Disease: Biology, Diagnosis, and Therapeutics*. Chichester, UK: John Wiley and Sons, 1997.

Prusiner, SB. Novel proteinaceous infectious particles cause scrapie. *Science*, 216:136–144, 1982.

Roses, AD. Apolipoprotein E affects the rate of Alzeimer disease expression. *Journal of Neuropathology and Experimental Neurology*, 53:4290437, 1994.

Vassar, R, Bennett, B, Babu-Khan, S, et al. Beta-secretase cleavage of Alzheimer's amyloid precursor protein by the transmembrane aspartic protease BACE. *Science*, 286(5440):735–41, (Oct 22) 1999.

Yan, R, Bienkowski, M, Shuck, M, Miao, H, Tory, M, Pauley, A, Brashier, J, Stratman, N, Mathews, W, Buhl, A, Carter, D, Tomasselli, A, Parodi, L, Heinrikson, R, Gurney, M. Membrane-anchored aspartyl protease with Alzheimer's disease beta-secretase activity. *Nature*, 402(6761):533–7, (Dec) 1999.

Chapter 11

Andreasen, NC. Post-traumatic stress disorder. In *Comprehensive Textbook of Psychiatry IV*, edited by Freedman, Kaplan, Sadock. Baltimore: Williams and Wilkins, Volume 1, pp. 918–924, 1984.

Bremner, JD, Marmar, CR. *Trauma, Memory, and Dissociation*. Washington, DC: American Psychiatric Press, 1998.

Gorman, JM, Kent, JM, Sullivan, GM, Coplan, JD. Neuroanatomical hypothesis of panic disorder, revised. *American Journal of Psychiatry*, 157:493–505, 2000.

LeDoux, J. *The Emotional Brain*. New York: Simon and Schuster, 1996.

Mazure, CM. *Does Stress Cause Psychiatric Illness?* Washington, DC: American Psychiatric Press, 1995.

McEwen, BS, Gould, EA, Sakai, RR. The vulnerability of the hippocampus to protective and destructive effects of glucocorticoids in relation to stress. *British Journal of Psychiatry*, 160:18–24, 1992.

Ursano, RJ, McMaughey, BG, Fullerton, CS. *Individual and Community Response to Trauma and Disaster*. Cambridge, UK: Cambridge University Press, 1994.

Chapter 12

Andreasen, NC. *The Broken Brain: The Biological Revolution in Psychiatry*. New York: Harper and Row, 1984.

Dawkins, R. *The Selfish Gene*. Second Edition. New York: Oxford University Press, 1989.

Huxley, A. *Brave New World*. New York: Harper Collins, 1932.

Kitcher, P. *The Lives to Come: The Genetic Revolution and Human Possibilities*. New York: Simon and Schuster, 1996.

Lyon, J, Gorner, P. *Altered Fates: Gene Therapy and the Retooling of Human Life*. New York: W. W. Norton, 1995.

Sober, E, Wilson, DS. *Unto Others: The Evolution and Psychology of Unselfish Behavior*. Cambridge, Mass: Harvard University Press, 1998.